Ionosphere Monitoring with Remote Sensing II

Ionosphere Monitoring with Remote Sensing II

Fabio Giannattasio

Basel • Beijing • Wuhan • Barcelona • Belgrade • Novi Sad • Cluj • Manchester

Fabio Giannattasio
Istituto Nazionale di
Geofisica e Vulcanologia
Rome
Italy

Editorial Office
MDPI AG
Grosspeteranlage 5
4052 Basel, Switzerland

This is a reprint of articles from the Special Issue published online in the open access journal *Remote Sensing* (ISSN 2072-4292) (available at: www.mdpi.com/journal/remotesensing/special_issues/824EVL2F4C).

For citation purposes, cite each article independently as indicated on the article page online and using the guide below:

Lastname, A.A.; Lastname, B.B. Article Title. *Journal Name* **Year**, *Volume Number*, Page Range.

ISBN 978-3-7258-1838-9 (Hbk)
ISBN 978-3-7258-1837-2 (PDF)
https://doi.org/10.3390/books978-3-7258-1837-2

Contents

About the Editor

Fabio Giannattasio

Fabio Giannattasio (Ph.D., II Level Master in Space Science in Technology) is a researcher at the Istituto Nazionale di Geofisica e Vulcanologia (INGV) in Rome, Italy. His research activity is focused on the study of Sun–Earth interactions and space weather. He has a particular interest in magnetosphere–ionosphere coupling, the complex nature of physical processes in the ionosphere, and some aspects inherent to solar physics.

remote sensing

Editorial

Ionosphere Monitoring with Remote Sensing Vol II

Fabio Giannattasio

Istituto Nazionale di Geofisica e Vulcanologia, Via di Vigna Murata 605, 00143 Roma, Italy;
fabio.giannattasio@ingv.it

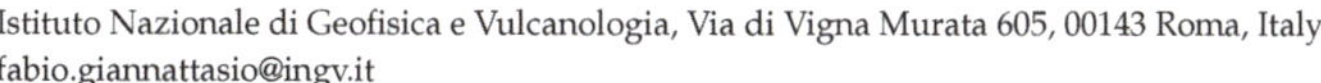

1. Introduction

Unveiling the physical properties of the Earth's ionosphere is crucial for the comprehension of the dynamic processes that occur within it across various spatial and temporal scales. This understanding is essential for comprehending numerous phenomena associated with space weather. The ionosphere, composed of ions and electrons, responds to the initiation, intensification, and evolution of magnetic and electric fields, potentially altering its physical properties and energy balance. These changes can significantly affect the propagation properties of electromagnetic signals passing through the ionosphere.

Thanks to a wealth of high-quality data, these ionospheric features can be thoroughly investigated across different scales using remote sensing and in situ instruments such as ionosondes, radars, satellites, and Global Navigation Satellite Systems (GNSS) receivers.

In this context, the Special Issue "Ionosphere Monitoring with Remote Sensing Vol. II" seeks to advance our understanding of the ionosphere by utilizing data from various facilities and established ionospheric models. In more detail, this Special Issue is primarily focused on: (1) the development or improvement of tools for ionospheric tomography reconstruction [1], signal propagation [2], the prediction of the critical frequency of the F2 layer [3,4], and the automatic identification of Spread-F (SF) on ionograms [5], and (2) the study of ionospheric features, anomalies, and irregularities associated with seasonal, geomagnetic and solar activity variations, solar eclipses, and volcanic eruptions [4,6–12].

2. Overview of Contributions and Future Perspectives

The following Special Issue contains 13 original research papers describing results obtained using different tools, data, models, and analysis techniques and is aimed at characterizing the properties of the ionosphere that are relevant, for instance, for signal transmission and space weather studies.

Significant attention has been paid to the development or improvement of algorithms and tools to increase our knowledge of the ionosphere. Wen et al. [1] proposed a truncated mapping singular value decomposition (TMSVD) method to improve the accuracy and computational efficiency of global navigation satellite system (GNSS)-based ionospheric tomography reconstruction, which is crucial for reconstructing the three-dimensional ionospheric electron density (IED) [13]. Ill-posedness of GNSS-based computerized ionospheric tomography represents a limitation in reconstructing the IED distribution and has been approached by the development of algorithms. Among them, it has been demonstrated that TMSVD methods represent a significant improvement and they have been successfully applied to reconstruct IED images based on actual GNSS observations. The new algorithm by Wen et al. [1] was used to reconstruct ionospheric structure, diurnal variations, and semiannual anomalies using GNSS observations in Huan province, China. Moreover, the method captured the ionospheric structure during an ionospheric storm, resulting in improvement with respect to the International Reference Ionosphere (IRI) model [14].

The path of radio waves propagating through the ionospheric medium can be retrieved by means of the ionospheric ray tracing technique [15]. This technique numerically integrates Hamilton's equations describing the propagation of electromagnetic waves in

check for
updates

Citation: Giannattasio, F. Ionosphere Monitoring with Remote Sensing Vol II. *Remote Sens.* **2024**, *16*, 2762.
https://doi.org/10.3390/rs16152762

Received: 22 July 2024
Accepted: 24 July 2024
Published: 29 July 2024

the high frequency (HF) band, namely between 3 and 30 MHz. The numerical solution provides the path of an HF radio wave propagating from a given transmitting point (and with a given azimuth of the ray) to an unknown receiving point on the Earth's surface. Pietrella et al. [2] upgraded the applicative software tool IONORT (IONOspheric Ray Tracing) developed at the Istituto Nazionale di Geofisica e Vulcanologia by Azzarone et al. [16]. The tool uses the electron density model provided by the IRI [17]; however, its use also provides the possibility to select any other model. The original Fortran 77 source code was rewritten in Fortran 90 and thoroughly tested. However, the most important novelties concern the extension from regional to global scales (such that a user can apply the tool anywhere on Earth) and the development of a new graphical user interface.

A crucial parameter in radio wave propagation is the critical frequency of the ionospheric F2 layer, i.e., foF2, which represents the maximum frequency of a signal reflected by the ionospheric F2 layer in long-distance radio communication. This parameter changes due to the different conditions of the ionospheric layer induced, for instance, by space weather events, and may affect the signal propagation and communication quality [18]. This is the reason as to why great efforts have been made to study foF2, with the ultimate aim of forecasting the variation of this parameter following geoeffective space weather events and thus being able to pursue a series of actions to mitigate possible failures, for example, of high-frequency communication and GNSS. In the last few decades, the development of artificial intelligence technology has given a boost to the emergence of new ionospheric prediction models based on artificial neural networks (ANN), statistical machine learning (ML), and deep learning (DL).

These models may perform better than traditional models [19–21]. Bi et al. [20] proposed a high-performance hybrid neural network with a quantile mechanism to model and forecast foF2 variations at low latitudes following space weather events during solar cycle 24. In particular, the model takes advantage of foF2 data together with the solar radio flux at 10.7 cm (F10.7), the sunspot number, the vertical component of the interplanetary magnetic field, and the geomagnetic indices ap and Dst from 2009 to 2014. They found that the proposed model performs better than other previous models such as IRI during both high and low solar activity levels and geomagnetic storm periods. Wang et al. [4] improved the prediction accuracy of foF2 by using an ML-based dynamic prediction method that takes into account the variation of foF2 with the solar cycle. The model uses dynamic training due to dynamic data updating and performs better than the last updated IRI model and the Asia Regional foF2 model.

Ionospheric plasma irregularities manifest in many ways. One of the most important is SF, which causes scintillation and degrades GNSS, communication, and radar signals. An easy and effective way to observe SF and study its diurnal and seasonal changes is through the use of ionosondes. However, identifying SF in ionogram traces is not always easy. Automatic scaling of ionograms has been extensively used for several decades [22–24]. Recently, in the machine learning era, Lan et al. [25] demonstrated that convolutional neural networks (CNNs) represent a promising method for SF automatic identification. Rao et al. [26] presented an auto-detection technique to identify Sporadic E and SF signatures, reaching a detection efficiency of up to 96.7%. Fen et al. [5] proposed a statistical analysis of SF in East Asia by using a Bayesian classifier. Their automatic identification algorithm reached an accuracy of up to 97% and was applied to five stations to expand the research of SF in that area.

Ionospheric anomalies, which include annual, semiannual, winter, equatorial, and mid-latitude summer nighttime anomalies, among others, are ionospheric variations mainly related to plasma dynamics and changes in the composition of the atmosphere [27]. Since its discovery [28], the winter anomaly has been extensively studied over the last century. It consists of daytime F2-layer peak electron density (NmF2) greater in winter than in summer at middle latitudes. The basic physical mechanism that is responsible for this anomaly is the change in the ratio O/N2 [29] resulting from global circulation in the thermosphere [30]. Conversely, the annual anomaly [31] consists of NmF2 from both hemispheres being greater during December than during June. Wang et al. [4] assessed the contribution of the winter anomaly to the annual anomaly based on regression analysis over 10 years of total electron content (TEC) and F10.7 data during solar cycle 24 conditioned to geomagnetic quietness. Among the other remarkable results, they found that TEC in the winter hemisphere is the main factor that determines the annual anomaly during high solar activity. Specifically, it contributes more than 50% to the annual anomaly when F10.7 exceeds 90 solar flux units (SFU), and up to 65% when F10.7 exceeds 130 SFU.

Ionospheric anomalies have also been studied in conjunction with natural phenomena such as earthquakes, volcanic eruptions, and nuclear blasts [32–36]. For example, Feng et al. [7] studied ionospheric anomalies before the submarine volcanic eruption of Hunga Tonga–Hunga Ha'apai in January 2022. The authors used GNSS data, global ionospheric maps, and radio occultation data to detect negative TEC anomalies starting from ten days before the eruption near its epicenter. Moreover, a few days earlier, the equatorial ionization anomaly (EIA) double peak showed relevant shifts and decreased until disappearing and merging in a single peak. The authors were able to rule out the effects of solar and geomagnetic activity, together with the lower atmospheric forcing.

A completely new work of its kind is that of Valdes Abreu et al. [8], who studied the ionospheric effects of two solar eclipses observed in South America and Antarctica during periods of disturbed geomagnetic activity and in different seasons. The rationale behind this interesting work is that solar eclipses provide a great opportunity to investigate the response of the ionosphere to sharp transitions from light to dark to light. During solar eclipses, the sudden decrease in photoionization and photoelectron heating results in the loss of plasma with a remarkable decrease in electron density. As a consequence, TEC fluctuates, and its variations may influence regions outside the moon's shadow due to a number of processes. While at high latitudes ionospheric effects induced by solar eclipses are mainly driven by magnetosphere–ionosphere coupling [37–39], at low latitudes, such effects are due to processes involving the EIA [40]. Valdes Abreu et al. [8] compared the effects of both eclipses on TEC over South America and Antarctica in different seasons by using data from 390 GNSS stations, the European Space Agency's Swarm A satellite, and the Defense Meteorological Satellite Program DMSP F18 satellite. They also investigated the effects of these eclipses on the EIA. They found that although the forcing was due to solar and geomagnetic activity, there were evident plasma depletions during the penumbra, with differential vertical TEC (DVTEC) values lower than 40%. Moreover, different values of DVTEC in the crests of the EIA were found during the two different eclipse periods.

Particular interest lies in the study of ionospheric plasma dynamics at mid and high latitudes, where physical processes are strongly influenced by coupling with the magnetosphere and solar wind. In fact, at high latitudes, the ionosphere quickly responds to solicitations of external origin and exhibits distinctive features [41]. In light of this, research aimed at ionospheric modeling in recent decades has benefited from the ever-increasing number and quality of available data, with the aim of reproducing the observed varying spatiotemporal structures, which depend not only on latitude, longitude, and season but also on solar and geomagnetic activity and solar wind conditions (primarily the physical properties of the plasma and interplanetary magnetic field). Such models are also crucial for identifying the various physical mechanisms at work. To achieve this, it is of primary importance to identify the scales at which these mechanisms operate. Lovati et al. [9] applied the Multivariate Empirical Mode Decomposition (MEMD) technique to electron density data acquired by the Swarm A satellite to detect the spatial scales impacting high-latitude ionospheric plasma dynamics. The modes extracted via MEMD contribute to different weights in the original data. The authors identified four modes that represent (i) noise, (ii) a constant frequency such as the satellite's orbit, (iii) seasonality, and (iv) geomagnetic activity dependence. Such modes are able to discern some specific processes, e.g., the ionospheric circulation typical of the polar cap and phenomena driven by sunlit, recombination, and neutral wind. Lovati et al. [9] de facto advanced our understanding of the spatiotemporal features in electron density, which future models should take into account for a better prediction of future ionospheric conditions.

The increase in ionization in the upper atmosphere leads to ionospheric disturbances driven, for example, by bursts of UV and X-ray radiation from the Sun. During solar flares, such bursts result in increased ionization velocity and the subsequent increase in electron density in the ionosphere. The variation in electron concentration may give rise to the absorption of HF radio waves, especially in the lower ionosphere [42–44]. However, strong solar flares may also impact the upper ionospheric layers [45,46]. Sergeeva et al. [10] studied the effect of two solar flares in 2022 at low latitudes, in Mexico, on both the topside and the lower ionosphere. They confirmed that the X-ray solar radiation increased electron density in the lower ionosphere more than in the upper layers. Moreover, the fainter flare (an M9.6 flare) resulted in irregularities at medium scales; in comparison, the stronger flare (an X1.3 flare) caused medium- and small-scale irregularities. The results of other studies [10] highlight the fact that the narrowing of the frequency operation range and signal amplitude decrease in ionograms exhibited a few minutes delay longer for the fainter flare than for the stronger one; the recovery of the ionosphere began before the flare ended; and the flare effects lasted longer for the stronger flare. The ionospheric response to flares depends more on the active region position on the Sun and the flare class than the zenith angle.

Information on the vertical distribution of plasma in the ionosphere–plasmasphere system and electron density profile can be retrieved using the slab thickness. This fundamental ionospheric parameter is also related to the plasma scale height and neutral temperature. In addition, it is an important component in ionospheric modeling and data assimilation methods. In fact, it is defined as the ratio between TEC and NmF2 and represents the equivalent depth in meters of the ionosphere given a uniform electron density of NmF2 [47–49]. Like other ionospheric parameters, the slab thickness has distinctive features varying with local time, latitude, longitude, season, solar, and geomagnetic activity. Due to the scarcity of ionospheric facilities, there are relatively few statistical studies of slab thickness at high latitudes. In previous works, it was found that nighttime mean values of the slab thickness are larger than in the daytime, with the highest and lowest mean values being in the daytime in summer and winter, respectively [50]. Different features have been found by other authors. Yadav and Bhawre [51] showed that the mean values of the slab thickness are greatest during the equinox and smallest during the summer. Pignalberi et al. [52] found that the largest slab thickness occurs during the nighttime and dawn hours in winter and the equinox. In addition, the pre-sunrise peak is more evident during low- and mid-solar activity periods, except for in the summer, while the post-sunset peak is evident only in winter except for the high solar activity. Feng et al. [6] investigated the variation in the slab thickness in the high latitudes of East Asia in order to improve the global empirical modeling of this parameter and study the effects of geomagnetic activity on it. They utilized TEC and ionosonde data at Yakutsk using 7 years' worth of data from 2010 to 2017 and found a dependence of diurnal and seasonal features of slab thickness on solar and geomagnetic activity. The slab thickness is greatest during winter and smallest during the summer. Moreover, during the summer and equinox, there are smaller diurnal variations and peaks with different strengths during sunrise and post-sunset times. The geomagnetic activity affects the slab thickness, enhancing it during geomagnetic storms due to intense particle precipitation and an expanded plasma convection electric field, as highlighted by examining a case study in June 2013.

Fluctuations in electron density are responsible for variations in the refractive index of the ionosphere, causing refraction and diffraction phenomena that underlie amplitude and phase variations in electromagnetic signals (i.e., scintillation) such as GNSS signals transmitted by navigation satellites. Phase scintillation can be quantified via an index based on the ensemble standard deviation of detrended phase measurements. Phase scintillation results in cycle slips or, at worst, in the complete loss of the lock of the receiver [53–55]. The former consists of the temporary disruption of phase tracking that can be detected, for instance, by changes in TEC [56]. Beeck et al. [11] statistically investigated the relationship between phase scintillation and cycle slips for GPS and the Galileo constellation in Greenland and performed a simulation of the effects on these signals. The authors found a high percentage of cycle slips for the Galileo constellation, while signal outages were the most relevant impact of phase scintillation on GPS data. They concluded that there was a different response to ionospheric disturbances for the two facilities due to the GPS signal underestimating TEC during large TEC changes.

Clumps of ionized plasma of various spatial scales are known as ionospheric field-aligned irregularities (FAIs) and are recognized to be sources of disturbance for navigation, communication, and radar systems. In particular, mid-latitude FAIs in the ionospheric E-region have been extensively studied since their detection with Arecibo radar more than four decades ago [57–59]. Yamamoto et al. [60] found two types of FAIs in the E layer: daytime continuous echoes at 90–100 km and nighttime quasi-periodic echoes at 100–130 km. The mechanisms of generation proposed include gravity waves [61], gradient drift instability [62], Kelvin–Helmohotz instability [63], and sporadic E instability [64]. To overcome the range resolution limitations of VHF radars that make it impossible to resolve small-scale structures in a radar-illuminated volume, multifrequency radar imaging (RIM) technology has been developed and successfully used. Chen et al. [12] applied the RIM technique and found that the spatial resolution of FAIs in the E-region greatly improved, and it was possible to obtain finer structures such as QP echoes. The features observed suggest that their generation is modulated by gravity waves. This technique was also applied for the observation of multiple layers of the E-region FAIs. The authors found that the multilayered FAI features at night may be due to gradient-drift instability acting on multiple ionized layers formed by tides or waves.

Funding: The above research received no external funding.

Conflicts of Interest: The author declares no conflicts of interest.

References

1. Wen, D.; Xie, K.; Tang, Y.; Mei, D.; Chen, X.; Chen, H. A New Algorithm for Ill-Posed Problem of GNSS-Based Ionospheric Tomography. *Remote Sens.* **2023**, *15*, 1930. [CrossRef]
2. Pietrella, M.; Pezzopane, M.; Pignatelli, A.; Pignalberi, A.; Settimi, A. An Updating of the IONORT Tool to Perform a High-Frequency Ionospheric Ray Tracing. *Remote Sens.* **2023**, *15*, 5111. [CrossRef]
3. Bi, C.; Ren, P.; Yin, T.; Xiang, Z.; Zhang, Y. Modeling and Forecasting Ionospheric foF2 Variation in the Low Latitude Region during Low and High Solar Activity Years. *Remote Sens.* **2022**, *14*, 5418. [CrossRef]
4. Wang, J.; Yu, Q.; Shi, Y.; Liu, Y.; Yang, C. An Explainable Dynamic Prediction Method for Ionospheric foF2 Based on Machine Learning. *Remote Sens.* **2023**, *15*, 1256. [CrossRef]
5. Feng, J.; Zhang, Y.; Gao, S.; Wang, Z.; Wang, X.; Chen, B.; Liu, Y.; Zhou, C.; Zhao, Z. Statistical Analysis of SF Occurrence in Middle and Low Latitudes Using Bayesian Network Automatic Identification. *Remote Sens.* **2023**, *15*, 1108. [CrossRef]
6. Feng, J.; Zhang, Y.; Xu, N.; Chen, B.; Xu, T.; Wu, Z.; Deng, Z.; Liu, Y.; Wang, Z.; Zhou, Y.; et al. Statistical Study of the Ionospheric Slab Thickness at Yakutsk High-Latitude Station. *Remote Sens.* **2022**, *14*, 5309. [CrossRef]
7. Feng, J.; Yuan, Y.; Zhang, T.; Zhang, Z.; Meng, D. Analysis of Ionospheric Anomalies before the Tonga Volcanic Eruption on 15 January 2022. *Remote Sens.* **2023**, *15*, 4879. [CrossRef]
8. Valdés-Abreu, J.C.; Díaz, M.; Bravo, M.; Stable-Sánchez, Y. IonosphericTotal Electron Content Changes during the 15 February 2018 and 30 April 2022 Solar Eclipses over South America and Antarctica. *Remote Sens.* **2023**, *15*, 4810. [CrossRef]
9. Lovati, G.; De Michelis, P.; Alberti, T.; Consolini, G. Unveiling the Core Patterns of High-Latitude Electron Density Distribution at Swarm Altitude. *Remote Sens.* **2023**, *15*, 4550. [CrossRef]
10. Sergeeva, M.A.; Maltseva, O.A.; Vesnin, A.M.; Blagoveshchensky, D.V.; Gatica-Acevedo, V.J.; Gonzalez-Esparza, J.A.; Chernov, A.G.; Orrala-Legorreta, I.D.; Melgarejo-Morales, A.; Gonzalez, L.X.; et al. Solar Flare Effects Observed over Mexico during 30–31 March 2022. *Remote Sens.* **2023**, *15*, 397. [CrossRef]
11. Beeck, S.S.; Mitchell, C.N.; Jensen, A.B.O.; Stenseng, L.; Pinto Jayawardena, T.; Olesen, D.H. Experimental Determination of the Ionospheric Effects and Cycle Slip Phenomena for Galileo and GPS in the Arctic. *Remote Sens.* **2023**, *15*, 5685. [CrossRef]
12. Chen, B.; Liu, Y.; Feng, J.; Zhang, Y.; Zhou, Y.; Zhou, C.; Zhao, Z. High-Resolution Observation of Ionospheric E-Layer Irregularities Using Multi-Frequency Range Imaging Technology. *Remote Sens.* **2023**, *15*, 285. [CrossRef]
13. Kunitsyn, V.E.; Andreeva, E.S.; Razinkov, O.G. Possibilities of the near-space environment radio tomography. *Radio Sci.* **1997**, *32*, 1953–1963. [CrossRef]

14. Bilitza, D.; Altadill, D.; Truhlik, V.; Shubin, V.; Galkin, I.; Reinisch, B.; Huang, X. International Reference Ionosphere 2016: From ionospheric climate to real-time weather predictions. *Space Weather* **2017**, *15*, 418–429. [CrossRef]

15. Haselgrove, J. Ray Theory and a New Method for Ray Tracing. In *The Physics of the Ionosphere: Report of the Physical Society Conference, Held at Cavendish Laboratory, Cambridge, September 1954*; Physical Society: London, UK, 1955; pp. 355–364.

16. Azzarone, A.; Bianchi, C.; Pezzopane, M.; Pietrella, M.; Scotto, C.; Settimi, A. IONORT: A Windows soft-ware tool to calculate the HF ray tracing in the ionosphere. *Comput. Geosci.* **2012**, *42*, 57–63. [CrossRef]

17. Bilitza, D.; Pezzopane, M.; Truhlik, V.; Altadill, D.; Reinisch, B.W.; Pignalberi, A. The International Reference Ionosphere model: A review and description of an ionospheric benchmark. *Rev. Geophys.* **2022**, *60*, e2022RG000792. [CrossRef]

18. Bilitza, D.; Obrou, O.; Adeniyi, J.; Oladipo, O. Variability of foF2 in the equatorial ionosphere. *Adv. Space Res.* **2004**, *34*, 1901–1906. [CrossRef]

19. Chen, C.; Wu, Z.S.; Xu, Z.W.; Sun, S.J.; Ding, Z.H.; Ban, P.P. Forecasting the local ionospheric foF2 parameter 1 hour ahead during disturbed geomagnetic conditions. *J. Geophys. Res. Space Phys.* **2010**, *115*, 135–146. [CrossRef]

20. Bai, H.; Feng, F.; Wang, J. A Combination Prediction Model of Long-Term Ionospheric foF2 Based on Entropy Weight Method. *Entropy* **2020**, *22*, 442. [CrossRef]

21. Maltseva, O. The Influence of Space Weather on the Relationship Between the Parameters TEC and foF2 of the Ionosphere. *IEEE J. Radio Freq. Identif.* **2021**, *5*, 261–268. [CrossRef]

22. Pezzopane, M.; Scotto, C. Automatic scaling of critical frequency foF2 and MUF(3000)F2: A comparison between Autoscala and ARTIST 4.5 on Rome data. *Radio Sci.* **2007**, *42*, RS4003. [CrossRef]

23. Pezzopane, M.; Pillat, V.; Fagundes, P. Automatic scaling of critical frequency foF2 from ionograms recorded at São José dos Campos, Brazil: A comparison between Autoscala and UDIDA tools. *Acta Geophys.* **2017**, *65*, 173–187. [CrossRef]

24. Scotto, C.; Ippolito, A.; Sabbagh, D. A method for automatic detection of equatorial spread-F in Ionograms. *Adv. Space Res.* **2019**, *63*, 337–342. [CrossRef]

25. Lan, T.; Hu, H.; Jiang, C.; Yang, G.; Zhao, Z. A Comparative Study of Decision Tree, Random Forest, and Convolutional Neural Network for Spread-F Identification. *Adv. Space Res.* **2020**, *65*, 2052–2061. [CrossRef]

26. Rao, T.V.; Sridhar, M.; Ratnam, D.V. Auto-detection of sporadic E and spread F events from the digital ionograms. *Adv. Space Res.* **2022**, *70*, 1142–1152.

27. Mendillo, M.; Huang, C.-L.; Pi, X.; Rishbeth, H.; Meier, R. The Global Ionospheric Asymmetry in Total Electron Content. *J. Atmos. Sol.-Terr. Phys.* **2005**, *67*, 1377–1387. [CrossRef]

28. Berkner, L.V.; Wells, H.W.; Seaton, S.L. Characteristics of the Upper Region of the Ionosphere. *Terr. Magn. Atmos. Electr.* **1936**, *41*, 173–184. [CrossRef]

29. Rishbeth, H.; Setty, C.S.G.K. The F-Layer at Sunrise. *J. Atmos. Terr. Phys.* **1961**, *20*, 263–276. [CrossRef]

30. King, G.A.M. The Dissociation of Oxygen and High Level Circulation in the Atmosphere. *J. Atmos. Sci.* **1964**, *21*, 231–237. [CrossRef]

31. Berkner, L.V.; Wells, H.W. Non-Seasonal Change of F2-Region Ion-Density. Terr. *Magn. Atmos. Electr.* **1938**, *43*, 15–36. [CrossRef]

32. Whitcomb, J.H.; Garmany, J.D.; Anderson, D.L. Earthquake Prediction: Variation of Seismic Velocities before the San Francisco Earthquake. *Science* **1973**, *180*, 632–635. [CrossRef] [PubMed]

33. Pulinets, S. Ionospheric Precursors of Earthquakes; Recent Advances in Theory and Practical Applications. *Terr. Atmos. Ocean. Sci.* **2004**, *15*, 413–435. [CrossRef]

34. Ke, F.; Wang, Y.; Wang, X.; Qian, H.; Shi, C. Statistical analysis of seismo-ionospheric anomalies related to Ms > 5.0 earthquakes in China by GPS TEC. *J. Seismol.* **2016**, *20*, 137–149. [CrossRef]

35. Iwata, T.; Umeno, K. Preseismic ionospheric anomalies detected before the 2016 Kumamoto earthquake. *J. Geophys. Res. Space Phys.* **2017**, *122*, 3602–3616. [CrossRef]

36. Xie, T.; Chen, B.; Wu, L.; Dai, W.; Kuang, C.; Miao, Z. Detecting Seismo-Ionospheric Anomalies Possibly Associated with the 2019 Ridgecrest (California) Earthquakes by GNSS, CSES, and Swarm Observations. *J. Geophys. Res. Space Phys.* **2021**, *126*, e2020JA028761. [CrossRef]

37. Davis, C.J.; Lockwood, M.; Bell, S.A.; Smith, J.A.; Clarke, E.M. Ionospheric measurements of relative coronal brightness during the total solar eclipses of 11 August, 1999 and 9 July, 1945. *Ann. Geophys.* **2000**, *18*, 182–190. [CrossRef]

38. Krankowski, A.; Shagimuratov, I.; Baran, L.; Yakimova, G. The effect of total solar eclipse of October 3, 2005, on the total electron content over Europe. *Adv. Space Res.* **2008**, *41*, 628–638. [CrossRef]

39. Chen, X.; Dang, T.; Zhang, B.; Lotko, W.; Pham, K.; Wang, W.; Lin, D.; Sorathia, K.; Merkin, V.; Luan, X.; et al. Global Effects of a Polar Solar Eclipse on the Coupled Magnetosphere-Ionosphere System. *Geophys. Res. Lett.* **2021**, *48*, e2021GL096471. [CrossRef]

40. Adekoya, B.; Chukwuma, V. Ionospheric F2 layer responses to total solar eclipses at low and mid-latitude. *J. Atmos. Sol. Terr. Phys.* **2016**, *138–139*, 136–160. [CrossRef]

41. Cowley, S.W.H. TUTORIAL: Magnetosphere-Ionosphere Interactions: A Tutorial Review. *Geophys. Monogr. Ser.* **2000**, *118*, 91.

42. Davies, K. *Ionospheric Radio Propagation*; Monograph 80; National Bureau of Standards: Gaithersburg, MD, USA, 1965; 487p.

43. Mitra, A.P. *Ionospheric Effect of Solar Flares*; Reidel: Norwell, MA, USA, 1974.

44. Hunsucker, R.D.; Hargreaves, J.K. *The High-Latitude Ionosphere and Its Effects on Radio Propagation*; Cambridge University Press: Cambridge, UK, 2003.

45. Dmitriev, A.V.; Yeh, H.-C.; Chao, J.-K.; Veselovsky, I.S.; Su, S.-Y.; Fu, C.C. Top-side ionosphere response to extreme solar events. *Ann. Geophys.* **2006**, *24*, 1469–1477. [CrossRef]
46. Mendillo, M.; Erickson, P.J.; Zhang, S.-R.; Mayyasi, M.; Narvaez, C.; Thiemann, E.; Chamberlain, P.; Andersson, L.; Peterson, W. Flares at Earth and Mars: An ionospheric escape mechanism? *Space Weather* **2018**, *16*, 1042–1056. [CrossRef]
47. Krankowski, A.; Shagimuratov, I.I.; Baran, L.W. Mapping of foF2 over Europe based on GPS-derived TEC data. *Adv. Space Res.* **2007**, *39*, 651–660. [CrossRef]
48. Gerzen, T.; Jakowski, N.; Wilken, V.; Hoque, M.M. Reconstruction of F2 layer peak electron density based on operational vertical total electron content maps. *Ann. Geophys.* **2013**, *31*, 1241–1249. [CrossRef]
49. Maltseva, O.A.; Mozhaeva, N.S.; Nikitenko, T.V. Validation of the Neustrelitz Global Model according to the low latitude ionosphere. *Adv. Space Res.* **2014**, *54*, 463–472. [CrossRef]
50. Jayachandran, B.; Krishnankutty, T.; Gulyaeva, T. Climatology of ionospheric slab thickness. *Ann. Geophys.* **2004**, *22*, 25–33. [CrossRef]
51. Yadav, R.; Bhawre, P. Ionospheric slab thickness over high latitude Antarctica during the maxima of solar cycle 23rd. *Int. J. Curr. Res.* **2020**, *12*, 10041–10046.
52. Pignalberi, A.; Pietrella, M.; Pezzopane, M.; Nava, B.; Cesaroni, C. The Ionospheric Equivalent Slab Thickness: A Review Supported by a Global Climatological Study Over Two Solar Cycles. *Space Sci. Rev.* **2022**, *218*, 37. [CrossRef]
53. Pi, X.; Mannucci, A.J.; Lindqwister, U.J.; Ho, C.M. Monitoring of global ionospheric irregularities using the worldwide GPS network. *Geophys. Res. Lett.* **1997**, *24*, 2283–2286. [CrossRef]
54. Jiao, Y.; Morton, Y.; Taylor, S.; Pelgrum, W. Characterization of high-latitude ionospheric scintillation of GPS signals. *Radio Sci.* **2013**, *48*, 698–708. [CrossRef]
55. Veettil, S.V.; Aquino, M.; Spogli, L. A statistical approach to estimate Global Navigation Satellite Systems (GNSS) receiver signal tracking performance in the presence of ionospheric scintillation. *J. Space Weather Space Clim.* **2018**, *8*, A51. [CrossRef]
56. Estey, L.; Wier, S. *Teqc Tutorial: Basic of Teqc Use and Teqc Products*; UNAVCO: Boulder, CO, USA, 2014.
57. Yamamoto, M.; Fukao, S.; Ogawa, T.; Tsuda, K.; Kato, S. A morphological study on mid-latitude E-region field-aligned irregularities observed with the MU radar. *J. Atmos. Sol. Terr. Phys.* **1992**, *54*, 769–777. [CrossRef]
58. Haldoupis, C.; Schlegel, K. Characteristic of midlatitude coherent backscatter from the ionospheric E region obtained with Sporadic E scatter experiment. *J. Geophys. Res.* **1996**, *101*, 13387–13397. [CrossRef]
59. Fukao, S.; Yamamoto, M.; Tsunoda, R.T.; Hayakawa, H.; Mukai, T. The SEEK (Sporadic-E Experiment over Kyushu) Campaign. *Geophys. Res. Lett.* **1998**, *25*, 1761–1764. [CrossRef]
60. Yamamoto, M.; Fukao, S.; Woodman, R.F.; Ogawa, T.; Tsuda, T.; Kato, S. Midlatitude E region field-aligned irregularities observed with the MU radar. *J. Geophys. Res.* **1991**, *96*, 15943–15949. [CrossRef]
61. Woodman, R.F.; Yamamoto, M.; Fukao, S. Gravity wave modulation of gradient drift instabilities in mid-latitude sporadic E irregularities. *J. Geophys. Res.* **1991**, *18*, 1197–1200. [CrossRef]
62. Maruyama, T.; Fukao, S.; Yamamoto, M. A possible mechanism for echo striation generation of radar backscatter from midlatitude sporadic E. *Radio Sci.* **2000**, *35*, 1155–1164. [CrossRef]
63. Larsen, M.F. A shear instability seeding mechanism for quasiperiodic radar echoes. *J. Geophys. Res.* **2000**, *105*, 24931–24940. [CrossRef]
64. Cosgrove, R.B.; Tsunoda, R.T. Simulation of the nonlinear evolution of the sporadic-E layer instability in the nighttime midlatitude ionosphere. *J. Geophys. Res.* **2003**, *108*, 1283. [CrossRef]

remote sensing

Technical Note

Experimental Determination of the Ionospheric Effects and Cycle Slip Phenomena for Galileo and GPS in the Arctic

S.S. Beeck [1,*], C.N. Mitchell [2], A.B.O. Jensen [1,3], L. Stenseng [1,4], T. Pinto Jayawardena [5] and D.H. Olesen [1]

1   Department of Space Research and Technology, Technical University of Denmark, Building 328 Elektrovej, 2800 Kongens Lyngby, Denmark
2   Department of Electronic and Electrical Engineering, University of Bath, Bath BA1 7AY, UK
3   AJ Geomatics, 4000 Roskilde, Denmark
4   Peak Wind, Mikkel Bryggers Gade 4, 1460 Copenhagen, Denmark
5   Athena Space Limited, Torquay TQ2 7TD, UK
*   Correspondence: saschu@space.dtu.dk

Abstract: The ionosphere can impair the accuracy, availability and reliability of satellite-based positioning, navigation and timing. The Arctic region is particularly affected by strong ionospheric gradients and phase scintillation, posing a safety issue for critical infrastructure and operations. Ionospheric warning and impact maps can provide support to Arctic operations, but to produce such maps threshold values have to be determined. This study investigates how such thresholds can be derived from the GPS and Galileo satellite signals. Rapid changes in total electron content (TEC) or scintillation-induced receiver tracking errors could result in cycle slips or even loss of lock. Cycle slips and data outages are used as a measure of impact on the receiver in this paper. For Galileo, 73.6% of the impacts were cycle slips and 26.4% were outages, while for GPS, 29.3% of the impacts were cycle slips and 70.7% were outages. Considering the sum of cycle slips and outages, it is worth noting that the sum of impacts for Galileo signals is larger than for GPS. A range of possible explanations have been examined through hardware-in-the-loop simulations. The simulations showed that the GPS L2 signal was not adequately tracked during rapid TEC changes and TEC changes were underestimated, thus the GPS cycle slips, derived from L1 and L2 derived TEC changes, were not all registered. These results are important in designing threshold values for TEC and for scintillation impact maps as well as for the operation of GNSS equipment in the Arctic. In particular, the results show that ionospheric changes could be underestimated if GPS L1 and L2 were used in isolation from other dual frequency combinations. It is the first time this analysis has been made for Greenland and the first time that the dual frequency derivation of ionospheric delay using GPS L1 and L2 has been shown to underestimate large TEC gradients. This has important implications for informing GNSS operations that rely on GPS to provide reliable estimates of the ionosphere.

Keywords: GNSS; Galileo; scintillation; sigma phi; ionosphere; cycle slip; signal tracking; Greenland; Arctic; space weather

Citation: Beeck, S.S.; Mitchell, C.N.; Jensen, A.B.O.; Stenseng, L.; Pinto Jayawardena, T.; Olesen, D.H. Experimental Determination of the Ionospheric Effects and Cycle Slip Phenomena for Galileo and GPS in the Arctic. *Remote Sens.* **2023**, *15*, 5685. https://doi.org/10.3390/rs15245685

Academic Editor: Fabio Giannattasio

Received: 1 October 2023
Revised: 15 November 2023
Accepted: 29 November 2023
Published: 11 December 2023

1. Introduction

The Global Navigation Satellite System (GNSS) signals travel more than 20,000 km through space towards our receivers on Earth. During the first $\sim$95% of the journey, they travel relatively unhindered through near vacuum. However, when they reach the last $\sim$5%, they encounter the Earth's atmosphere. One layer of the atmosphere that has a significant impact on radio signals is the ionosphere. The GNSS signals interact with free electrons that have resulted from solar extreme ultraviolet radiation, and impact ionisation from precipitation, ionising the upper atmosphere. The amount of free electrons the signals encounter is quantified by the Total Electron Content (TEC), defined as the total number of free electrons in an imaginary cylinder with a cross-section area of one square metre around the signal path. It is measured in TEC Units (TECU), which is 10^{16} electron per square metre.

GNSS signals travelling through the ionosphere will experience refraction as a function of the electron density along the signal path. When there are rapid fluctuations in the electron density or TEC structures with sizes around the first Fresnel radius, the signal can be subject to changes in the refractive index and diffraction. This can lead to fluctuations on the phase and amplitude of the signal [1,2]. These fluctuations of the phase and amplitude are termed scintillation. In the Arctic region phase, scintillation is the dominant type of scintillation [3], occurring primarily in the night side of the auroral oval and in the cusp [4,5].

In order to estimate phase scintillation of a received signal the fluctuations on the GNSS signal phase can be quantified using a scintillation index. The σ_ϕ index is a widely used phase scintillation index, based on the ensemble standard deviation of the detrended phase measurements. An increase in σ_ϕ indicates increased high-frequency fluctuations of the GNSS signal phase.

When GNSS signals arrive at the receiver, the effect of the phase scintillation on the positioning solution will depend on the receiver tracking architecture and the bandwidth of the carrier tracking loop [6]. This means that the effect of scintillation can differ depending on the receiver type, and this has been investigated in other studies [6–8]. However, it is widely acknowledged that phase scintillation on GNSS signals may cause cycle slips, or in the worst case result in a complete loss of lock in the receiver [3,9,10]. A cycle slip is a momentary disruption of phase tracking which does not cause loss of lock, and hence does not require re-acquisition. Several methods for detecting cycle slips exist. In the analysis in this paper, the UNAVCO approach is followed [11] where cycle slips are detected by changes in TEC above a certain threshold. In [12,13], a clear relation between σ_ϕ and cycle slips in the Canadian Arctic region was shown. Furthermore, ref. [14] found σ_ϕ to be a good indicator of loss of lock for GNSS receivers located within 20–100 km of a GNSS station tracking the scintillation activity.

Precise and reliable satellite-based positioning infrastructure is a critical capability to ensure safety in the Arctic. Ionospheric scintillation poses a threat to this infrastructure at high latitudes due to the possible degradation of GNSS reliability and positioning accuracy [3,10,12,14,15]. As a step towards local GNSS scintillation impact maps and warnings in Greenland, this study aims to statistically investigate the relations of phase scintillation with cycle slips and data outages for GPS and Galileo in Greenland. The signals chosen for this study are L1(C/A) (1575.42 MHz) and L2(P) (1227.6 MHz) for GPS and E1(OS) (1575.42 MHz) and E5a (1176.45 MHz) for Galileo, from now on referred to as L1, L2, E1 and E5a. These are chosen because they are the most widely used in the region, and there are still receivers in the Arctic that are not compatible with the newer GPS L2C and L5 signals. Previous studies have investigated the effect of scintillation on different signals; however, according to [16], the majority have been focused on the low latitudes, i.e., [17–19]. In [20], Rate of TEC index (ROTI) and ROT were investigated in the low latitude region and used to define a threshold for cycle slip event detection. While [21] relates TEC measurements to scintillation indices in the equatorial region using multiple constellations.

The novelty of this study is the analysis of the effects of phase scintillation on both GPS and Galileo frequencies related to cycle slips in signal monitoring for a full year of data from Greenland, combined with a simulation of the effects on GPS and Galileo signals. In the experimental analyses of the study, data from the Space Weather Forecasting for Arctic Defence Operations (SWADO) network has been used. The SWADO network has gathered scintillation data since the Autumn of 2021. The location of the stations in the network, which is currently being extended with more stations, can be seen in Figure 1.

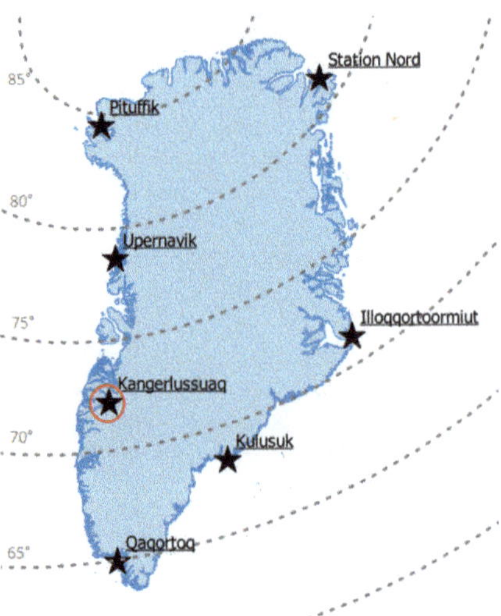

Figure 1. Location of the SWADO stations along the coast of Greenland marked with stars, and the magnetic latitudes shown by dotted lines. The station used in this study is marked by a red circle.

2. Data and Methods

2.1. GNSS Stations and Data Used

The stations in the SWADO network are equipped with Septentrio PolaRx5S receivers, which are categorised as GNSS Ionospheric Scintillation and TEC Monitor (GISTM) receivers. The output files used in this study are in ionospheric scintillation monitoring record (ISMR) format and contain 1 min σ_ϕ indices computed from 100 Hz GNSS data. The indices are computed at the stations to lower the amount of data that is transmitted from Greenland. This setup of the SWADO stations was chosen with a future near real-time service in mind.

For this study, the scintillation data are obtained from one of the stations in the SWADO network, located on the west coast of Greenland in Kangerlussuaq (KLQ2). The data for cycle slip detection is obtained from one of the stations (klsq) in the Greenland GNSS Network (GNET) with a PolaRx5 receiver. The use of a non-specialised (non-scintillation) receiver (such as the PolaRx5) for cycle slip detection allows for assessment of the impact on that type of receiver; this is more relevant for applied users. The two stations are co-located, sharing a Leica LEIAR25.R4 leit antenna with the geographic coordinates (66.9956°N, −50.6202°E). The station site will be referred to by its SWADO station code, KLQ2, from now on. The KLQ2 station was chosen because it is located in an area with frequent auroral activity. This is expected to provide more scintillation impact events during the data period, improving the foundation for statistical analysis. The KLQ2 station is also located on top of a mountain structure, providing good visibility in all directions. Furthermore, the fact that the GNET and SWADO receivers share an antenna ensures that the two GNSS receivers at the KLQ2 site collect data under the exact same conditions.

The study is based on one year of GPS and Galileo data in the format of ISMR, Receiver Independent Exchange format (RINEX) and navigation files. The analyses of the study utilise twenty-seven GPS satellites: G01–09, G12–21, G23–26, G29–32; and twenty Galileo satellites: E02–05, E07–09, E11–13, E15, E21, E24–27, E30–31, E33 and E36.

The SWADO network, shown in Figure 1, is a GNSS network established in the Autumn of 2021. There had not previously been a network of GNSS receivers in Greenland sampling at a high enough frequency to compute σ_ϕ. The data period for this study starts on the 8th of November 2021 and ends one year later on the 7 November 2022. The choice of exactly one year was taken to avoid any time of the year being over- or under-represented.

This precludes any yearly variations from affecting the analysis. The phase scintillation in the night side auroral oval and cusp region have different seasonal patterns [5] and, given the location of KLQ2, the data set could contain scintillation of both regions. This is supported by the most northern ionospheric pierce point being at 83 degrees magnetic latitude and the most southern at 68 degrees magnetic latitude at an ionospheric shell height of 300 km when an elevation mask of 15 degrees is applied. However, despite the precautions to represent each season equally, there will still be an overall increase in solar activity throughout the period as a result of the 11-year solar cycle.

Signals from low-elevation satellites can be subject to non-ionospheric disturbances such as multipath. To filter out these effects, an elevation mask was utilised in the analyses. Due to the geometry of both the GPS and Galileo constellations, the satellite-tracks peak at lower elevations in the sky when observed from higher latitudes. To include as much data as possible in the analyses, it was chosen to investigate the distribution of cycle slips at the station site to find the optimal cutoff angle for this particular site. This was performed by computing a 2D histogram of cycle slips as a function of satellite elevation and azimuth angles, with North at 0 degrees azimuth increasing towards East [22]. The 2D histogram is computed based on data from the entire one-year data period. Any low-elevation obstructions of the GNSS signals at the station are assumed to be due to permanent structures like houses or antenna towers and therefore constant in time. Thus, a set elevation mask is determined from the data. The 2D cycle slip histogram is shown in Figure 2.

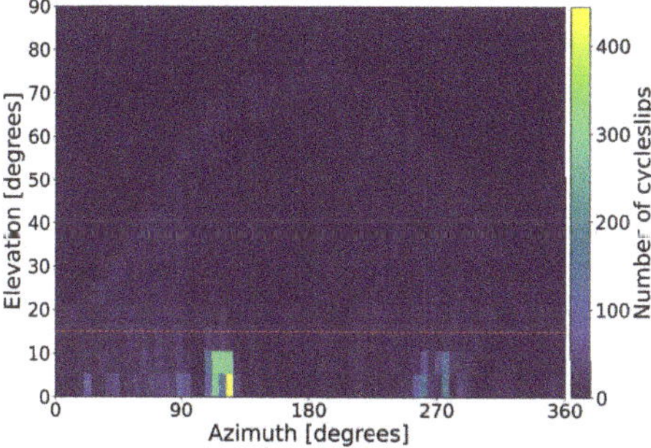

Figure 2. 2D histogram of the number of GPS and Galileo cycle slips seen from KLQ2 throughout the data period as a function of azimuth and elevation of the satellite transmitting the affected signal. The red line illustrates a 15-degree elevation mask.

Figure 2 displays a higher number of cycle slips in specific azimuth directions seen from KLQ2. The directional increases in cycle slips are attributed to permanent structures on the ground, which can be filtered out using an elevation mask of 15 degrees. This elevation cutoff angle is lower than what is used in many other studies of the ionosphere [9,10,23]. However, this is justified since the station is placed on high ground in an area with little disturbance, with the intention of being a stationary permanent GNSS station. The 15-degree elevation cutoff angle is therefore used to filter the data throughout the rest of the study.

2.2. Estimation of Phase Scintillation

The σ_ϕ index is computed as the standard deviation of the detrended carrier phase averaged over a defined time interval. The detrending is performed using a sixth-order Butterworth filter with a cutoff frequency of 0.1 Hz. The cutoff frequency can affect the σ_ϕ index [10,24], and 0.1 Hz is chosen because it is the default in the PolaRx5S and has been commonly used [5,25], which ensures the results of the study are widely applicable. The σ_ϕ index is generally computed over 60 s; however, there are several types of σ_ϕ indices in the ISMR files: Phi60, Phi30, Phi10, Phi03 and Phi01. The difference is the intervals over which σ_ϕ is computed. As an example, Phi03 is the average of the 20 standard deviations computed over 3 s intervals during the last minute, while Phi60 is the standard deviation over the entire last minute [25]. In this study, it is chosen to consistently use Phi60 and it will henceforth be referred to as σ_ϕ.

In this study, σ_ϕ for GPS is a measure of the phase fluctuations on L1, and for Galileo it is on E1. In the first analysis, which investigates the probability of cycle slips as a function of σ_ϕ for GPS and Galileo, the mean of the σ_ϕ values is computed within 15 min intervals for each satellite. The data stream from the SWADO network provides 1 min indices inside 15 min files. The number of cycle slips and the σ_ϕ values were therefore investigated in 15 min intervals. This is consistent with the applied aims of this work, where we want to associate elevated scintillation activity with impacts on receivers with a future view to producing scintillation impact maps. The mean was chosen because it is representative of all the σ_ϕ values in the interval and more robust to outliers in the data than the maximum would be. In the second analysis, which investigates cycle slips and data outages as a function of time of day, the mean σ_ϕ is computed based on σ_ϕ from all satellites binned by the minute of day of the measurement. The mean is taken of each bin, giving 1440 mean σ_ϕ values.

2.3. Identification of Cycle Slips and Data Outages

In this study, impacts are identified through an examination of cycle slip and data outages. The cycle slips are found by the software 'Translation, Editing and Quality Checking' (TEQC) (version 2019Feb25) [11] as ionospheric delay (IOD) slips. The TEQC programme computes the IOD from the phase observation equation shown in Equation (1) [26]

$$L_f = R + c(dt_r + dt_s) - I_f + N + m_f + n_f\lambda_f \tag{1}$$

where L_f is the phase observable, R is the geometric distance between the satellite and receiver, c is the speed of light, dt_r and dt_s are the receiver and satellite clock errors, I_f is the ionospheric delay, N is the tropospheric delay, m_f is the phase multipath, and $n_f\lambda_f$ is the phase ambiguity in wavelengths. The subscript f is used to denote the frequency. Since the ionosphere is dispersive to radio frequencies the ionospheric delay is frequency-dependent. The phase delay for GNSS signals is given in Equation (2).

$$I_{L_f} = \frac{-40.3}{f^2} TEC \tag{2}$$

If it is assumed that the signals at two frequencies follow approximately the same path through the ionosphere then the TEC is the same for the two frequencies, giving the following relation, in Equation (3), between the ionospheric delays at two frequencies.

$$I_1 f_1^2 = I_2 f_2^2 \tag{3}$$

Since TEC is unknown, two frequencies are needed to determine the ionospheric delays. The phase observations at two frequencies are therefore combined in Equations (4)–(6) in order to isolate the ionospheric delay, where $\alpha = \dfrac{f_1^2}{f_2^2}$.

$$L_1 - L_2 = -I_1 + I_2 + m_1 - m_2 + n_1\lambda_1 - n_2\lambda_2 \tag{4}$$

$$\frac{1}{\alpha - 1}L_1 - L_2 = -I_1 + \frac{1}{\alpha - 1}(m_1 - m_2 + n_1\lambda_1 - n_2\lambda_2) \tag{5}$$

$$I_1 = \frac{1}{\alpha - 1}(L_1 - L_2 - m_1 + m_2 - n_1\lambda_1 + n_2\lambda_2) \tag{6}$$

The IOD is the time derivative of the ionospheric delay. It is assumed that the change in phase multipath is minimal over two consecutive epochs and the time derivative of the ionospheric delay on frequency 1 can therefore be expressed as in Equation (7).

$$I_{(1)}OD = \frac{1}{\alpha - 1}[(L_1 - L_2)_j - (L_1 - L_2)_{j-1}]/(t_j - t_{j-1}) \tag{7}$$

The above equations are presented as the basis of the TEQC IOD computations in [26]. The software defines an IOD slip as a change in Equation (7) that exceeds 400.0 cm/min. This is chosen to allow a certain degree of temporal variation in the ionosphere and satellite and receiver motion. The RINEX data are cut into 15 min intervals and the sum of IOD slips in each interval is found for every GPS and Galileo satellite to receiver link. The first analysis uses this cycle slip count, to statistically estimate the probability of cycle slips as a function of mean σ_ϕ.

To gain a more complete view of the potential impact of GNSS scintillation, episodes with total loss of lock, in this study referred to as outages, were also identified. This was performed by computing the satellite positions from the navigation files and comparing the satellites that should be visible above 15 degrees elevation with the satellite observations that are actually present above 15 degrees in the ISMR files. The data has a 1 min resolution of satellite outages for the full data period, which is used together with the cycle slips in the second analysis.

As with all data collected in an uncontrolled environment, data are subject to error sources, some of which are challenging to identify and mitigate. Therefore, there can be GNSS signal outages which could be caused by other sources than the ionosphere. To support the analysis, a GNSS simulator was therefore used to investigate further.

2.4. Scintillation Simulation

The data set of this study is not large enough to completely isolate the effects of the constellations and signals. Therefore, the study is enhanced with a separate third analysis using a GNSS simulator to test GPS and Galileo signals under the exact same conditions in a fully controlled environment. The simulator is a Spirent GSS7000 unit [27], which can be used to emulate the effects of the same ionospheric scenario on the tracking of GPS and Galileo satellite signals in a PolaRx5S receiver. This is the same type of receiver that is used in the SWADO network.

To simulate the ionospheric conditions, a user command define (UCD) file was created and imported into the simulator module. The file consists of 13 columns where the last two define the carrier offset and code offset for specific satellites at specific times defined in the previous columns. A description of the UCD format and use of the simulator for ionospheric research at low latitudes can be found in [28].

The relationship between ionospheric TEC and the code propagation delay of a radio signal at GPS frequencies is given by Equation (8), while the phase propagation delay was given in Equation (2).

$$I_{R_f} = \frac{40.3}{f^2}TEC \tag{8}$$

In order to simulate TEC changes in the ionosphere, and thus changes in code delay and phase advance at the receiver, the code and carrier offsets are changed in accordance with their relationship to the frequency of the signal. These numbers for code and carrier offsets are input at the signal simulation rate using the input format of the UCD file.

To create a more realistic scenario, a snip of real carrier offset, observed during a scintillation event, was superimposed on a decrease in carrier offset, which together with an increase in code offset imitated an increase in TEC. The mix of scintillation and TEC increase was selected to simulate a realistic scintillation scenario with both rapid fluctuations and plasma drift. The scintillation profile and TEC are both scaled with respect to the radio frequency. The scintillation and TEC ramping scenario is repeated three times in the UCD files with increasing amplitude, to test where the GPS and Galileo links will break.

In the GSS7000 interface, a time was chosen when GPS satellite G16 and Galileo satellite E27 were both at high elevations for the duration of the simulation. Using the orbit parameters, the GSS7000 simulated the signal from one of the satellites for each constellation while imposing the carrier and code offsets from the UCD file on the respective signal. The simulated RF signal was fed into a PolaRx5S receiver, which computed ISMR and RINEX data files. The RINEX data were used to estimate relative TEC.

3. Results

3.1. Probability of Cycle Slips as a Function of σ_ϕ

The 15 min intervals with cycle slips were binned according to the mean σ_ϕ of the interval. The probability of cycle slips for each bin was found by dividing the number of intervals containing cycle slips by the total number of intervals. The result can be seen in Figure 3.

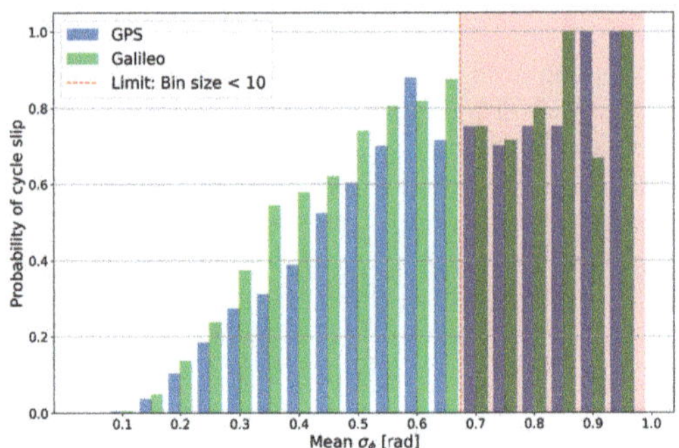

Figure 3. Probability of cycle slips as function of σ_ϕ for GPS and Galileo. The red area marks the bins containing less than ten values.

Figure 3 shows that the maximum chance of cycle slips lies for σ_ϕ from around 0.6 radians and up. For times with a mean σ_ϕ under 0.6 radians, the chance of cycle slips is seen to be consistently larger for Galileo than GPS. There can be several reasons for this which will be discussed in Section 4. The bins marked in red contain less than ten of the 15 min intervals. In some of the cases with high σ_ϕ values, the small amount of data in the bins results in a 100% probability of cycle slips. This illustrates a limitation of the data, namely that the data set does not cover a solar maximum, and therefore has a lack of very high scintillation events. The lack of events with high σ_ϕ can also be a result of losses of lock in the receiver during intense phase scintillation. Therefore, the number of outages in the data set is also investigated.

3.2. Cycle Slips and Outages in Numbers

Table 1 lists the number of cycle slips detected, the number of 15 min intervals that contain one or more cycle slips, and the number of outages for GPS and Galileo. As part of the analysis, the observation time for GPS and Galileo was also investigated. There are more GPS than Galileo satellites in the data set, but the Galileo satellites are on average visible for a slightly longer time. Overall the Galileo observation time was 77% of the total GPS observation time. This is partly due to the selection of satellites as discussed in Section 2.1. The correction was applied in the middle column of Table 1, to improve the comparability of GPS and Galileo. The correction involved multiplying the event number by the ratio of total events for Galileo divided by the total events for GPS. Hence, the middle column shows the expected number of cycle slips from GPS over a comparable observation interval to the Galileo results in column 3. The occurrence of cycle slips for GPS and Galileo shows that there are 3.6 times as many cycle slips for Galileo as for GPS when including the correction for differences in Galileo and GPS observation time. The results therefore indicate that the Galileo satellite signals experience more cycle slips (8183) than GPS satellite signals (2197). However, the Galileo satellite cycle slips occur over a similar number of 15 min intervals (1149) as that of GPS (921) over a comparable observation time.

Table 1. Number of cycle slips, number of 15 min intervals containing cycle slips, and number of outages for GPS and Galileo. The middle column is the number for GPS corrected for the difference in observation time for GPS and Galileo seen from the station KLQ2.

	GPS	GPS Corrected	Galileo
Number of cycle slips	2862	2197	8183
Intervals with cycle slips	1200	921	1149
Number of outages	6895	5292	2929
Sum of cycle slips and outages	9757	7489	11,112

As seen in Table 1 there are around 1.2 times as many intervals that contained cycle slips for Galileo as for GPS. Galileo has 7.1 cycle slips per 15 min interval containing cycle slips, while GPS only has 2.4 cycle slips per interval containing cycle slips. Although the GPS cycle slips are fewer in number, the Galileo cycle slips are spread across a similar number of time intervals. This indicates that there are more cycle slips occurring around the same scintillation events for Galileo than for GPS. As a result, the outages for GPS and Galileo were investigated. It is seen from Table 1 that there are 1.8 times more outages for GPS than Galileo over the entire one-year data period when taking the observation time into account. This indicates distinct ratios between cycle slips and outages for the two constellations. For Galileo 73.6% of the impacts are cycle slips and 26.4% are outages, while for GPS 29.3% of the impacts are cycle slips and 70.7% are outages. Considering the sum of cycle slips and outages in Table 1, it is worth noting that the sum of impacts for Galileo signals is larger than for GPS, but the impacts on GPS are generally more severe since a higher percentage of the impacts are signal outages.

3.3. Daily Variations in Cycle Slips and Outages

In Figure 4, the number of cycle slips and outages for GPS and Galileo are shown together with the mean σ_ϕ as a function of the time of day to investigate further the relation between phase scintillation and the receiver impacts. The difference in observation time for GPS and Galileo is taken into account also in Figure 4.

Figure 4 shows daily variations in the number of cycle slips and outages for both GPS and Galileo. Referring to Figure 4a,c, the cycle slip distribution trend for GPS and Galileo is seen to be similar to each other and both are following the trend of σ_ϕ throughout the day. Both GPS and Galileo have peaks in the number of cycle slips in the afternoon around 4 pm local time and again later just before midnight, with the latter peak being higher. The peaks at certain times of day are not likely to be the result of the rise and descent of specific satellites, since the visibility of the satellites shifts every day. Referring to Figure 4b,d, there may be a relationship between phase scintillation and outages for GPS, since there is evidence of peaks at 05–09, 14–17 and 20–23 in both outages and σ_ϕ; however, looking at Galileo, the relationship between σ_ϕ and outages is not evident. This suggests that there are some outages for Galileo which are not related to phase scintillation.

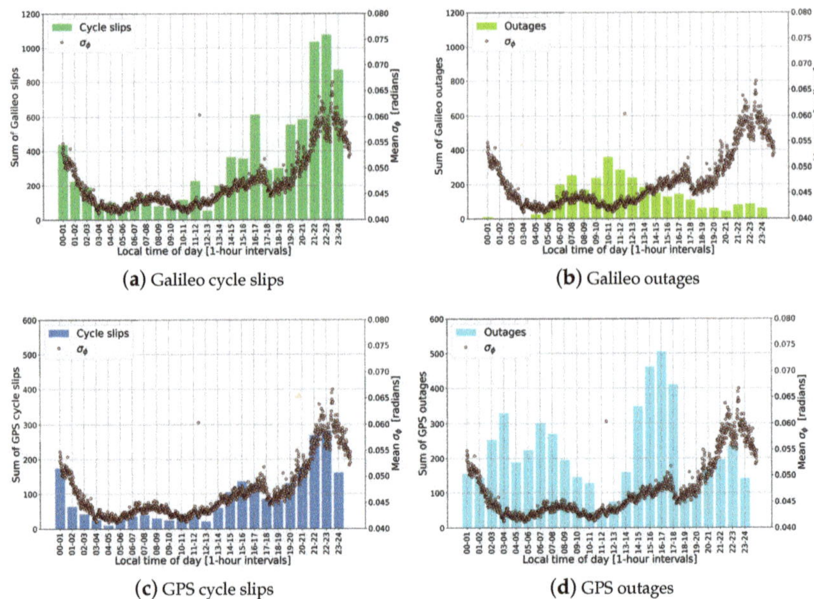

(**a**) Galileo cycle slips

(**b**) Galileo outages

(**c**) GPS cycle slips

(**d**) GPS outages

Figure 4. Sum of outages and cycle slips for GPS and Galileo for each hour of the day throughout the investigation period in local time (UTC-3 h). Note that the y-axis of (**c**,**d**) ranges from 0 to 600, while for (**a**,**b**) it reaches 1200. The mean σ_ϕ for each minute is overlaid, to show the relation between cycle slips, satellite outages, and phase scintillation.

3.4. Simulation of Scintillation

The results so far have indicated different responses to scintillation on GPS and Galileo signals. A controlled experiment was therefore carried out to investigate receiver response for these two signals, using a Spirent GSS7000 simulator. This allowed for identical ionospheric effects to be applied to different signals, while keeping all other signal effects benign, and to examine the receiver records. To set up this controlled test a simple ramp was applied to the TEC, with options to add in scintillation to the signals. The ramp in TEC was set to be close to the cycle slip detection levels as discussed in Section 2.3, and the receiver response to the simulated TEC ramps was investigated.

The input to the simulator is the UCD file described in Section 2.4. The time series of the TEC in the file is plotted in Figure 5a. The figure shows the constant TEC for the first two minutes and then the ramping up of TEC for two minutes, resulting in a TEC change of around 24 TECU per minute for the first two minutes. The subsequent two ramps are at 1.3 times higher gradients, so they should result in cycle slip detection.

The first TEC ramp is where the L1 and E1 signals have been subject to an increase in the code offset of 0.0013 m per 0.02 s. Corresponding higher values of code delays and phase advances resulting from the TEC ramp were applied to the L2 and E5a because these experience a larger effect due to their lower frequencies (see Equations (2) and (8)). Further, the scintillation profiles were also applied to the carrier phase offsets.

Figure 5b,c show the relative TEC computed for GPS L1 & L2 and Galileo E1 & E5a from the output RINEX files that were recorded from the simulated signals. Note that the initial value of TEC has an offset and the receiver takes time to gain lock. Nevertheless, the results show that the Galileo signals have been tracked correctly because they show the same TEC changes as were input (compare the TEC changes in Figure 5a–c). However, comparing the TEC changes in Figure 5b,c, it can be seen that the GPS signals result in an underestimation of the TEC that was simulated. Further investigation showed that the GPS receiver had a problem tracking the L2 signals through these large TEC changes and further, that even the TEC changes alone without scintillation applied created tracking issues for L2. This may well relate to the codeless tracking requirement on L2, which is not an issue for E5a. Consequently, the cycle slip detections on GPS are not going to be triggered because the TEC changes are being underestimated, thereby showing fewer cycle slips for GPS compared with Galileo.

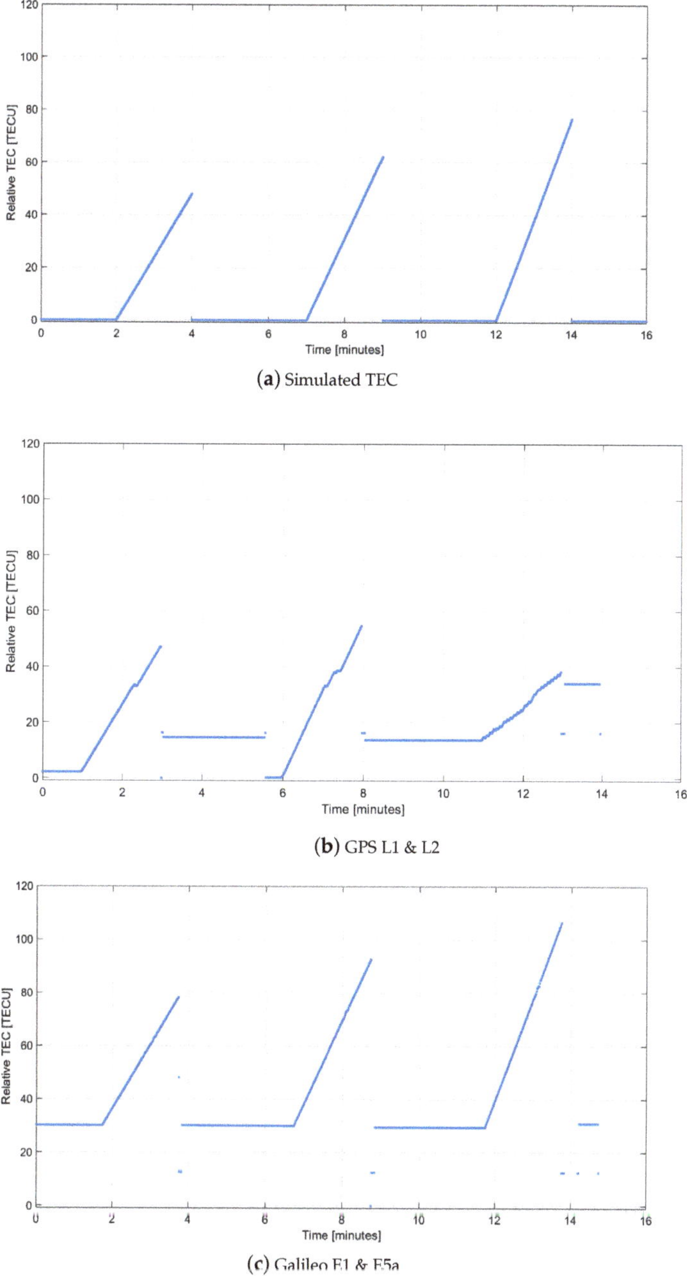

(**a**) Simulated TEC

(**b**) GPS L1 & L2

(**c**) Galileo E1 & E5a

Figure 5. Relative TEC based on L1 & L2 for GPS and E1 & E5a for Galileo.

4. Discussion

When using GNSS data for determining the impact of scintillation in the Arctic, it is a challenge that the geometry of the satellites seen from the GNSS stations is sub-optimal. Multiple GNSS constellations should therefore be considered in order to increase the amount of data. This is especially relevant in Greenland, where limited access to power and the internet restricts the possible locations of ground stations. This affects the spatial distribution of the data, resulting in spatial data gaps or regions covered mainly by low elevation data [29]. However, GNSS signals from different constellations cannot be assumed to be affected by scintillation in the same way.

This study has investigated the importance of considering the constellations used when relating σ_ϕ values to a scintillation impact level. Figure 3 showed that the probability of cycle slips increases at low scintillation levels for Galileo compared to GPS. Table 1 and Figure 4 further revealed that the tracking of signals from Galileo experiences more cycle slips, and that these cycle slips are more clustered in time than for GPS. In contrast, Table 1 showed a significantly higher number of outages of signal tracking for GPS than Galileo. These results indicate a more rapid succession of cycle slips for Galileo than for GPS. It is stated in [30] that a rapid succession of multiple cycle slips is a challenge with some cycle slip detection methods. However, the same cycle slip detection program was used for both constellations in this study, so this should not be the reason why fewer cycle slips per 15 min interval are detected for GPS compared with Galileo. It is more likely that the difference in the number of cycle slips detected for GPS and Galileo is inherent to the signal design and the tracking architecture of the receiver. This was evidenced by the results of the simulator analysis in Section 3.4.

The fact that the impact of scintillation on GNSS reliability is dependent on receiver signal tracking [31] means that preferably every receiver type used in the Arctic should be investigated to find the relationship between scintillation and its impact on receiver performance. In this study, the cycle slips were identified in data from a PolaRx5 receiver. This means that the cycle slip detection was based on data from a high-quality geodetic GNSS station, which is not necessarily representative of the receiver types applied by end users of an impact warning service. Even though the tracking architecture is not necessarily consistent between different receivers, the signal designs of the received signals are the same for all receivers capable of utilising the same frequencies.

Considering the signal designs, the Galileo E1 signal has a broader bandwidth than GPS L1, which could make E1 more vulnerable to cycle slips than L1. Furthermore, the frequency of E5a is lower than for L2, resulting in larger signal fluctuations on E5a [18,32]. Both arguments could point to the conclusion that there are more cycle slips detected for Galileo because the signals are more susceptible to cycle slips. However, it could also be a result of a higher signal strength for Galileo than for GPS, making the Galileo signal less prone to experiencing a complete loss of lock.

The simulations indicate that the increased amount of loss of lock on GPS compared to Galileo stems from L2 tracking issues. This is in line with previous studies [14,31,33]. In [14], it was concluded that, for the most severe of the high-latitude storms they examined, the tracking performance of the codeless receiver was significantly degraded, although the effect on the L1 tracking performance for both types of receivers was negligible. Likewise, it was found in a global scintillation study that the number of cycle slips on L1 was at least one level of magnitude lower than for L1–L2 [34]. A study on the effects of TEC jumps on GPS L1 and L2 in the auroral region also stated that failures in phase tracking on L2 can occur due to lower signal power or low signal-to-noise ratio on receivers for civil use since these have to utilise codeless tracking methods [35].

The GPS constellation does have alternative frequencies available today where codeless tracking techniques like cross-correlation are not needed. However, the two alternative frequencies L2C and L5 are not being transmitted from all GPS satellites yet. As of 12 April 2023, L2C is transmitted from 25 satellites and L5 from 18 satellites [36]. The complete use of these signals should be possible for L2C by 2023 and L5 by 2027 [37]. Even though

L2C and L5, once fully operational, could obviate the need for codeless or semi-codeless tracking techniques, it would require the end user segment to transition into using receivers compatible with L2C and L5. The impact of scintillation on the newer GPS signals L2C and L5 were investigated in a statistical analysis of GPS tracking during ionospheric scintillation and it was found that the L2C and L5 tracking seems to be less robust to scintillation than L1 [18].

The analyses of the study are limited to the GNSS signals utilised by the PolaRx5S and PolaRx5 receivers. The study is also limited by the selection of receiver type. Different receiver design and configuration will have an effect on the response to the ionospheric changes and the tracking response through scintillation events. The results of the study suggest that high precision GNSS users who depend on a high resilience should consider including the Galileo signals or the modern GPS signals rather than relying solely on GPS L1 and L2. The results also point to a need for weighting σ_ϕ measurements from GPS and Galileo separately if they are used in combination for evaluating the impact on the reliability of GNSS positioning. An example of this could be a future service where scintillation measurements are translated into impact maps or other types of local warning systems.

In the future, a simulator, like the one used in this study, can be useful in testing the response of different receiver types when imposing various ionospheric disturbance scenarios. To test a receiver thoroughly without a simulator, the receiver should be deployed in the region of interest, preferably for one or more solar cycles. With a simulator, a thorough testing of receivers and new signals types can be performed in significantly less time. Furthermore, the test is carried out in a controlled environment where the receiver is not left unprotected. In [7], a simulator was also used in order to test the impact of simulated scintillation on different receivers.

Further investigation into the relationship between cycle slips and the ionospheric plasma irregularities is an intriguing topic. It is well established that ionospheric plasma can cause rapid TEC changes that can result in phase jitter, and that irregularities can also cause diffractive effects that cause phase jitter and amplitude scintillation. The ionospheric state causing a scintillation event can be partly inferred using GPS observations but other instrumentation needs to be used to gather additional information about the underlying physical mechanisms. Interesting work by authors such as [38–41] have made significant advances in this topic. In future, it would be interesting to advance the understanding of cycle slip events by relating it to a model of the ionospheric irregularity structure together with a forward model of refractive and diffractive processes.

In the development towards an operational local warning system for Greenland, the next step would be to discuss the tolerance for false positives versus missing warnings with the user segment, and also to include more signals in the analysis. With time, it will only become more relevant to include the new GPS signals L2C and L5 in this type of analysis and perhaps other GNSS constellations to gain a better data coverage over Greenland.

5. Conclusions

Analyses pointing to differences in the impact of ionospheric scintillation on GPS and Galileo signals were presented in this paper. The impact was evaluated using cycle slips in a PolaRx5 receiver from GNET and outages in a PolaRx5S receiver from SWADO, both located in Kangerlussuaq ($66.9956°$N, $-50.6202°$E). For Galileo, a high percentage of the impacts were cycle slips as opposed to signal outages. This is in contrast to GPS behaviour, where the opposite is true. The Galileo phase tracking experienced a significantly larger number of cycle slips than for GPS, while GPS overall experienced more outages than Galileo did. Collectively, Galileo experienced the highest number of scintillation impacts as a result of the large number of cycle slips. Several possible explanations were investigated, and the results point towards the reason being issues in tracking the L2 signal, leading to fewer cycle slips being detected for GPS. Inconsistencies between ROTI based on L2 and

L2C were found in [42] and further discussed in [43], which provides supporting evidence for the results found in this paper. The conclusion of this study is therefore that there is a different response to ionospheric disturbances for GPS and Galileo due to the GPS signals underestimating the TEC during large TEC changes, as a result of the codeless tracking requirement on L2, which is not an issue for Galileo E5a.

Author Contributions: Conceptualisation, S.S.B., A.B.O.J. and L.S.; methodology, All authors; software, S.S.B. and C.N.M.; validation, S.S.B.; formal analysis, All authors; investigation, All authors; resources, All authors; writing—original draft preparation, S.S.B.; writing—review and editing, All authors; visualisation, S.S.B.; supervision, L.S., A.B.O.J., D.H.O., C.N.M. All authors have read and agreed to the published version of the manuscript.

Funding: C.N.M. has received funding from NERC grant number NE/W003074/1.

Data Availability Statement: The GNET dataset from SDFI which was analysed in this study is publicly available and can be found here: https://dataforsyningen.dk/data/4804, accessed on 8 November 2022. The SWADO data presented in this study are available on request from the corresponding author.

Acknowledgments: C.N.M. acknowledges support from a Royal Society Industry Fellowship. S.S.B. acknowledges DALO for funding part of her PhD studies. This study was made possible by access to data from the SWADO network which is operated by DTU on behalf of the Danish Defence and from the GNET owned by SDFI. Also important was access to the Spirent GSS7000 simulator at the University of Bath which made it possible to test the response of the GPS and Galileo signals to the same ionospheric disturbance under the exact same environment conditions. UNAVCO is acknowledged for developing and making the TEQC software version 2019Feb25 available.

Conflicts of Interest: A.B.O. Jensen was employed by the AJ Geomatics, L. Stenseng was employed by the AJ Geomatics and T. Pinto Jayawardena was employed by the Athena Space Limited. The remaining authors declare that the research was conducted in the absence of any commercial or financial relationships that could be construed as a potential conflict of interest.

Abbreviations

The following abbreviations are used in this manuscript:

GISTM	GNSS Ionospheric Scintillation and TEC Monitor
GLONASS	Global Navigation Satellite System
GNET	Greenland GPS Network
GNSS	Global Navigation Satellite System
GPS	Global Positioning System
IOD	Ionospheric Delay
ISMR	Ionospheric Scintillation Monitoring Record
RF	Radio Frequency
RINEX	Receiver Independent Exchange format
SWADO	Space Weather Forecasting for Arctic Defence Operations
TEC	Total Electron Content
TECU	TEC Units
TEQC	Translating, Editing and Quality Checking
UCD	User Command Define
UTC	Universal Time Coordinated

References

1. Titheridge, J.E. The diffraction of satellite signals by isolated ionospheric irregularities. *J. Atmos. Terr. Phys.* **1971**, *33*, 47–69. [CrossRef]
2. Kintner, P.M.; Ledvina, B.M.; de Paula, E.R. GPS and ionospheric scintillations. *Space Weather* **2007**, *5*, S09003. [CrossRef]
3. Jiao, Y.; Morton, Y.; Taylor, S.; Pelgrum, W. Characterization of high-latitude ionospheric scintillation of GPS signals. *Radio Sci.* **2013**, *48*, 698–708. [CrossRef]
4. Aarons, J. Global positioning system phase fluctuations at auroral latitudes. *J. Geophys. Res.* **1997**, *102*, 17219–17231. [CrossRef]
5. Prikryl, P.; Jayachandran, P.T.; Mushini, S.C.; Chadwick, R. Climatology of GPS phase scintillation and HF radar backscatter for the high-latitude ionosphere under solar minimum conditions. *Ann. Geophys.* **2011**, *29*, 377–392. [CrossRef]

6. Humphreys, T.E.; Psiaki, M.L.; Kintner, P.M. GPS Carrier Tracking Loop Performance in the presence of Ionospheric Scintillation. In Proceeding of the 18th International Technical Meeting of the Satellite Division of the Institute of Navigation, ION GNSS 2005, Long Beach, CA, USA, 13–16 September 2005.

7. Hinks, J.C.; Humphreys, T.E.; O'Hanlon, B.; Psiaki, M.L.; Kintner, P.M. Evaluating GPS receiver robustness to ionospheric scintillation. In Proceeding of the 21st International Technical Meeting of the Satellite Division of the Institute of Navigation, Ion GNSS 2008, Savannah, GA, USA, 16–19 September 2018; Volume 3, pp. 1390–1401.

8. Taylor, S.; Morton, Y.; Marcus, R.; Bourne, H.; Pelgrum, W.; Van Dierendonck, A.J. Ionosphere Scintillation Receivers Performance Based on High Latitude Experiments. In Proceeding of the ION 2013 Pacific PNT Meeting, Honolulu, HI, USA, 23–25 April 2013.

9. Pi, X.; Mannucci, A.J.; Lindqwister, U.J.; Ho, C.M. Monitoring of global ionospheric irregularities using the worldwide GPS network. *Geophys. Res. Lett.* **1997**, *24*, 2283–2286. [CrossRef]

10. Veettil, S.V.; Aquino, M.; Spogli, L. A statistical approach to estimate Global Navigation Satellite Systems (GNSS) receiver signal tracking performance in the presence of ionospheric scintillation. *J. Space Weather Space Clim.* **2018**, *8*, A51. [CrossRef]

11. Estey, L.; Wier, S. *Teqc Tutorial: Basic of Teqc Use and Teqc Products*; UNAVCO: Boulder, CO, USA, 2014.

12. Prikryl, P.; Jayachandran, P.T.; Mushini, S.C.; Pokhotelov, D.; MacDougall, J.W.; Donovan, E.; Spanswick, E.; St.-Maurice, J.-P. GPS, TEC, scintillation and cycle slips observed at high latitudes during solar minimum. *Ann. Geophys.* **2010**, *28*, 1307–1316. [CrossRef]

13. Prikryl, P.; Jayachandran, P.T.; Mushini, S.C.; Richardson, I.G. High-latitude GPS phase scintillation and cycle slips during high-speed solar wind streams and interplanetary coronal mass ejections: A superposed epoch analysis. *Earth Planets Space* **2014**, *66*, 62. [CrossRef]

14. Meggs, R.W.; Mitchell, C.N. GPS scintillation over the European Arctic during the November 2004 storms. *GPS Solut.* **2008**, *12*, 281–287. [CrossRef]

15. Datta-Barua, S.; Doherty, P.H.; Delay, S.H.; Dehel, T.; Klobuchar, J.A. Ionospheric Scintillation Effects on Single and Dual frequency GPS Positioning. In Proceedings of the 16th International Technical Meeting of the Satellite Division of the Institute of Navigation ION GPS/GNSS 2003, Portland, OR, USA, 9–12 September 2003; pp. 336–346.

16. Romero, R.; Linty, N.; Cristodaro, C.; Dovis, F.; Alfonsi, L. On the Use and Performance of new Galileo Signals for Ionospheric Scintillation Monitoring. In Proceeding of the Institute of Navigation Technical International Meeting (ITM 2017), Monterey, CA, USA, 30 January–2 February 2017.

17. Carrano, C.S.; Groves, K.M.; McNeil, W.J.; Doherty, P.H. Scintillation Characteristics Across the GPS Frequency Band. In Proceedings of the 25th International Technical Meeting of the Satellite Division of the Institute of Navigation (ION GNSS 2012), Nashville, TN, USA, 17–21 September 2012; pp. 1972–1989.

18. Delay, S.H.; Carrano, C.; Groves, K. A Statistical Analysis of GPS L1, L2 and L5 Tracking Performance During Ionospheric Scintillation. In Proceeding of the 2015 ION Pacific PNT Conference, Honolulu, HI, USA, 20–23 April 2015.

19. Hlubek, N.; Berdermann, J.; Wilken, V.; Gewies, S.; Jakowski, N.; Wassaie, M.; Damtie, B. Scintillation of the GPS, GLONASS, and Galileo signals at equatorial latitude. *J. Space Weather Space Clim.* **2014**, *4*, A22. [CrossRef]

20. Sinha, S.; Bhardwaj, S.C.; Vidyarthi, A.; Jassal, B.S. Ionospheric Scintillation analysis using ROT and ROTI for Slip Cycle Detection. In Proceedings of the 4th International Conference on Information Systems and Computer Networks (ISCON), Mathura, India, 21–22 November 2019.

21. Vankadara, R.K.; Jamjareegulgarn, P.; Seemala, G.K.; Siddiqui, M.I.H.; Panda, S.K. Trailing Equatorial Plasma Bubble Occurrences at a Low-Latitude Location through Multi-GNSS Slant TEC Depletions during the Strong Geomagnetic Storms in the Ascending Phase of the 25th Solar Cycle. *Remote Sens.* **2023**, *15*, 4944. [CrossRef]

22. Septentrio. *PolaRx5S Reference Guide*; Septentrio: Leuven, Belgium, 2020.

23. Yang, Z.; Morton, Y. Time lags in Ionospheric Scintillation Response to Geomagnetic Storms; Alaska Observations. In Proceedings of the 33rd International Technical Meeting of the Satellite Division of the Institute of Navigation (ION GNSS+ 2020), Online, 21–25 September 2020; pp. 3494–3501.

24. Forte, B.; Radicella, S.M. Problems in data treatment for ionospheric scintillation measurements. *Radio Sci.* **2002**, *37*, 6. [CrossRef]

25. Septentrio. *PolaRx5S User Manual*; Version 2.3; Septentrio: Leuven, Belgium, 2021.

26. Estey, L.H.; Meertens, C.M. TEQC: The Multi-Purpose Toolkit for GPS/GLONASS Data. *GPS Solut.* **1999**, *3*, 42–49. [CrossRef]

27. Spirent.com. Spirents Official Website. Available online: https://www.spirent.com/products/gnss-simulator-gss7000 (accessed on 29 August 2023).

28. Mohd Ali, A. GNSS in Aviation: Ionospheric Threats at Low Latitudes. Ph.D. Thesis, University of Bath, Bath, UK, 2018.

29. Beeck, S.S.; Jensen, A.B.O. ROTI maps of Greenland using kriging. *J. Geod. Sci.* **2021**, *11*, 83–94. [CrossRef]

30. Breitsch, B.; Morton, Y.J. Triple-Frequency GNSS Cycle Slip Detection Performance in Presence of Diffractive Ionosphere Scintillation. In Proceedings of the ION Position, Location and Navigation Symposium (PLANS), Portland, ON, USA, 20–23 April 2020.

31. Skone, S.; Knudsen, K.; de Jong, M. Limitations in GPS Receiver Tracking Performance Under Ionospheric Scintillation conditions. *Phys. Chem. Earth (A)* **2001**, *26*, 613–621. [CrossRef]

32. Hong, J.; Chung, J.; Kim, Y.H.; Park, J.; Kwon, H.; Kim, J.; Chol, J.; Kwak, Y. Characteristics of Ionospheric Irregularities Using GNSS Scintillation Indices Measured at Jang Bogo Station, Antarctica (74.62°S, 164.22°E). *Space Weather* **2020**, *18*, e2020SW002536. [CrossRef]

33. Nichols, J.; Hansen, A.; Walter, T.; Enge, P. High-Latitude Measurements of Ionospheric Scintillation Using the NSTB. *Navig. J. Inst. Navig.* **2000**, *47*, 2. [CrossRef]
34. Afraimovich, E.L.; Lesyuta, O.S.; Ushakov, I.I.; Voeykov, S.V. Geomagnetic storms and the occurrence of phase slips in the reception of GPS signals. *Ann. Geophys.* **2002**, *45*, 55–72. [CrossRef]
35. Chernyshov, A.A.; Miloch, W.J.; Jin, Y.; Zakharov, V.I. Relationship between TEC jumps and auroral substorms in the high-latitude ionosphere. *Sci. Rep.* **2020**, *10*, 6363. [CrossRef]
36. GPS.gov. Official U.S. Government Information about the Global Positioning System (GPS) and Related Topics. Available online: https://www.gps.gov/systems/gps/modernization/civilsignals/ (accessed on 6 July 2023).
37. Esper, M.; Chao, E.L.; Wolf, C.F. *Federal Radio Navigation Plan*; United States Department of Defense, Department of Transportation, Department of Homeland Security: Washington, DC, USA, 2019.
38. Deshpande, K.B.; Bust, G.S.; Clauer, C.R.; Rino, C.L.; Carrano, C.S. Satellite-beacon Ionospheric-scintillation Global Model of the upper Atmosphere (SIGMA) I: High-latitude sensitivity study of the model parameters. *J. Geophys. Res. Space Phys.* **2014**, *119*, 4026–4043. [CrossRef]
39. Deshpande, K.B.; Bust, G.S.; Clauer, C.R.; Scales, W.A.; Frissell, N.A.; Ruohoniemi, J.M.; Spogli, L.; Mitchell, C.; Weatherwax, A.T. Satellite-beacon Ionospheric-scintillation Global Model of the upper Atmosphere (SIGMA) II: Inverse modeling with high-latitude observations to deduce irregularity physics. *J. Geophys. Res. Space Phys.* **2016**, *121*, 9188–9203. [CrossRef]
40. Deshpande, K.B.; Zettergren, M.D. Satellite-Beacon Ionospheric-Scintillation Global Model of the upper Atmosphere (SIGMA) III: Scintillation simulation using a physics-based plasma model. *Geophys. Res. Lett.* **2019**, *46*, 4564–4572. [CrossRef]
41. Zernov, N.N.; Driuk, A.V. Coherence properties of high-frequency wave field propagating through inhomogeneous ionosphere with anisotropic random irregularities of electron density: 1. Theoretical background. *J. Atmos. Sol.-Terr. Phys.* **2020**, *205*, 105313. [CrossRef]
42. Yang, Z.; Liu, Z. Investigating the inconsistency of ionospheric ROTI indices derived from GPS modernized L2C and legacy L2 P(Y) signals at low-latitude regions. *GPS Solut.* **2017**, *21*, 783–796. [CrossRef]
43. McCaffrey, A.M.; Jayachandran, P.T.; Langley, R.B.; Sleewaegen, J.-M. On the accuracy of the GPS L2 observable for ionospheric monitoring. *GPS Solut.* **2018**, *22*, 23. [CrossRef]

Article

An Updating of the IONORT Tool to Perform a High-Frequency Ionospheric Ray Tracing

Marco Pietrella [1,*], Michael Pezzopane [1], Alessandro Pignatelli [1], Alessio Pignalberi [1] and Alessandro Settimi [2]

1   Istituto Nazionale di Geofisica e Vulcanologia, Via di Vigna Murata 605, 00143 Rome, Italy; michael.pezzopane@ingv.it (M.P.); alessandro.pignatelli@ingv.it (A.P.); alessio.pignalberi@ingv.it (A.P.)
2   Istituto di Istruzione Superiore Statale (I.I.S.S.) "J. von Neumann" RMIS022001, 00143 Rome, Italy; alessandro.settimi1@posta.istruzione.it
*   Correspondence: marco.pietrella@ingv.it

Abstract: This paper describes the main updates characterizing the new version of IONORT (IONOsperic Ray Tracing), a software tool developed at Istituto Nazionale di Geofisica e Vulcanologia to determine both the path of a high frequency (HF) radio wave propagating in the ionospheric medium, and the group time delay of the wave itself along the path. One of the main changes concerns the replacement of a regional three-dimensional electron density matrix, which was previously taken as input to represent the ionosphere, with a global one. Therefore, it is now possible to carry out different ray tracings from whatever point of the Earth's surface, simply by selecting suitable loop cycles thanks to the new ray tracing graphical user interface (GUI). At the same time, thanks to a homing GUI, it is also possible to generate synthetic oblique ionograms for whatever radio link chosen by the user. Both ray tracing and homing GUIs will be described in detail providing at the same time some practical examples of their use for different regions. IONORT software finds practical application in the planning of HF radio links, exploiting the sky wave, through an accurate and thorough knowledge of the ionospheric medium. HF radio waves users, including broadcasting and civil aviation, would benefit from the use of the IONORT software (version 2023.10).

Keywords: ionospheric HF ray tracing; homing; three-dimensional electron density; IRI model; ISP model; oblique sounding; space weather

Citation: Pietrella, M.; Pezzopane, M.; Pignatelli, A.; Pignalberi, A.; Settimi, A. An Updating of the IONORT Tool to Perform a High-Frequency Ionospheric Ray Tracing. *Remote Sens.* **2023**, *15*, 5111. https://doi.org/10.3390/rs15215111

Academic Editor: Yunbin Yuan

Received: 13 September 2023
Revised: 12 October 2023
Accepted: 20 October 2023
Published: 25 October 2023

1. Introduction

Through the ray tracing technique, it is possible to establish the travel path of radio waves propagating in an anisotropic and inhomogeneous medium [1]. When the medium is the terrestrial ionosphere, then we talk about *ionospheric ray tracing*. Nevertheless, what do we really mean with the terminology *ionospheric ray tracing*.

Since the mid 1950s, a set of differential equations implementing Hamilton's equations were formulated to analyze the propagation in the ionosphere of electromagnetic waves in the high frequency (HF) band (3–30 MHz) [2]. These equations cannot be solved analytically; thus, they were reformulated, and their solution was found via numerical integration techniques first by Haselgrove and Haselgrove [3] and then by Haselgrove [4]. The solutions of Haselgrove's equations provide the paths of dominant energy flow, which are referred to as *ionospheric ray tracing*.

In other terms, the numerical solution of Haselgrove's equations allows determining, once the initial point and the azimuth of the ray have been specified, the path of a HF radio wave propagating through the ionosphere from a well specified transmitting point (Tx) to an unknown landing point (Rx), being Tx and Rx any pair of points on the Earth's surface.

The numerical approach provides a very accurate ionospheric ray tracing when a three-dimensional (3-D) electron density matrix and a geomagnetic field model are considered. In this case, we speak about a 3-D magnetoionic ray tracing. Nevertheless, the numerical ray tracing for the 3-D magnetoionic case requires very high computational costs. This is why simplified software packages, including two-dimensional (2-D) and 3-D numerical ray tracing neglecting geomagnetic field effects, and analytical ray tracings have been developed over the years.

The first codes written for the ionospheric ray tracing computation date back to the 1960s, and they provided just a numerical output with runtimes relatively long [5–7]. In particular, it is worth mentioning the numerical ray tracing algorithm developed by Jones [8], which was improved and modified by Jones and Stephenson [9], that became one of the best-known numerical ionospheric ray tracing codes readily available to be executed on a computer. Later on, other ray tracing algorithms were developed. Reilly [10], after a substantial revision of an earlier 3-D ionospheric ray tracing program, investigated the performance of upgraded versions of different numerical integration techniques, comparing measured oblique ionograms with synthetic ones obtained for near-vertical incidence sky wave (NVIS) conditions. In the late 1990s, the numerical Homing Ray Tracing software package developed by Norman et al. [11] was compared with the Segmented Method for 2-D Analytic Ray Tracing [12]. The latter, based on the Fully Analytic Ionospheric Model [13], outputs a reliable ray tracing through complicated horizontal gradients varying with altitude along the ray path direction. Another 2-D numerical ray tracing formulation proposed by Coleman [14] has been shown to be effective in simulating oblique and backscatter ionograms as well as the output of some ionosondes, supporting Over-The-Horizon Radar (OTHR) operations. Generally speaking, HF direction finding systems (HFDF) can provide valuable information about the angles of arrival of signals of a given frequency; this can be suitably exploited as input by a ray tracing program to determine the signal source (i.e., the transmitting site), once the electron density distribution between the HFDF site and the transmitter is known. To this regard, Huang and Reinisch [15], based on ionograms and sky-map measurements taken at a HFDF site, reconstructed in real time the 3-D electron density distribution between the HFDF site and the transmitter locations, and then used these data as input for a ray tracing program. So doing, from the measured arrival angles at the HFDF site, footprints for both the ordinary and extraordinary ray tracing in the frequency range 7–11 MHz were found to cluster around the actual position of the transmitter within an error of $\approx$15–55 km, depending on whether the measured ionospheric tilt is considered. More recently, Haselgrove's equations were properly optimized and adapted for numerical integration on powerful computers, and corresponding results about ray tracing, ray homing, HF propagation-channel modeling, coordinate registration for OTHR, and the detrimental effects that travelling ionospheric disturbances (TIDs) have on an OTHR system were discussed [16]. A numerical and stepped ray tracing method, working on the empirical 3-D electron density model TWIM [17], was developed on the basis of ionospheric radio occultation data, and was successfully evaluated for oblique transmissions comparing synthetic ionograms with measured ones at Chung-Li (24.97°N, 121.19°E) [18]. The numerical 3-D ray tracing program developed by Jones and Stephenson [9] was also applied to calculate the ray paths through a polar ionosphere including patches and arcs of enhanced electron density; corresponding simulations showed satisfactory results about the ground coverage for a transmitter located at Cambridge Bay (69.1°N, 105.1°E) [19].

The Provision of High-frequency Raytracing Laboratory for Propagation studies (PHaRLAP) tool developed by Cervera and Harris [20] for studying and modelling the propagation of HF radio waves in the Earth's ionosphere, proved to be effective in reproducing a small and a larger imposed ionospheric tilt, as well as in modeling the signatures of a TID observed on both ordinary and extraordinary traces of synthetic ionograms obtained under NVIS propagation conditions [20]. PHaRLAP was also used to model the propagation of HF signals in studies turned to the assessment of climatological models to be used for the OTHR [21], as well as to model the interference environment in the HF band, which is an important factor for both the design and the development of OTHR systems [22,23]. A modified version of the Jones and Stephenson [9] ray tracing code, taking as input different profiles of electron density derived by the International Reference

Ionosphere (IRI) model ([24] and reference therein), was used to simulate the Super Dual Auroral Radar Network (SuperDARN) at African equatorial latitudes [25]. The corresponding analysis of ray tracing results obtained for an azimuth angle ranging between 1° and 360°, an elevation angle ranging between 1° and 90°, and frequencies equal to 12, 16, 20, and 24 MHz, has shown that ionospheric backscatter, as well as ground scatter, can be achieved by a SuperDARN-type radar at equatorial latitudes, in the east–west azimuthal direction [26].

When both the Tx and the Rx are known a priori, we talk about a *homing* approach. This method consists in sending out signals characterized by a wide range of frequencies (f), elevation angles (Δ), and azimuth angle (α); the arrival point of the ray (characterized by a definite triplet: f, α, Δ) can be approximated with Rx only whether it is relatively close to the Rx position. How close the landing point is to the receiver is established evaluating whether some logical conditions have been fulfilled. However, as an alternative to the homing method, a form of variational principle could be used as the basis for calculating the ionospheric ray tracing when the landing points are known, as clearly demonstrated by Coleman [27] for several ionospheric scenarios. The homing method is, at the current time, the main approach to perform a point-to-point ionospheric ray tracing. Strangeways [28] presented a homing method capable of obtaining a precise bending of the ray paths, their phases and group delays, for both terrestrial HF links and Earth-to-satellite paths, taking into account also the effects of horizontal electron density gradients. Kashcheyev et al. [29] used a numerical homing ray tracing algorithm to estimate the impact that the second- and third-order ionospheric residuals errors have on the Global Navigation Satellite System (GNSS) positioning. They found that the greatest errors are observed when the ray paths cross the crest of the equatorial ionization anomaly. Abdullah et al. [30] considered a reference and a mobile station spaced 10 km apart to estimate the ionospheric differential error associated to ray paths of Global Positioning System (GPS) signals using the Jones and Stephenson [9] 3-D ray tracing program properly modified, and including an accurate homing procedure. They found that the differential delay is of the order of 1–5 cm at low elevation angles for both north–south and east–west directions; an amount that, even if usually neglected by the standard differential GPS method, can be important for improving the positioning results that are obtained by the single-frequency GPS users. We conclude this excursus by saying that the ionospheric ray tracing finds its main applications where a detailed knowledge of the ray path is requested; for instance, when there is a need either to predict operating frequencies or to manage HF radio communications, as well as for experimental studies on OTHR systems, single station location, and HFDF [31]. The upgrade of a whatever synoptic model with real-time ionospheric data recorded for example by a network of either ground-based ionosondes or GNSS receivers (e.g., [32–38] and reference therein), provides a more reliable specification of the ionospheric conditions. This is essential to enhance, through the ray tracing tool, the operation and reliability of sky-wave radio communications and radar systems beyond the line-of-sight, especially in case of Space Weather events.

At the Istituto Nazionale di Geofisica e Vulcanologia, an applicative software tool named IONORT (IONOsperic Ray Tracing) was proposed by Azzarone et al. [39] and used for more than a decade. Since then, the reliability of IONORT has been tested comparing synthetic and measured oblique ionograms [40,41], considering as the ionospheric background that provided by the assimilative ISP (IRI-SIRMUP-PROFILE) regional model [33,34]. In this paper, after describing the mathematical background behind IONORT, also giving a short overview of the numerical integration methods embedded in the code, we present the main upgrades recently included in IONORT. First of all, the original Fortran 77 source code was completely rewritten in Fortran 90, and carefully debugged. Secondly, the applicability of IONORT has been extended from a regional to a global scale. Finally, we present the new graphical user interfaces (GUIs) of IONORT, including a novel GUI for homing applications.

Section 2 describes the basic mathematical background of the integration computational code of IONORT. Section 3 discusses the consistency checks of the IONORT integration algorithm. Section 4 presents and discusses the main upgrades of the IONORT software package. Sections 5 and 6 present and discuss some ray tracing and homing results. Conclusive remarks are the subject of Section 7.

2. IONORT Mathematical Background

The core of IONORT is based on the versatile computer ray tracing program developed by Jones and Stephenson [9] to trace the path of radio waves propagating in the ionosphere, that is an anisotropic, dispersive, and varying in time medium, whose index of refraction changes continuously in space. A short description of fundamental mathematical aspects on which the IONORT algorithm is based will be given. This description is, however, not exhaustive, and the most interested reader is referred to the work by Jones and Stephenson [9] for further information.

Jones and Stephenson [9] adopted the Haselgrove's equations [3,4] for the development of their 3-D ray tracing computer program. These equations consist of a system of six coupled first-order differential equations with Hamiltonian formalism. For a geocentric spherical coordinate system (Figure 1), the first-order differential equations system assumes the following form:

$$
\begin{cases}
\dfrac{d\rho}{dP'} = -\dfrac{1}{c}\dfrac{\partial H/\partial k_\rho}{\partial H/\partial \omega}, \\[2mm]
\dfrac{d\theta}{dP'} = -\dfrac{1}{\rho \cdot c}\dfrac{\partial H/\partial k_\theta}{\partial H/\partial \omega}, \\[2mm]
\dfrac{d\varphi}{dP'} = -\dfrac{1}{\rho \cdot c \sin\theta}\dfrac{\partial H/\partial k_\varphi}{\partial H/\partial \omega}, \\[2mm]
\dfrac{dk_\rho}{dP'} = \dfrac{1}{c}\dfrac{\partial H/\partial \rho}{\partial H/\partial \omega} + k_\theta \dfrac{d\theta}{dP'} + k_\varphi \sin\theta \dfrac{d\varphi}{dP'}, \\[2mm]
\dfrac{dk_\theta}{dP'} = \dfrac{1}{\rho}\left(\dfrac{1}{c}\dfrac{\partial H/\partial \theta}{\partial H/\partial \omega} - k_\theta \dfrac{d\rho}{dP'} + k_\varphi \cdot \rho \cos\theta \dfrac{d\varphi}{dP'}\right), \\[2mm]
\dfrac{dk_\varphi}{dP'} = \dfrac{1}{\rho \sin\theta}\left(\dfrac{1}{c}\dfrac{\partial H/\partial \varphi}{\partial H/\partial \omega} - k_\varphi \cdot \sin\theta \dfrac{d\rho}{dP'} - k_\varphi \cdot \rho \cos\theta \dfrac{d\theta}{dP'}\right);
\end{cases}
\tag{1}
$$

where ρ is the distance between the origin O of the geocentric reference system and the point M (see Figure 1); θ and φ are the colatitude and the longitude in spherical polar coordinates, respectively; k_ρ, k_θ, and k_φ are the components of the wave vector $\mathbf{k}$ orthogonal to the wave front in the ρ, θ, and φ direction, respectively; H is the Hamiltonian, $P' = c \cdot t$ is the group path length, being c the speed of an electromagnetic wave in the vacuum, and t the group time delay; $\omega = 2\pi f$ is the angular wave frequency, being f the wave frequency.

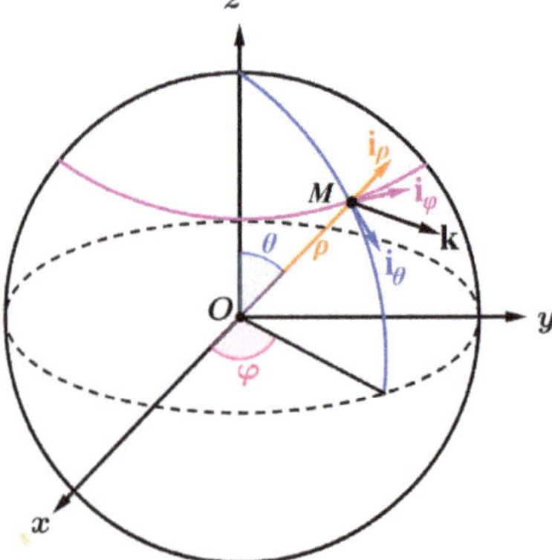

Figure 1. Geocentric reference system in spherical (ρ, θ, φ) and Cartesian (x, y, z) coordinates. $\mathbf{k}$ is the wave vector, with corresponding projections along the versors $\mathbf{i}_\rho$ (orange), $\mathbf{i}_\theta$ (blue), and $\mathbf{i}_\varphi$ (purple).

The system (1) is the one considered for calculations. Its solution allows determining the coordinates (ρ, θ, φ) reached by the wave vector, its three components (k_ρ, k_θ, k_φ), and the group time delay (t). In (1), the group path $P' = c \cdot t$ is the independent variable, because the derivatives with respect to P' do not depend on the choice of the Hamiltonian function. This is convenient because it allows the ray tracing program taking smaller steps near the reflection point where the calculations are more critical, and a major accuracy is requested [9]. Since analytical solutions of the system (1) are not possible, numerical solutions are found by means of numerical integration techniques.

2.1. The Hamiltonian

With regard to the choice and calculation of H some issues arise. In a dissipative medium such as the ionosphere, the dispersion relation is complex, which means that H is in general complex and consequently the solutions of the system (1), i.e., the coordinates of the ray path are complex. In this case, H in the reference system of Figure 1 should be written as:

$$H = H\left(t, \rho, \theta, \varphi, k_\rho, k_\theta, k_\varphi, n^2\right) = \left\{ \frac{1}{2}\left[\frac{c^2}{\omega^2}\left(k_\rho^2 + k_\theta^2 + k_\varphi^2\right) - n^2 \right] \right\}, \tag{2}$$

where k_ρ, k_θ, k_φ, and n are complex quantities.

n^2 is the square of the phase refractive index for an electromagnetic wave propagating in a cold magnetized plasma with collisions, which is calculated by the following Appleton–Hartree equation:

$$n^2 = 1 - \frac{X}{1 - iZ - \frac{Y_T^2}{2(1-iZ-X)} \pm \sqrt{\frac{Y_T^4}{4(1-iZ-X)^2} + Y_L^2}}, \tag{3}$$

being $X = (f_N/f)^2$ where f_N is the plasma frequency, $Z = \nu/\omega$ where ν is the electron collision frequency, $Y_T = (f_H/f)\sin\psi$ and $Y_L = (f_H/f)\cos\psi$, where f_H is the gyrofrequency and ψ the angle between the normal to the wave front and the geomagnetic field, and $i = (-1)^{1/2}$ is the imaginary unit.

Due to the birefringence of the ionospheric plasma, Equation (3) provides two refractive indices: n_{ord} for the ordinary ray and n_{ext} for the extraordinary ray, where both of them are complex quantities (being $n_{\text{ord}} = \mu_{\text{ord}} - i\chi_{\text{ord}}$ and $n_{\text{ext}} = \mu_{\text{ext}} - i\chi_{\text{ext}}$) [1]. The two refractive indices are obtained from Equation (3) through the choice of either the positive or the negative sign in the denominator, which must be decided applying the so-called Booker's rule [42]. Once the Booker's critical frequency $\omega_c = (|e|B/2m_e)\sin^2(\psi)/\cos(\psi)$ is defined, where B is the geomagnetic field magnitude, e is the electron charge, and m_e is the electron mass, Booker's rule states that if $\omega_c/\nu < 1$, to achieve the continuity of μ_{ord} and χ_{ord}, in Equation (3) the positive sign must be adopted for both $X < 1$ and $X > 1$, while to achieve the continuity of μ_{ext} and χ_{ext}, in Equation (3) the negative sign must be adopted for both $X < 1$ and $X > 1$. Instead, if $\omega_c/\nu > 1$, to achieve the continuity of μ_{ord} and χ_{ord}, in Equation (3) the positive and the negative sign must be adopted, respectively, for $X < 1$ and for $X > 1$; while to achieve the continuity of μ_{ext} and χ_{ext}, in Equation (3) the negative and the positive sign must be adopted, respectively, for $X < 1$ and for $X > 1$ (e.g., [43] and reference therein).

The ray tracing in a complex space is necessary when low-frequency radio waves propagate in the ionospheric D layer, where they suffer significant energy losses. In case of HF radio waves, these losses are smaller, and their effect is only an attenuation of the signal. This is why it is desirable that the system (1) have real solutions. This is accomplished by considering H as:

$$H = H(t, \rho, \theta, \varphi, k_\rho, k_\theta, k_\varphi, n^2) = \left\{ \frac{1}{2}\left[\frac{c^2}{\omega^2}(k'_\rho + k'_\theta + k'_\varphi) - \text{Re}(n^2) \right] \right\}, \tag{4}$$

where the real part of the complex phase refractive index is obtained assuming $\nu << \omega$, which means that the electron collision frequency is much smaller than the wave angular frequency. In this case, in Equation (3) the term Z is negligible. Moreover, by expliciting in Equation (3) the magnetoionic parameters X, Y_T, and Y_L, i.e., writing them in terms of the plasma frequency $f_N = (N_e e^2/4\pi^2 \varepsilon_0 m_e)^{1/2}$ and the gyrofrequency $f_H = |e|B/2\pi m_e$, the real part of the phase refractive index assumes the following form:

$$
n^2 = 1 - \cfrac{\dfrac{N_e(h)e^2}{4\pi^2 \varepsilon_0 m_e f^2}}{1 - \dfrac{\left(\dfrac{|e|B}{2\pi m_e f}\sin\psi\right)^2}{2\left(1 - \dfrac{N_e(h)e^2}{4\pi^2 \varepsilon_0 m_e f^2}\right)} \pm \sqrt{\dfrac{\left(\dfrac{|e|B}{2\pi m_e f}\sin\psi\right)^4}{4\left(1 - \dfrac{N_e(h)e^2}{4\pi^2 \varepsilon_0 m_e f^2}\right)^2} + \left(\dfrac{|e|B}{2\pi m_e f}\cos\psi\right)^2}} ,
\tag{5}
$$

where the signs (+) and (−) are associated to the ordinary and extraordinary propagation mode, respectively; $N_e(h)$ is the electron density, which is function of the real height h, and ε_0 is the dielectric constant in the vacuum.

Equation (5) clearly highlights how a 3-D electron density specification of the ionosphere, as well as an Earth's magnetic field model, are fundamental elements to obtain accurate ray tracing and homing results in the hypothesis of propagation in a collisionless ionospheric plasma.

Anyway, it must be noted that the Hamiltonian of Equation (4), which is function of the phase refractive index given by Equation (5), cannot be used when a "spitze" phenomenon occurs. This phenomenon consists of a cusp characterizing the ray path at the reflection point. It occurs when the elevation angles of the electromagnetic waves are very large, to such an extent to form a critical angle depending on the propagating wave frequency, the geomagnetic field intensity, and its inclination [44]. In this case, the phase refractive index, along with some of its derivatives, cannot be determined at the "spitze" and the Hamiltonian of Equation (4) is replaced by another Hamiltonian (see Equation (22) in [9]) which is not based on the phase refractive index.

2.2. Numerical Integration Methods Embedded in the IONORT Code

Generally, finding an analytical solution of an ordinary first-order differential equation is not always possible and, consequently, it is necessary to rely on numerical integration methods. The purpose of numerical methods is to find an approximate solution for a certain finite time interval. To solve numerically the system (1), the Runge-Kutta (R-K) method of the fourth order (R-K 4) [45] has been implemented in the IONORT source code, because it provides the best accuracy/computational cost ratio, ensuring an error for each integration step of $\approx o(h^5)$. Another numerical integration technique that has been implemented in the IONORT source code to solve numerically the system (1) is the Adams–Bashforth (A-B)/Adams–Moulton (A-M) predictor–corrector method, where the four-step A-B 4 method [46] is the predictor and the three-step A-M 3 method [47] is the corrector. A-B 4 and A-M 3 form a predictor–corrector pair that is a particular case of a broad category of multistep predictor–corrector methods exploiting the information from the previous steps to obtain a better approximation in the next step.

3. Consistency of the IONORT Integration Algorithm

The group time delay t is the parameter that usually a ray tracing program outputs when the numerical integration stops, and the ray has reached the ground after being reflected in the ionosphere. In principle, to test the goodness of any ray tracing algorithm, t could be compared to the group time delay as measured by OTHR applications, synchronized oblique sounding campaigns, or backscattering ionospheric soundings. Unfortunately, these kinds of measurements are uncommon. Nevertheless, taking into account the secant law, and the Breit–Tuve and Martyn propagation equivalent path theorems (e.g., [31]), an escamotage can be used. To this aim, Figure 2 shows the oblique path Tx-B_i-Rx of a HF wave propagating through the ionosphere with a frequency $f_{ob,i}$, at an elevation angle Δ, reaching the receiver after being reflected at the real height h_i, i.e., at B_i. ϕ is the incidence angle of the wave on the ionosphere that, in the approximation of a flat Earth, is simply the complementary angle of Δ. The subscript i varies from 1 to N, with N indicating different oblique frequencies reflected at different heights; consequently, also the apex of the real oblique path (B_i) and virtual triangular equivalent path (C_i) will be different, depending on how much the frequency $f_{ob,i}$ penetrates the ionosphere.

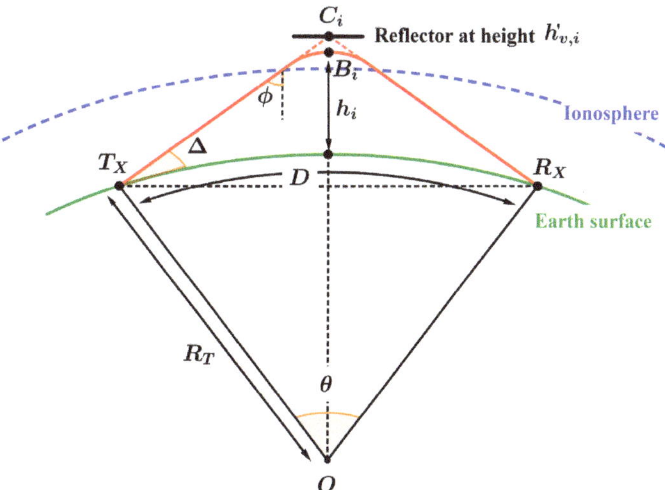

Figure 2. Sketch illustrating the geometry of an oblique circuit. Tx-B_i-Rx is the real oblique path. Tx-C_i-Rx is the equivalent virtual path. Δ and ϕ are, respectively, the elevation angle and the incidence angle on the ionosphere of the wave. h_i is the real height of reflection, while C_i denotes the height of the virtual reflector $h'_{v,i}$. D is the distance Tx-Rx, which is the ground range. R_T is the Earth's radius. θ is the angle at the center of the Earth subtended by the path.

The secant law states that $f_{ob,i} = f_{v,i} \sec\phi_i$, where $f_{v,i}$ is the frequency that is vertically reflected at the height h_i in the middle point B_i of the radio link Tx-Rx.

The Breit–Tuve theorem states that, assuming a monotonic increasing trend of the electron density profile, the group time $t_{i,\text{Tx-Bi-Rx}}$ taken by the wave of frequency $f_{ob,i}$ in travelling the real oblique path is equal to the time that such wave takes in travelling the virtual triangular equivalent path with speed c, that is $t_{i,\text{Tx-Bi-Rx}} = t_{i,\text{Tx-Ci-Rx}}$. Therefore, if N different oblique frequencies ($f_{ob,i}$ with $i = 1, \ldots, N$) are sent out with the same elevation angle Δ, N different real oblique paths and corresponding virtual triangular equivalent paths will be obtained so as to have $t_{i,\text{Tx-Bi-Rx}} = t_{i,\text{Tx-Ci-Rx}}$.

This relationship suggests that assuming as real measurements the times $t_{i,\text{Tx-Ci-Rx}}$, we can compare these to the times $t_{i,\text{Tx-Bi-Rx}}$ output by IONORT, to evaluate its performance. It is clear that to do this we need to know the position of the various reflectors, which is the height of the virtual reflection points C_i. This is possible thanks to the Martyn's theorem which establishes that, assuming valid the secant law, the virtual vertical height of reflection $h'_{v,i}$ is equal to the height $h'_{ob,i}$ of the virtual equivalent triangular path for the oblique signal, i.e., $h'_{v,i} = h'_{ob,i}$.

If a numerical electron density profile $N_e = N_e(h_i)$ is known in the middle point of the considered radio link Tx-Rx, and neglecting the geomagnetic field in Equation (5), it is possible to calculate the group refractive index n_G as $1/n$:

$$n_G = \frac{1}{\left(1 - \frac{N_e(h_i)e^2}{4\pi^2\varepsilon_0 m_e f_{v,i}^2}\right)^{1/2}};\qquad(6)$$

the value of $h'_{v,i}$ is then calculated by solving numerically the following integral:

$$h'_{v,i}(f_{v,i}) = h_b + \int_{h_b}^{h_i} n_G(f_{v,i}, N_{e,i}(h_i))\mathrm{d}h = \int_{h_b}^{h_i} \frac{1}{\left(1 - \frac{N_e(h_i)e^2}{4\pi^2\varepsilon_0 m_e f_{v,i}^2}\right)^{1/2}}\mathrm{d}h, \tag{7}$$

where h_b and h_i are, respectively, the base of the ionosphere and the real height of reflection of the vertical path from which the wave of frequency $f_{v,i}$ reflects coming back to the ground. In Equation (7), $f_{v,i}$ is a parameter during the integration; therefore, the knowledge of N_e allows calculating the heights $h'_{v,i}$ of N reflectors. Once the values of $h'_{v,i}(f_{v,i})$ are known, it is then possible to calculate $t_{i,\text{Tx-Ci-Rx}}$ as follows:

$$t_{i,\text{Tx}-Ci-\text{Rx}} = \frac{2\sin\left(\frac{D}{2R_T}\right) \cdot \left[R_T + h'_{v,i}\right]}{c \cdot \cos(\Delta)}, \tag{8}$$

where D, R_T and Δ are known.

Specifically, in order to fully test the IONORT algorithm, a numerical 3-D electron density background was considered, and three runs have been carried out, for three different values of Δ: 18°, 30°, and 45°. For each run, IONORT takes as input parameters values of $f_{ob,i}$ ranging between 3 and 30 MHz. In this way, three different datasets of IONORT times are obtained. Subsequently, each dataset was compared with the corresponding one calculated through Equation (8). According to the Breit–Tuve theorem, datasets corresponding to the same elevation angle should be equal. As a consequence, the following comparison $t_{i,\text{Tx-Bi-Rx}}$ vs. $t_{i,\text{Tx-Ci-Rx}}$ has been carried out for each of the three chosen elevation angles.

An example of this analysis is shown in Figure 3 for $\Delta = 30°$, where the percentage relative error:

$$\Delta t_i = \frac{\left|t_{i,\text{Tx}-B_i-\text{Rx}} - t_{i,\text{Tx}-C_i-\text{Rx}}\right|}{t_{i,\text{Tx}-C_i-\text{Rx}}} \cdot 100 \tag{9}$$

is also reported.

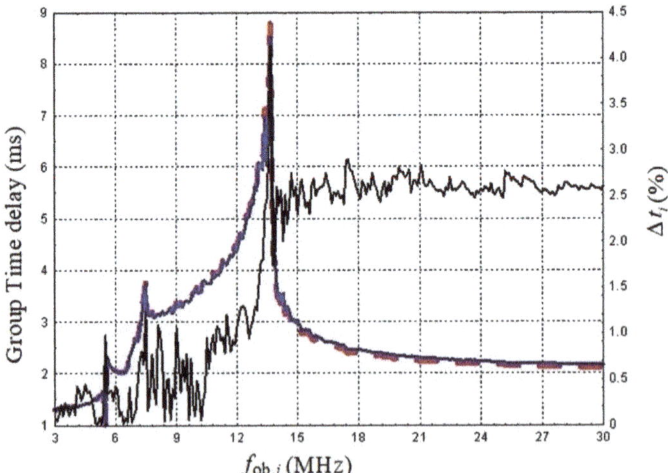

Figure 3. Trends of group time delays calculated by the IONORT algorithm (blue) and through Equation (8) (dashed red), and corresponding percentage relative error (black) given by Equation (9), for an elevation angle of 30°.

The trend of Δt_i given by Equation (9) represents the performance of the IONORT algorithm, and shows that for low frequencies, from 3 to about 12 MHz, the percentage relative error is below 1%. Smaller errors at lower frequencies are those related to waves travelling for relatively short paths; in this case, the Earth's curvature can be neglected and the approximation of flat ionospheric layers is more appropriate, leading to a reduction in the error that can be ascribed only to the integration procedure used by IONORT. The errors increase when getting closer to the maximum of electron density and then decrease again showing a fluctuating trend that practically does not ever exceed 3%.

The outcome of Figure 3 holds also for elevation angles equal to 18° and 45°, for which the results are not shown here. The interested reader can refer to the work by Bianchi et al. [48] where additional details about this issue can be found.

4. Main IONORT Upgrades

Compared to Azzarone et al. [39], the IONORT tool was upgraded from both the source code point of view and the graphical design point of view.

Concerning the code, this was completely rewritten from Fortran 77 to Fortran 90 programming language, several bugs have been fixed and, at the same time, some pointless parts of the code have been removed and the existing routines have been optimized. As a result, an improvement of the running speed was obtained.

The conversion from Fortran 77 to Fortran 90 programming language has allowed to solve the significant problem affecting the previous version that forced the user to work with a regional electron density background, limited in both latitude and longitude. Now the user can work with a global electron density background, for instance that given by the IRI model [24], and generate both numerical ray tracing data, for whatever transmitting point on the Earth's surface, and numerical homing data for any Tx-Rx pair in the world.

The Matlab programming language has been employed for developing a new graphical design. In fact, there are many updates regarding the GUI. The first one concerns the possibility, when launching IONORT, to choose what procedure to launch: ray tracing or homing (Figure 4).

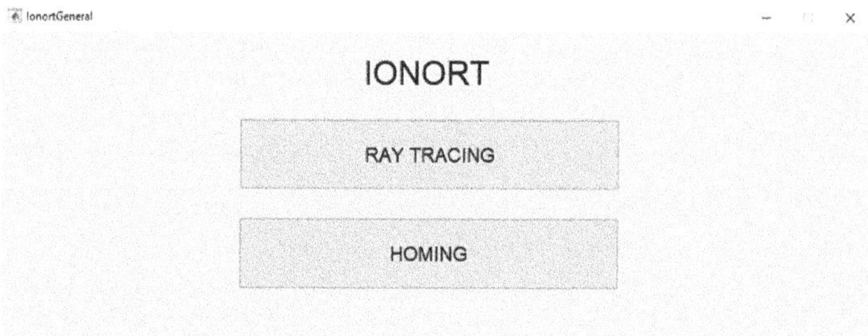

Figure 4. The initial GUI of IONORT.

4.1. The Ray Tracing GUI

A new version of the ray tracing GUI was developed (Figure 5), that gives the user the possibility to select nested loop cycles and visualize the ray tracing results, whatever is the transmitting point chosen on the Earth's surface. On the left side, the user has the possibility to insert the coordinates of the transmitting point and the desired loop cycles in frequency, elevation, and azimuth angles, as well as to choose the integration algorithm. In the central part, the user can include the geomagnetic field and choose both the kind of ray (ordinary or extraordinary) and the integration step. The results of the ray tracing are shown both in the lower part of the GUI as an altitude vs. ground range plot, and in the right side of the GUI as rays on the Earth's surface.

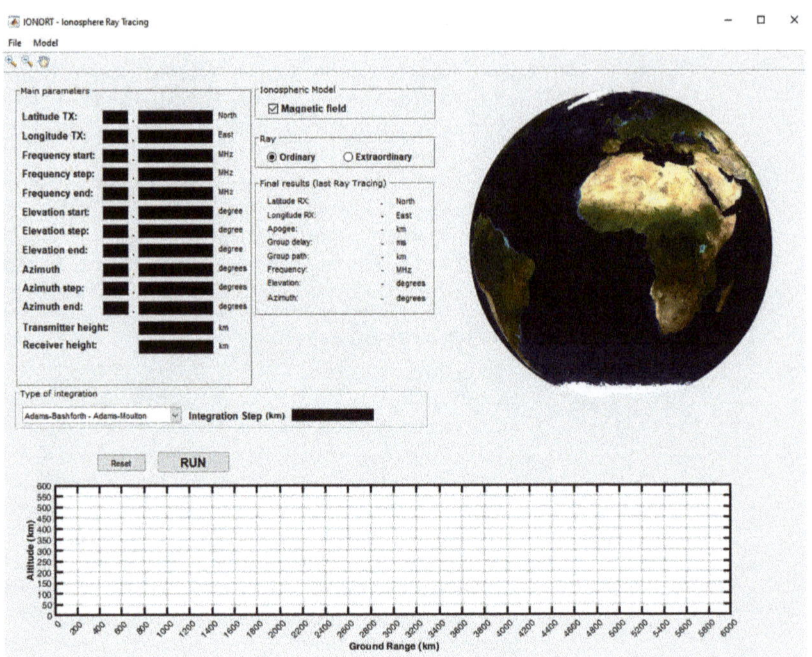

Figure 5. The ray tracing GUI appearing after clicking the ray tracing button shown in Figure 4.

4.2. The Homing GUI

Another novelty is the inclusion of a homing GUI (Figure 6), which allows visualizing a synthetic oblique ionogram for whatever radio link the user wants to simulate on the terrestrial surface. As for the ray tracing GUI, the user can insert the loop cycles in frequency, elevation angle, and azimuth angles, and choose the kind of ray to be processed (ordinary or extraordinary), the type of integration and the corresponding step, and whether the run has to be carried out with or without the geomagnetic field. The right side of the GUI shows the list of the "winner triplets", while the frequencies and group time delays values are automatically plotted to visualize the corresponding synthetic oblique ionogram. The winner triplets are the values of frequency, elevation, and azimuth angle, embedded in the loop cycles, thanks to which the electromagnetic wave sent out from the Tx reaches the Rx, on the basis of the chosen accuracy values. On the left side, the user, besides inserting the coordinates of both the transmitting and receiving points, can also insert the accuracy values requested along the directions north–south and east–west. It is worth saying a few words about the meaning of parameters *Latitude Accuracy* and *Longitude Accuracy*, which are chosen for running the homing procedure, because they are fundamental for the verification of the homing. The values *Latitude Accuracy* and *Longitude Accuracy* are those for which it is requested that the following logical conditions are fulfilled:

$$\left[\frac{\lambda_{\text{Rx}} - \lambda_{\text{LandingPoint}}}{(\pi/2) - \lambda_{\text{Rx}}} \right] \cdot 100 \leq \textit{Latitude Accuracy}$$

$$\text{AND} \tag{10}$$

$$\left[\frac{\varphi_{\text{LandingPoint}} - \varphi_{\text{Rx}}}{\varphi_{\text{Rx}}} \right] \cdot 100 \leq \textit{Longitude Accuracy}$$

where λ_{Rx} and φ_{Rx} are the latitude and longitude of the receiver, while $\lambda_{\text{LandingPoint}}$ and $\varphi_{\text{LandingPoint}}$ are the latitude and longitude of the landing point. Whenever the condition (10) is fulfilled, the homing subroutine assumes that the position of the landing point can be approximated with the position of the receiver. In this case, the homing is considered successful. When this circumstance is verified for a given "triplet" (f, α, Δ), then a point, whose coordinates are the frequency and the time delay, is drawn so as to obtain at the end of the nested loop cycle a synthetic oblique ionogram, as shown in Section 6.

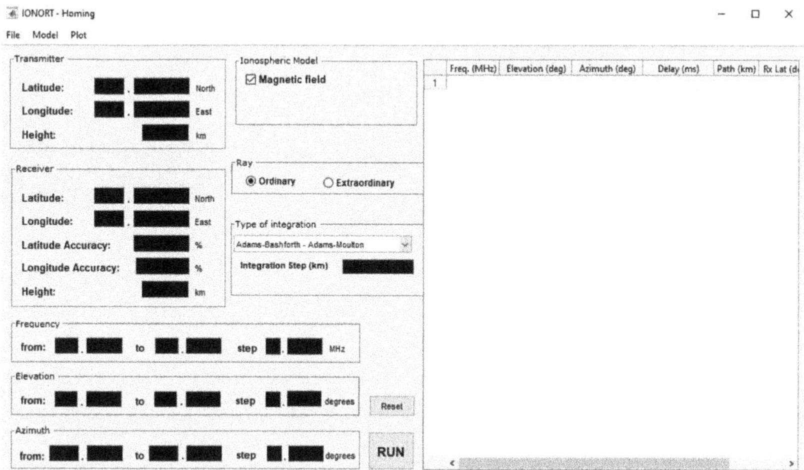

Figure 6. The homing GUI appearing after clicking the homing button shown in Figure 4.

It is worth highlighting that the input settings to be inserted in both GUIs can be saved as a Matlab file. This is a very useful feature whenever the user wants to repeat the same ray tracing or homing analysis, because this file can be loaded by opening the menu "File" and clicking "Load Input data", thus avoiding rewriting each time the same settings.

5. The Ray Tracing Feature: Some Results

In this section, two ray tracing cases obtained by placing the transmitter at high and middle latitude in the Northern hemisphere will be shown. The electron density background will be that provided by the IRI model [24], based on both the 12-month smoothed ionospheric index (IG_{12}) [49] and the 12-month smoothed sunspot number index (R_{12}) [50,51]. The ray tracing simulations have been performed using as integration method that of A-B/A-M, setting the ordinary (O) mode of propagation, and considering as geomagnetic field model the International Geomagnetic Reference Field (IGRF-13) [52]. To highlight the global potentialities of IONORT, additional examples can be found in the Supplementary Material.

Figure 7 shows the ray tracing results for the O propagation mode for the epoch 6 January 2022 at 09:00 UT, simulating a transmitter positioned at Qasigiannguit (68.8°N, 51.2°W, Greenland), and setting loop cycles in frequency from 2 to 12 MHz (with a step of 2 MHz), and in elevation angle from 2.5° to 12.5° (with a step of 5°). Two values of the azimuth angle equal to 130° and 260° have been considered. For α = 130° the minimum length path ($\approx$992 km, group delay $\approx$ 3.3112 ms) occurs for f = 2 MHz and Δ = 7.5°, while the maximum length path ($\approx$2448 km, group delay $\approx$ 8.1672 ms) occurs for f = 8 MHz and Δ = 2.5°. For α = 260° the minimum length path ($\approx$917 km, group delay $\approx$ 3.0685 ms) occurs for f = 2 MHz and Δ = 12.5°, while the maximum length path ($\approx$3013 km, group delay $\approx$ 10.0525 ms) occurs for f = 8 MHz and Δ = 2.5°.

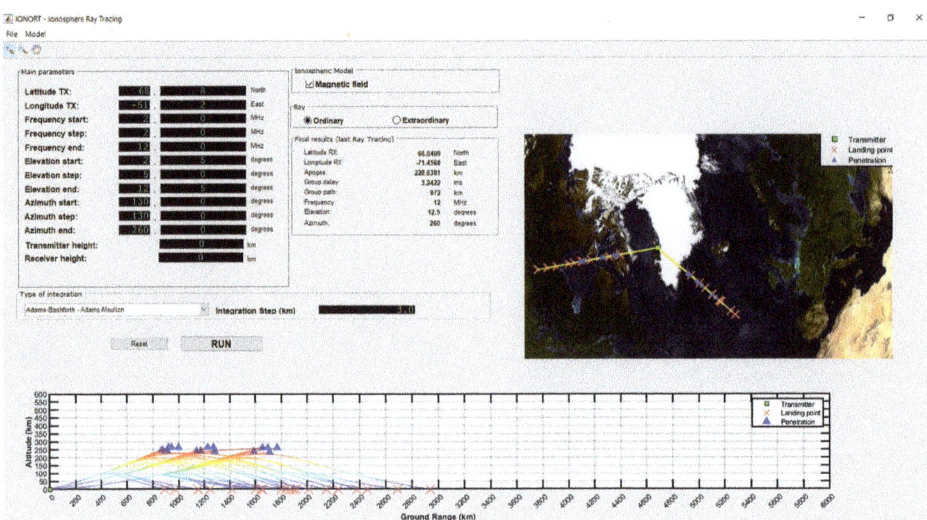

Figure 7. Ray tracing results in the Arctic region obtained for the O ray in presence of the geomagnetic field, simulating a transmitter positioned in Qasigiannguit for the epoch 6 January 2022 at 09:00 UT. The geographical map shows that the farthest landing points reach the Atlantic Ocean and Canada for $\alpha = 130°$ and $\alpha = 260°$, respectively. Looking at the altitude vs. ground range plot some penetration cases are observed. Colors are indicative of the different altitudes reached by the electromagnetic wave.

The same settings used to obtain the results shown in Figure 7 for the O propagation mode, were used to run IONORT for both the extraordinary (X) propagation mode and without including the geomagnetic field. The corresponding results are not shown here. Nevertheless, an example of the obtained ray paths for the O and X modes, as well as in absence of the geomagnetic field, are shown in Figure 8 for $f = 6$ MHz, $\Delta = 12.5°$ and $\alpha = 130°$.

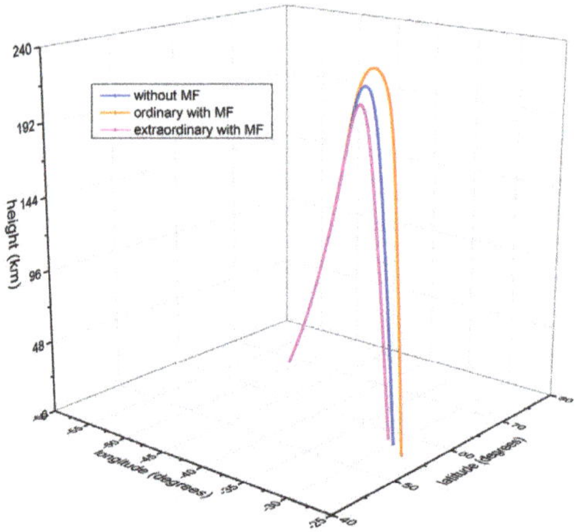

Figure 8. Example of 3-D ray tracing calculated by IONORT for $f = 6$ MHz, $\Delta = 12.5°$, and $\alpha = 130°$, for the epoch 6 January 2022 at 09:00 UT, for the O (orange) and X (magenta) propagation modes, and in the absence of the geomagnetic field (blue). The azimuthal displacement of the propagation plane due to the anisotropy introduced by the presence of the Earth's magnetic field is, respectively, $\Delta\alpha \approx 0.396°$ and $\Delta\alpha \approx 0.014°$ for the O and X ray.

In addition, the azimuthal displacement due to the anisotropy introduced by the presence of the Earth's magnetic field has been calculated also for some of ray tracings drawn in the 2-D plot shown in Figure 7. Corresponding results are shown in Table 1, where α_{start} is the initial azimuth angle given as input through the GUI.

Table 1. Azimuthal displacement showed by the O propagation mode due to the presence of the geomagnetic field, for some of ray tracings of Figure 7, whose frequency and elevation angle are written in the first column.

f (MHz), Δ (°)	$\alpha_{start} = 130°$ Δα° (O)	$\alpha_{start} = 260°$ Δα° (O)
4, 2.5	0.059	0.123
4, 12.5	0.280	0.211
6, 2.5	0.099	0.085
6, 7.5	0.161	0.132
6, 12.5	0.396	0.226
8, 2.5	0.090	0.085

Figure 9 shows the ray tracing results for the O propagation mode for the epoch 24 May 2022 at 15:00 UT simulating a transmitter positioned in Rome (41.9°N, 12.5°E; Italy), with the same loop cycles of Figure 7. Azimuth angles equal to 110° and 260° have been chosen. For $\alpha = 110°$ the minimum length path ($\approx$729 km, group delay $\approx$ 2.4319 ms) occurs for f = 2 MHz and Δ = 12.5°, while the maximum length path ($\approx$2054 km, group delay $\approx$ 6.8546 ms) occurs for f = 12 MHz and Δ = 12.5°. For $\alpha = 260°$ the minimum length path ($\approx$728 km, group delay $\approx$ 2.4293 ms) occurs for f = 2 MHz and Δ = 12.5°, while the maximum length path ($\approx$1768 km, group delay $\approx$ 5.9003 ms) occurs for f = 12 MHz and Δ = 2.5°.

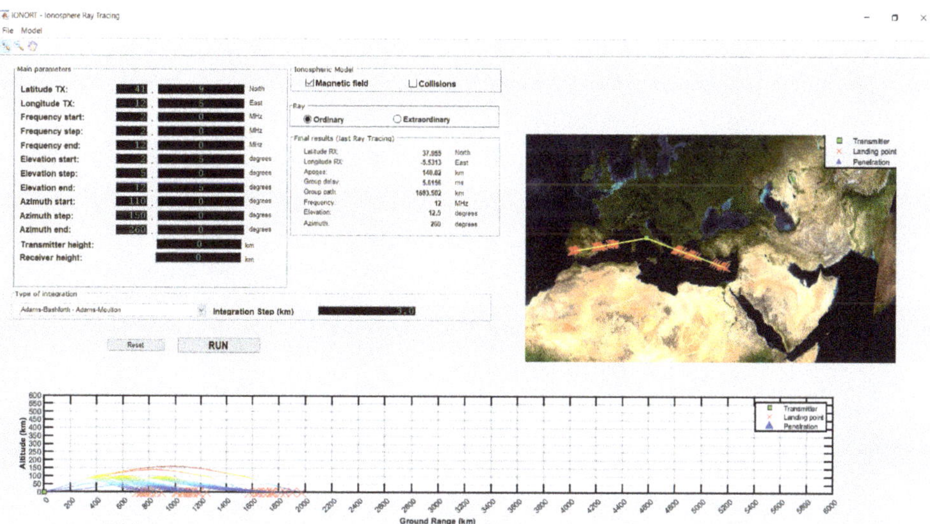

Figure 9. Ray tracing results in the middle latitude European region obtained for the O ray in presence of the geomagnetic field, simulating a transmitter positioned in Rome for the epoch 24 May 2022 at 15:00 UT. The geographical map exhibits landing points on the Mediterranean Sea along the direction $\alpha = 110°$, while along the direction $\alpha = 260°$ landing points on the southern part of Spain and Portugal are observed. Looking at the altitude vs. ground range plot, no penetration case is observed. Colors are indicative of the different altitudes reached by the electromagnetic wave.

As carried out for the previous epoch, also in this case the same settings used to obtain the results shown in Figure 9 for the O mode of propagation, were used to run IONORT for both the X propagation mode and without including the geomagnetic field. An example of the obtained ray paths calculated for the O and X modes, as well as in absence of the geomagnetic field, are shown in Figure 10 for f = 2 MHz, Δ = 2.5 °, and α = 260°.

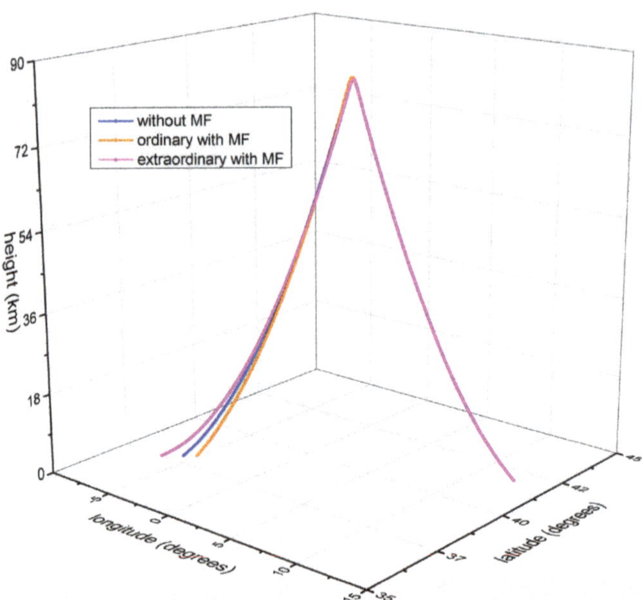

Figure 10. Example of 3-D ray tracing calculated by IONORT for f = 2 MHz, Δ = 2.5 °, and α = 260°, for the epoch 24 May 2022 at 15:00 UT, for the O (orange) and X (magenta) propagation modes, and in the absence of the geomagnetic field (blue). The azimuthal displacement of the propagation plane due to the anisotropy introduced by the presence of the Earth's magnetic field is, respectively, $\Delta\alpha \approx 0.039°$ and $\Delta\alpha \approx 0.036°$ for the O and X ray.

In addition, the azimuthal displacement due to the anysotropy introduced by the presence of the Earth's magnetic field has been calculated also for some of ray tracings drawn in the 2-D plot shown in Figure 9. Corresponding results are shown in Table 2, where α_{start} is the initial azimuth angle given as input through the GUI.

Table 2. Azimuthal displacement showed by the O propagation mode due to the presence of the geomagnetic field, for some of ray tracings of Figure 9, whose frequency and elevation angle are written in the first column.

f (MHz), Δ (°)	α_{start} = 110° $\Delta\alpha°$ (O)	α_{start} = 260° $\Delta\alpha°$ (O)
4, 2.5	0.071	0.038
4, 12.5	0.075	0.040
6, 2.5	0.070	0.038
6, 7.5	0.072	0.039
6, 12.5	0.075	0.039
8, 2.5	0.070	0.038

Comparing Table 1 with Table 2 it is clear how the effect of the geomagnetic field is more important at high latitudes than at middle latitudes. A larger value of the geomagnetic field at high latitudes causes a larger azimuthal displacement of the plane of propagation than at middle latitudes.

6. The Homing Feature: Some Results

In this section, two homing cases obtained by placing the transmitter and the receiver both in the Northern Europe and Middle East will be shown. As it was carried out in the previous section, the electron density background will be that output by the IRI model [24]. The homing simulations will be performed using as integration method the A-B/A-M, setting both modes of propagation, O and X, and considering as geomagnetic field model the International Geomagnetic Reference Field (IGRF-13) [52]. Moreover, different loop cycles in frequency, elevation angle, and azimuth angle are tried to find the "winner triplets", namely those combinations of (f, Δ, α) that allow the homing. The synthetic oblique ionograms resulting from the homing will be shown superimposed to the list of the "winner triplets". To highlight the global potentialities of IONORT, additional examples can be found in the Supplementary Material.

Figure 11 shows the homing result obtained for the epoch 9 August 2022 at 23:00 UT, for the O mode, simulating a transmitter located in London (51.5°N, 359.8°E; England) and a receiver located in Stockholm (59.3°N, 18.0°E; Sweden). The corresponding oblique ionogram shows a Maximum Usable Frequency (MUF) for the simulated radio link (ground range $\approx$ 1433 km) of 8.7 MHz.

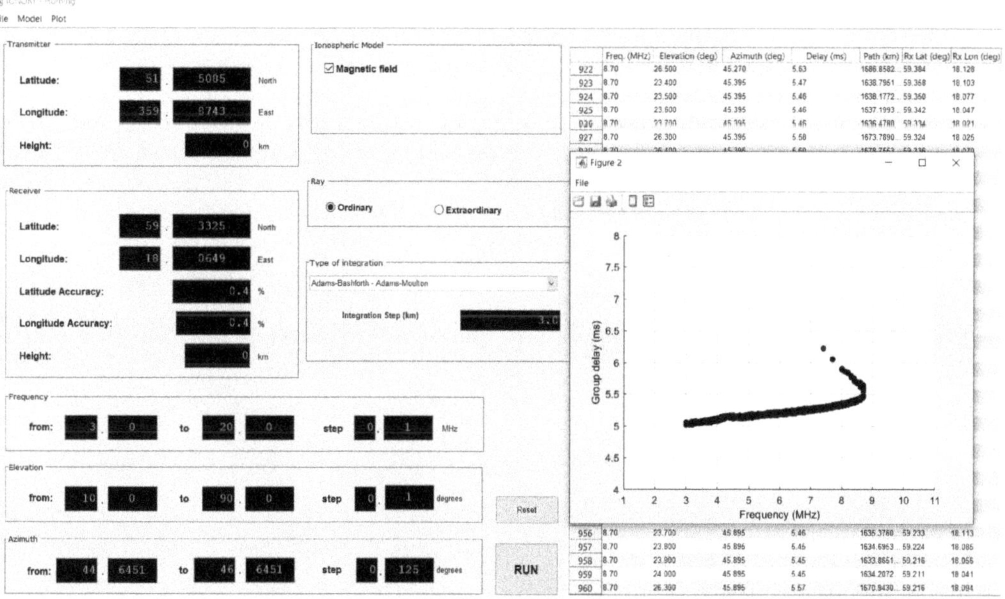

Figure 11. Example of homing simulation obtained in the Northern Europe for the epoch 9 August 2022 at 23:00 UT, for the O propagation mode. The left side of the GUI shows the main input parameters that are: the geographic coordinates of both the transmitter (London) and the receiver (Stockholm); the loop cycles in frequency (3–20 MHz, step = 0.1 MHz), elevation (10–90°, step = 0.1°), and azimuth (44.6–46.6°, step = 0.125°). The right side of the GUI shows the list of the "winner triplets", that in the specific case are 960. The synthetic oblique ionogram obtained plotting the 960 pairs (frequency vs. group time delay) is also shown.

Figure 12 shows the homing result for the same radio link considered in Figure 11, for the X propagation mode. The corresponding oblique ionogram shows that the MUF is now 9.3 MHz.

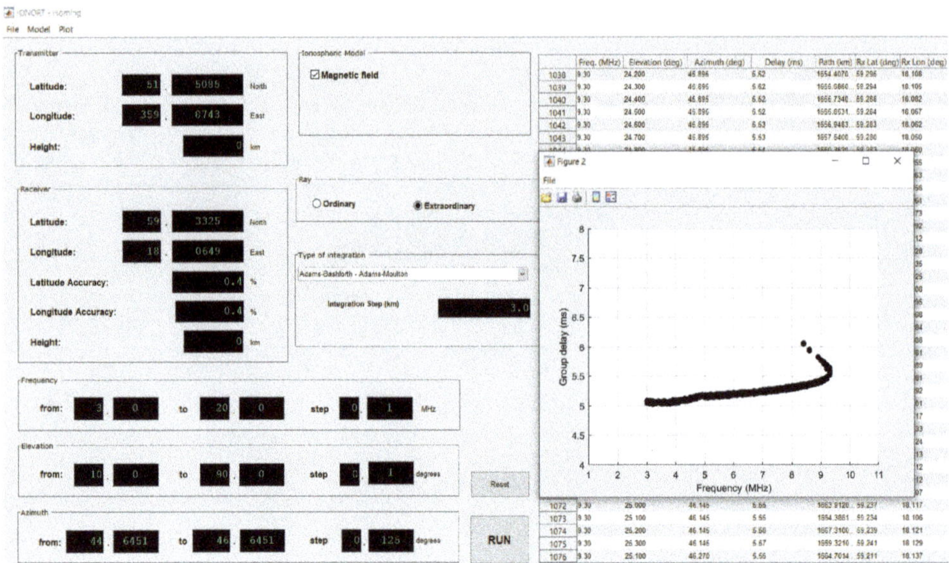

Figure 12. Example of homing simulation obtained in the Northern Europe for the epoch 9 August 2022 at 23:00 UT, for the X propagation mode. The left side of the GUI shows the main input parameters that are: the geographic coordinates of both the transmitter (London) and the receiver (Stockholm); the loop cycles in frequency (3–20 MHz, step = 0.1 MHz), elevation (10–90°, step = 0.1°), and azimuth (44.6–46.6°, step = 0.125°). The right side of the GUI shows the list of the "winner triplets", that in the specific case are 1076. The synthetic oblique ionogram obtained plotting the 1076 pairs (frequency vs. group time delay) is also shown.

In Figure 13 the O and X trace have been plotted together to better visualize their differences.

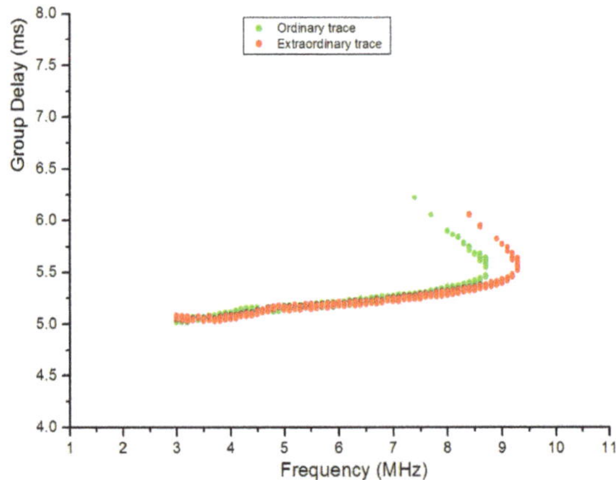

Figure 13. The O (green) and X (red) oblique ionograms shown in Figures 11 and 12 are here plotted together. The MUF value of the X trace exceeds that of the O trace by 0.6 MHz.

Figure 14 shows the homing result obtained for the epoch 29 July 2022 at 09:00 UT, for the O mode, simulating a transmitter located in Ryadh (24.6°N, 46.7°E; Saudi Arabia) and a receiver located at Teheran (35.6°N, 51.4°E; Iran). The oblique ionogram corresponding to the simulated radio link (ground range ≈ 1294 km) is constituted by two well distinct traces: the lower trace is related to the F1-layer, characterized by a MUF of about 16.0 MHz; the upper trace is related to the F2-layer, characterized by a MUF of 16.5 MHz.

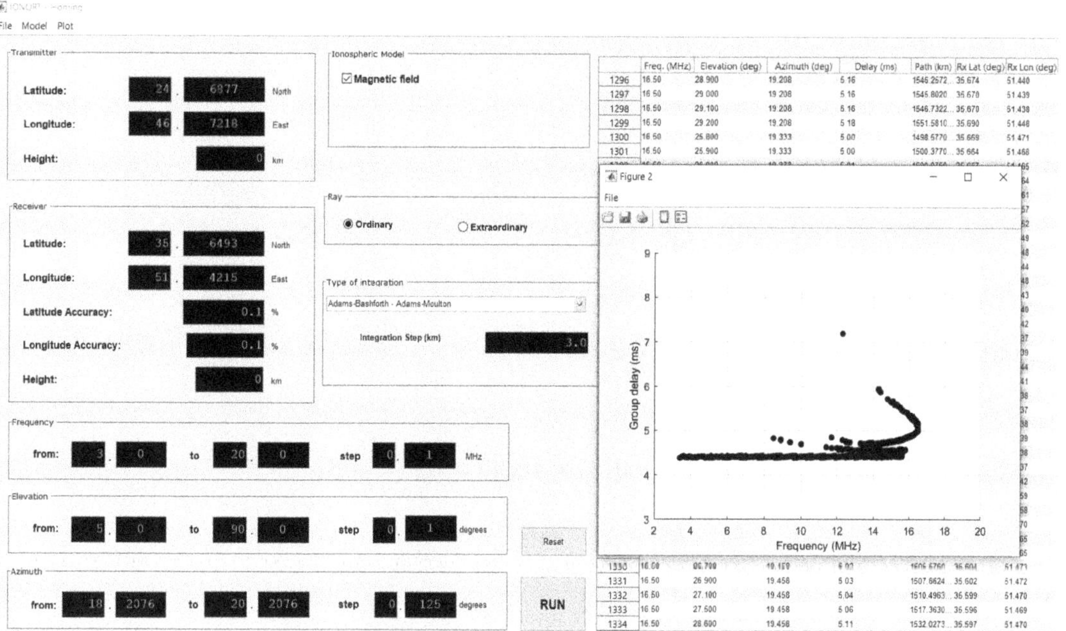

Figure 14. Example of homing simulation obtained in the Middle East for the epoch 29 July 2022 at 09:00 UT, for the O propagation mode. The left side of the GUI shows the main input parameters that are: the geographic coordinates of both the transmitter (Ryadh) and the receiver (Teheran); the loop cycles in frequency (3–20 MHz, step = 0.1 MHz), elevation (5–90°, step = 0.1°), and azimuth (18.2–20.2°, step = 0.125°). The right side of the GUI shows the list of the "winner triplets", that in the specific case are 1334. The synthetic oblique ionogram obtained plotting the 1334 pairs (frequency vs. group time delay) is also shown.

Figure 15 shows the homing result for the same radio link considered in Figure 14, for the X propagation mode. The corresponding oblique ionogram shows a lower trace related to the F1-layer, characterized by a MUF slightly larger than 16.0 MHz, and an upper trace, related to the F2-layer, showing a MUF of 16.8 MHz.

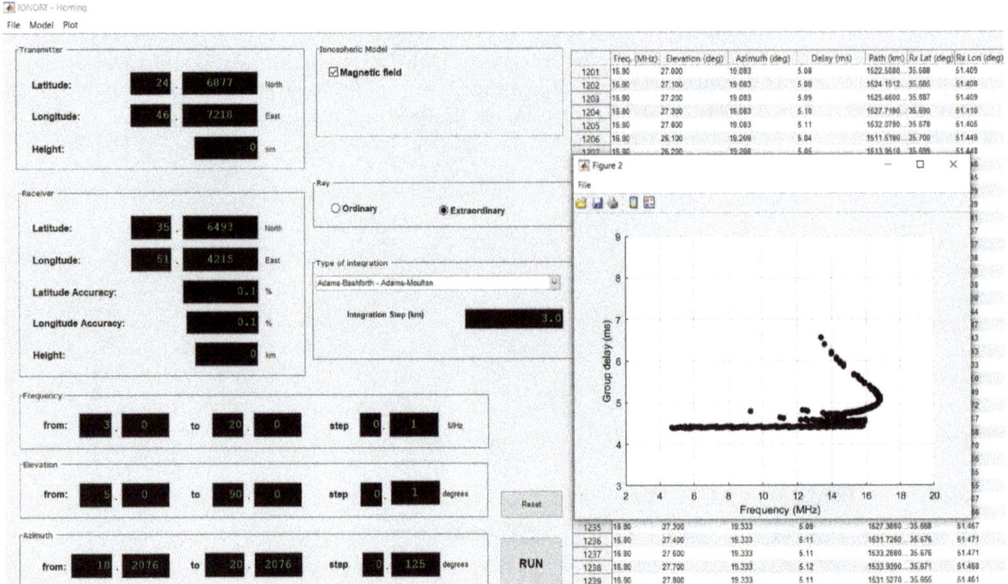

Figure 15. Example of homing simulation obtained in the Middle East for the epoch 29 July 2022 at 09:00 UT, for the X propagation mode. The left side of the GUI shows the main input parameters that are: the geographic coordinates of both the transmitter (Ryadh) and the receiver (Teheran); the loop cycles in frequency (3–20 MHz, step = 0.1 MHz), elevation (5–90°, step = 0.1°), and azimuth (18.2–20.2°, step = 0.125°). The right side of the GUI shows the list of the "winner triplets", that in the specific case are 1239. The synthetic oblique ionogram obtained plotting the 1239 pairs (frequency vs. group time delay) is also shown.

As it was carried out for the previous case, in Figure 16 the O and X trace have been plotted together to better visualize their differences.

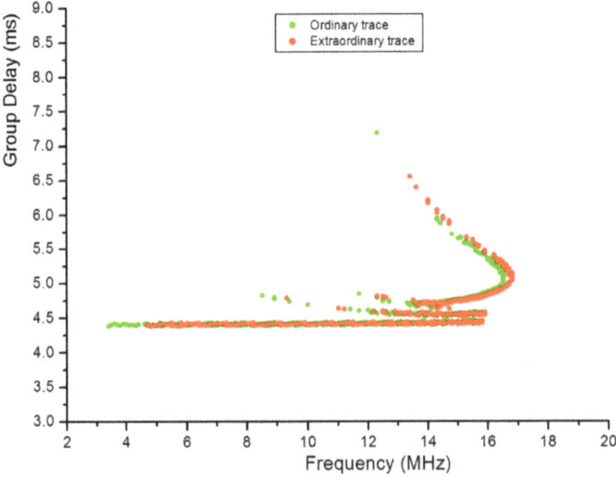

Figure 16. The O (green) and X (red) oblique ionograms shown in Figures 14 and 15 are here plotted together. The MUF value of the X trace exceeds that of the O trace by 0.3 MHz.

Comparing Figure 13 with Figure 16 it is clear how a larger value of the geomagnetic field at higher latitudes causes a larger difference between the O and X MUF values.

7. Conclusive Remarks

In this paper, the main novelties characterizing the new version of IONORT (version 2023.10), with respect to the previous one proposed by Azzarone et al. [39], have been shown and discussed. In particular, the most important improvement is that now the users can perform their own simulations anywhere in the world. Moreover, besides a ray tracing simulation, now a user has also the possibility to perform a homing simulation, obtaining as output all the triplets (frequency, elevation angle, and azimuth angle) that satisfy the initial conditions chosen in terms of accuracy values requested along the directions north–south and east–west. It is worth highlighting that all the simulations described in the paper have been performed using as electron density background that given by the IRI model [24]. Moreover, a user can use any other ionospheric model; obviously, the important thing is to respect the electron density matrix format currently in force. In this regard, one thing that needs to be underlined is that the IRI model is a climatological model, and consequently its use in the Space Weather context could be limiting [53]. With regard to this, the consideration of nowcasting methods, as for instance the IRI UPdate [35] or the IRI Real Time Assimilative Model [32], could be more appropriate. With regard to future developments, it is currently under study the calculation of the nondeviative absorption due to the propagation through the D layer. A further possible work could be based on testing IONORT under severe Space Weather events and comparing the results obtained with both a climatological and a nowcasting background.

Supplementary Materials: The following supporting information can be downloaded at: https://www.mdpi.com/article/10.3390/rs15215111/s1, Figure S1: Ray tracing results for the O propagation mode over the equatorial zone of the South-East Asia; Figure S2: Ray tracing results for the O propagation mode over the equatorial zone of the South America; Figure S3: Example of homing simulation obtained in the Southern Africa for the O propagation mode; Figure S4: Example of homing simulation obtained in the Southern Africa for the X propagation mode; Figure S5: O trace of Figure S3 and X trace of Figure S4 plotted together; Figure S6: Example of homing simulation obtained in the Central Asia for the O propagation mode; Figure S7: Example of homing simulation obtained in the Central Asia for the X propagation mode; Figure S8: O trace of Figure S6 and X trace of Figure S7 plotted together.

Author Contributions: Writing—original draft preparation M.P. (Marco Pietrella); revision—original draft M.P. (Michael Pezzopane); Conceptualization, M.P. (Marco Pietrella) and M.P. (Michael Pezzopane); methodology M.P. (Michael Pezzopane) and M.P. (Marco Pietrella); investigation and validation, M.P. (Marco Pietrella) and M.P. (Michael Pezzopane); software, A.P. (Alessandro Pignatelli) A.S. (Alessandro Settimi) and A.P. (Alessio Pignalberi). All authors have read and agreed to the published version of the manuscript.

Funding: This research received no external funding.

Data Availability Statement: International Reference Ionosphere (IRI) Fortran code is available at the IRI website (http://irimodel.org/, accessed on 24 May 2023).

Acknowledgments: The IRI team is acknowledged for developing and maintaining the IRI model and for giving access to the corresponding Fortran code via the IRI website (http://irimodel.org/, accessed 24 May 2023).

Conflicts of Interest: The authors declare no conflict of interest.

References

1. Budden, K.G. *The Propagation of Radio Waves: The Theory of Radio Waves of Low Power in the Ionosphere and Magnetosphere*; Cambridge University Press: Cambridge, UK, 1988; p. 688.
2. Haselgrove, J. Ray Theory and a New Method for Ray Tracing. In *The Physics of the Ionosphere: Report of the Physical Society Conference, Held at Cavendish Laboratory, Cambridge, September 1954*; Physical Society: London, UK, 1955; pp. 355–364.
3. Haselgrove, C.B.; Haselgrove, J. Twisted ray paths in the ionosphere. *Proc. Phys. Soc.* **1960**, *75*, 357–363. [CrossRef]

4.   Haselgrove, J. The Hamilton ray path equations. *J. Atmos. Terr. Phys.* **1963**, *25*, 397–399. [CrossRef]

5.   Dudziak, W.F. *Three-Dimensional Ray Trace Computer Program for Electromagnetic Wave Propagation Studies*; Technical Military Planning Operation RM 61TMP-32, Defense Atomic Support Agency DASA 1232; General Electrical Company: Santa Barbara, CA, USA, 1961; p. 170.

6.   Lawrence, R.S.; Posakony, D.J. A digital ray tracing program for ionospheric research. In Proceedings of the Second International Space Science Symposium, Florence, Italy, 10–14 April 1961; van de Hulst, H.C., de Jager, C., Moore, A.F., Eds.; North-Holland Publishing Co.: Amsterdam, The Netherlands; New York, NY, USA, 1961; Volume 2, pp. 258–276.

7.   Croft, T.A.; Gregory, L. *Fast, Versatile Ray-Tracing Program for IBM 7090 Digital Computers*; Defense Technical Information Center, Ed.; Rept. SEL-63-107, TR 82, Contract No. 225 (64); Office of Naval Research, Advanced Research Projects Agency: Stanford, CA, USA, 1963; p. 28.

8.   Jones, R.M. *A Three Dimensional Ray Tracing Computer Program*; ESSA Technical Report, IER 17-ITSA 17; Government Printing Office: Washington, DC, USA, 1966.

9.   Jones, R.M.; Stephenson, J.J. *A Versatile Three-Dimensional Ray Tracing Computer Program for Radiowaves in the Ionosphere*; OT Report, 75–76; U.S. Department of Commerce, Office of Telecommunication, U.S. Government Printing Office: Washington, DC, USA, 1975.

10.   Reilly, M.H. Upgrades for efficient three-dimensional ionospheric ray-tracing: Investigation of HF near vertical incidence sky wave effects. *Radio Sci.* **1991**, *26*, 971–980. [CrossRef]

11.   Norman, R.J.; Bennett, J.A.; Dyson, P.L.; Nguyen, L. *HIRT: Homing-in Ray Tracing Program*; Research report; School of Physics, La Trobe University: Bundoora, VIC, Australia, 1994.

12.   Norman, R.J.; Cannon, P.S. A two-dimensional analytic ray tracing technique accommodating horizontal gradients. *Radio Sci.* **1997**, *32*, 387–396. [CrossRef]

13.   Anderson, D.N.; Forbes, J.M.; Codrescu, M. A fully analytic, low- and middle-latitude ionospheric model. *J. Geophys. Res.* **1989**, *9*, 1520–1524. [CrossRef]

14.   Coleman, C.J. A ray-tracing formulation and its application to some problems in over-the-horizon radar. *Radio Sci.* **1998**, *33*, 1187–1197. [CrossRef]

15.   Huang, X.; Reinisch, B. Real Time HF ray tracing through a tilted ionosphere. *Radio Sci.* **2006**, *41*, RS5S47. [CrossRef]

16.   Nickisch, L.J. Practical applications of Haselgrove's equations for HF systems. *Radio Sci. Bull.* **2008**, *325*, 36–48.

17.   Tsai, L.C.; Liu, C.H.; Hsiao, T.Y.; Huang, J.Y. A near real-time phenomenological model of ionospheric electron density based on GPS radio occultation data. *Radio Sci.* **2009**, *44*, RS5002. [CrossRef]

18.   Tsai, L.C.; Liu, C.H.; Huang, J.Y. Three-dimensional numerical ray tracing on a phenomenological ionospheric model. *Radio Sci.* **2010**, *44*, RS5017. [CrossRef]

19.   Warrington, E.M.; Zaalov, N.Y.; Naylor, J.S.; Stocker, A.J. HF propagation modeling within the polar ionosphere. *Radio Sci.* **2012**, *47*, RS0L13. [CrossRef]

20.   Cervera, M.A.; Harris, T.J. Modeling ionospheric disturbance features in quasi-vertically incident ionograms using 3-D magnetoionic ray tracing and atmospheric gravity waves. *J. Geophys. Res. Space Phys.* **2014**, *119*, 431–440. [CrossRef]

21.   Cervera, M.A.; Francis, D.B.; Frazer, G.J. Climatological Model of Over-the-Horizon Radar. *Radio Sci.* **2018**, *53*, 988–1001. [CrossRef]

22.   Pederick, L.K.; Cervera, M.A. Modeling the interference environment in the HF band. *Radio Sci.* **2016**, *51*, 82–90. [CrossRef]

23.   Pederick, L.K.; Cervera, M.A. A directional HF noise model: Calibration and validation in the Australian region. *Radio Sci.* **2016**, *51*, 25–39. [CrossRef]

24.   Bilitza, D.; Pezzopane, M.; Truhlik, V.; Altadill, D.; Reinisch, B.W.; Pignalberi, A. The International Reference Ionosphere model: A review and description of an ionospheric benchmark. *Rev. Geophys.* **2022**, *60*, e2022RG000792. [CrossRef]

25.   Greenwald, R.A.; Baker, K.B.; Dudeney, J.R.; Pinnock, M.; Jones, T.B.; Thomas, E.C.; Villain, J.-P.; Cerisier, J.-C.; Senior, C.; Hanuise, C.; et al. DARN/SuperDARN: A global view of the dynamics of high-latitude convection. *Space Sci. Rev.* **1995**, *71*, 761–796. [CrossRef]

26.   Michael, C.M.; Yeoman, T.K.; Wright, D.M.; Milan, S.E.; James, M.K. A Ray Tracing Simulation of HF Ionospheric Radar Performance at African Equatorial Latitudes. *Radio Sci.* **2020**, *55*, e2019RS006936. [CrossRef]

27.   Coleman, C.J. Point-to-point ionospheric ray tracing by a direct variational method. *Radio Sci.* **2011**, *46*, RS5016. [CrossRef]

28.   Strangeways, H.J. Effects of horizontal gradients on ionospherically reflected or transionospheric paths using a precise homing-in method. *J. Atmos. Sol. Terr. Phys.* **2000**, *62*, 1361–1376. [CrossRef]

29.   Kashcheyev, A.; Nava, B.; Radicella, S.M. Estimation of higher-order ionospheric errors in GNSS positioning using a realistic 3-D electron density model. *Radio Sci.* **2012**, *47*, RS4008. [CrossRef]

30.   Abdullah, M.; Strangeways, H.J.; Zulkifli, S.S.N. Ionospheric differential error determination using ray tracing for a short baseline. *Adv. Space Res.* **2010**, *46*, 1326–1333. [CrossRef]

31.   McNamara, L.F. *The Ionosphere: Communications, Surveillance, and Direction Finding*; Krieger Pub Co., Technical Book Store: Malabar, FL, USA, 1991; p. 248.

32.   Galkin, I.A.; Reinisch, B.W.; Huang, X.; Bilitza, D. Assimilation of GIRO data into a real-time IRI. *Radio Sci.* **2012**, *47*, RS0L07. [CrossRef]

33. Pezzopane, M.; Pietrella, M.; Pignatelli, A.; Zolesi, B.; Cander, L.R. Assimilation of autoscaled data and regional and local ionospheric models as input sources for real-time 3-D International Reference Ionosphere modeling. *Radio Sci.* **2011**, *46*, RS5009. [CrossRef]
34. Pezzopane, M.; Pietrella, M.; Pignatelli, A.; Zolesi, B.; Cander, L.R. Testing the three-dimensional IRI-SIRMUP-P mapping of the ionosphere for disturbed periods. *Adv. Space Res.* **2013**, *52*, 1726–1736. [CrossRef]
35. Pignalberi, A.; Pezzopane, M.; Rizzi, R.; Galkin, I. Effective solar indices for ionospheric modeling: A review and a proposal for a real-time regional IRI. *Surv. Geophys.* **2018**, *39*, 125–167. [CrossRef]
36. Pignalberi, A.; Pietrella, M.; Pezzopane, M.; Rizzi, R. Improvements and validation of the IRI UP method under moderate, strong, and severe geomagnetic storms. *Earth Planets Space* **2018**, *70*, 180. [CrossRef]
37. Pignalberi, A.; Habarulema, J.B.; Pezzopane, M.; Rizzi, R. On the development of a method for updating an empirical climatological ionospheric model by means of assimilated vTEC measurements from a GNSS receiver network. *Space Weather* **2019**, *17*, 1131–1164. [CrossRef]
38. Pietrella, M.; Pezzopane, M.; Zolesi, B.; Cander, L.R.; Pignalberi, A. The Simplified Ionospheric Regional Model (SIRM) for HF Prediction: Basic Theory, Its Evolution and Applications. *Surv. Geophys.* **2020**, *41*, 1143–1178. [CrossRef]
39. Azzarone, A.; Bianchi, C.; Pezzopane, M.; Pietrella, M.; Scotto, C.; Settimi, A. IONORT: A Windows soft-ware tool to calculate the HF ray tracing in the ionosphere. *Comput. Geosci.* **2012**, *42*, 57–63. [CrossRef]
40. Settimi, A.; Pezzopane, M.; Pietrella, M.; Bianchi, C.; Scotto, C.; Zuccheretti, E.; Makris, J. Testing the IONORT-ISP system: A comparison between synthesized and measured oblique ionograms. *Radio Sci.* **2013**, *48*, 167–179. [CrossRef]
41. Pietrella, M.; Pezzopane, M.; Settimi, A. Ionospheric response under the influence of the solar eclipse occurred on 20 March 2015: Importance of autoscaled data and their assimilation for obtaining a reliable modeling of the ionosphere. *J. Atmos. Sol. Terr. Phys.* **2016**, *146*, 49–57. [CrossRef]
42. Booker, H.G. The application of the magnetoionic theory to the ionosphere. *Proc. R. Soc. Lond. Ser. A Math. Phys. Sci.* **1935**, *150*, 267–286. [CrossRef]
43. Settimi, A. An in-depth analysis on the Quasi-Longitudinal approximations applied to ionospheric ray-tracing, oblique and vertical sounding, and absorption. *Ann. Geophys. Italy* **2022**, *65*, PA003. [CrossRef]
44. Elias, A.G.; Fagre, M.; Zossi, B.S.; Amite, H. Spitze Angle Changes during Rapid Geomagnetic Core Field Variation. *Geomagn. Aeron.* **2021**, *61*, 134–142. [CrossRef]
45. Bachvalov, N.S. *Metodi Numerici*; Editori Riuniti University Press: Rome, Italy, 2012; p. 620. (In Italian)
46. Butcher, J.C. *Numerical Methods for Ordinary Differential Equations*; John Wiley: Hoboken, NJ, USA, 2003; p. 103. [CrossRef]
47. Hairer, E.; Nørsett, S.P.; Wanner, G. *Solving Ordinary Differential Equations I: Nonstiff Problems*, 2nd ed.; Springer: Berlin, Germany, 1993.
48. Bianchi, C.; Settimi, A.; Scotto, C.; Azzarone, A.; Lozito, A. A method to test HF ray tracing algorithm in the ionosphere by means of the virtual time delay. *Adv. Space Res.* **2011**, *48*, 1600–1605. [CrossRef]
49. Liu, R.Y.; Smith, P.A.; King, J.W. A new solar index which leads to improved foF2 predictions using the CCIR Atlas. *Telecommun. J.* **1983**, *50*, 408–414.
50. Clette, F.; Svalgaard, L.; Vaquero, J.M.; Cliver, E.W. Revisiting the Sunspot Number. *Space Sci. Rev.* **2014**, *186*, 35–103. [CrossRef]
51. Clette, F.; Lefèvre, L. The New Sunspot Number: Assembling All Corrections. *Sol. Phys.* **2016**, *291*, 2629–2651. [CrossRef]
52. Thébault, E.; Beggan, C.D.; Amit, H.; Aubert, J.; Baerenzung, J.; Bondar, T.N.; Brown, W.J.; Califf, S.; Chambodut, A.; Chulliat, A.; et al. International Geomagnetic Reference Field: The thirteenth generation. *Earth Planets Space* **2021**, *73*, 49. [CrossRef]
53. Pignalberi, A.; Pezzopane, M.; Tozzi, R.; De Michelis, P.; Coco, I. Comparison between IRI and preliminary Swarm Langmuir probe measurements during the St. Patrick storm period. *Earth Planets Space* **2016**, *68*, 93. [CrossRef]

Article

Analysis of Winter Anomaly and Annual Anomaly Based on Regression Approach

Kaixin Wang [1,2], Jiandi Feng [1,2,3,*], Zhenzhen Zhao [1] and Baomin Han [1]

1 School of Civil Engineering and Geomatics, Shandong University of Technology, Zibo 255000, China; happysdut@163.com (K.W.); zzzhao@whu.edu.cn (Z.Z.); hanbm@sdut.edu.cn (B.H.)
2 State Key Laboratory of Space Weather, Chinese Academy of Sciences, Beijing 100190, China
3 State Key Laboratory of Geo-Information Engineering, Xi'an 710054, China
* Correspondence: jdfeng@whu.edu.cn; Tel.: +86-189-5442-5161

Abstract: Studying the temporal and spatial dependence of ionospheric anomalies using total electron content (TEC) can provide an important reference for developing empirical ionospheric models. In this study, winter anomaly, annual anomaly, and the contributions of winter anomaly to annual anomaly were investigated during solar cycle 24 (2008–2018) by using the global ionosphere maps of the Center for Orbit Determination in Europe during the geomagnetic activity quiet period (Kp $\leq$ 5) based on a regression approach. Our detailed analysis shows the following: (1) Winter anomaly is more significant at 11:00–13:00 local time (LT), and the region of winter anomaly extends from North America to the Far East with increasing solar activity levels. (2) The minimum level of solar activity corresponding to the occurrence of winter anomaly was calculated at each grid point, which can provide a reference for single-point ionospheric modeling. (3) The annual anomaly reaches its maximum at 12:00 LT when the TEC in December is 34.4% higher than in June. (4) At 12:00 LT, the winter anomaly contributes up to 32% to the annual anomaly (at this time, the winter hemisphere contributes 57% to the annual anomaly).

Keywords: total electron content; ionospheric anomalies; global ionosphere maps; solar activity

Citation: Wang, K.; Feng, J.; Zhao, Z.; Han, B. Analysis of Winter Anomaly and Annual Anomaly Based on Regression Approach. *Remote Sens.* **2023**, *15*, 4968. https://doi.org/10.3390/rs15204968

Academic Editor: Fabio Giannattasio

Received: 18 September 2023
Revised: 10 October 2023
Accepted: 11 October 2023
Published: 15 October 2023

1. Introduction

Behaviors of the ionospheric F2 layer that depart from the prediction of Chapman's model are considered anomalies [1]. These ionospheric anomalies are related to the plasma motion of the ionosphere, variations in the composition of the atmosphere, etc., [2]. Ionospheric anomalies include annual anomaly, semiannual anomaly, winter anomaly, equatorial anomaly, and midlatitude summer nighttime anomaly, etc., which show regular temporal and spatial variation characteristics. The annual anomaly indicates that the F2-layer peak electron density (NmF2) combined from both hemispheres is greater (about 20%) during December than during June [3]. The semiannual anomaly indicates that NmF2 is greater during equinoxes than during either solstice [4]. The winter anomaly indicates that daytime NmF2 is greater in winter than in summer at middle latitudes at approximately the same solar activity level [5]. The equatorial anomaly indicates the occurrence of NmF2 peaks in the region 15°–20° north and south of the magnetic equator in magnetic coordinates [6]. The midlatitude summer nighttime anomaly indicates that the peak of summer NmF2 occurs around midnight [7]. Among the above ionospheric anomalies, winter anomaly is coupled with the annual anomaly, with both attracting extensive attention from scholars.

In 1936, Berkner et al. [5] discovered winter anomaly for the first time in both hemispheres, and since then, winter anomaly has been extensively investigated. Many scholars have studied the relationship between winter anomaly and solar activity. For example, Torr and Torr [8] constructed a global distribution of NmF2 at low-, medium-, and high-solar activity levels. They found winter anomaly in the middle and high latitudes of the

Northern Hemisphere and the Australian region of the Southern Hemisphere. Moreover, the winter anomaly is more intense and widespread with increasing solar activity levels. Based on global ionosphere maps (GIMs) from the Jet Propulsion Laboratory (JPL) for the years spanning 1998–2015, Yasyukevich et al. [9] analyzed the relationship between winter anomaly and solar activity using regression. They found that winter anomaly was most significant in North America, the Far East, and Australia and showed a positive correlation with solar activity. Azpilicueta and Nava [10] calculated the minimum level of solar activity corresponding to the occurrence of winter anomaly at 66 Global Positioning System (GPS) single stations. They found that the proportion of stations with winter anomaly in the northern and southern hemispheres was 1/33 and 21/33, respectively. In addition, Yasyukevich et al. [9] studied the relationship between winter anomaly and geomagnetic activity for the first time. They found that the growth of geomagnetic disturbance facilitates the intensity of the winter anomaly. Pavlov and Pavlova [11] found a higher correlation between winter anomaly and geomagnetic latitude than geographical latitude. Many scholars investigated the relationship between winter anomaly and magnetic pole positions and found that winter anomaly is more obvious near the poles [12–14]. But this conclusion is not absolute. Huo et al. [15] studied the geographical distribution of winter anomaly by using GPS total electron content (TEC) in 2002. They found that winter anomaly is more significant in the East Siberian region (far pole region) than in Europe. Many scholars also studied the altitudinal extent of winter anomaly and found that the winter anomaly exists only over a limited height range (about 150–450 km) around the peak height [13,16,17].

The physical mechanisms of winter anomaly have been extensively studied. Rishbeth and Setty [18] suggested that changes in neutral composition cause the winter anomaly, specifically the ratio of O/N_2. King [19] suggested that the changes are the results of global circulation in the thermosphere. This mechanism has been the most commonly accepted and confirmed in many publications [2,4,9,12,20,21]. However, longitudinal variations of the winter anomaly cannot be attributed to O/N_2 behavior. Yasyukevich et al. [9] concluded that it may be related to longitudinal variations of the vertical plasma transport induced by the thermosphere wind in local winter. Other influence factors of winter anomaly have also been put forward, such as the effects of temperature on the recombination coefficient between O^+ and the molecular neutral gas [20], the interruptions of the equatorial anomaly for the solar-heating induced summer-to-winter meridional wind [22], and the influences of vibrationally excited molecules [23].

In 1938, Berkner and Wells [3] discovered annual anomaly for the first time. Subsequently, Yonezawa [24] further studied this anomaly by using the NmF2 data from pairing ionosonde measurements. Rishbeth [4] and Rishbeth et al. [25] provided a good overview of the work on annual anomaly variations in the ionosphere. Rishbeth and Müller-Wodarg [26] combined paired NmF2 data from both hemispheres to describe annual anomaly by defining an annual asymmetry index. They found annual anomaly at noon and midnight, which was positively correlated with the level of solar activity. Mendillo et al. [2] demonstrated that annual anomaly does not have a longitude effect but is a global-scale phenomenon by using global TEC data. Su et al. [27] found more significant annual anomaly in the top ionosphere than in the bottom. Gowtam and Ram [28] calculated the annual asymmetry index at different altitudes and found that the index reached the maximum at an altitude of 300–500 km during periods of low solar activity and at 600 km as the solar activity increased. Currently, the physical mechanism of the annual anomaly remains controversial, and is listed as one of the top scientific goals concerning the ionosphere by Rishbeth [29].

Since 1998, the ionospheric analysis centers of the International Global Navigation Satellite System Service have been releasing global ionosphere map data products that span more than two solar activity cycles [30]. Many researchers have focused on the study of ionospheric anomaly by using GIMs. For example, Zhao et al. [12] and Yasyukevich et al. [9] studied the winter anomaly by using the GIMs provided by the JPL. Mendillo et al. [2] also used these data to study the annual anomaly. Feng et al. [31] studied the midlatitude summer nighttime anomaly by using the GIMs released by the Center for

Orbit Determination in Europe (CODE). Feng et al. [32] studied anomalies produced by the Tonga volcano using CODE GIMs. Their results can provide an important reference for establishing empirical ionospheric models. For example, the TECM-GRID model and TEC model of multi-source fusion introduce midlatitude summer nighttime anomaly correction terms [33,34], and the Global Neustrelitz TEC Model introduces equatorial anomaly correction terms [35,36]. These corrections significantly improve the ionospheric model accuracy.

In summary, it is of great significance to study the temporal and spatial variation characteristics of ionospheric anomalies based on TEC data, and some of the current works need to be further developed. For example, at each grid point, calculating the minimum solar activity level corresponding to the occurrence of winter anomaly can provide a reference for single-point ionospheric modeling. Winter anomaly is coupled with annual anomaly, and a quantitative analysis of the contribution of the winter anomaly to the annual anomaly can help deepen the understanding of these two anomalies. The main purpose of this study is to show the maps for the solar activity levels at which the winter anomaly appears and to assess the contribution of winter anomaly on the annual anomaly. Additionally, we verified the reliability of the regression approach.

2. Data and Approach

2.1. Data

In 1999, the CODE developed a global ionospheric model by using a spherical harmonics expansion up to degree and order 15, which has significant advantages for fitting global TEC [37]. From 3 November 2002 to 18 October 2014, the data started at 00:00 universal time (UT) and ended at 24:00 UT, with a time resolution of 2 h for a total of 13 global ionosphere maps per day. From 19 October 2014 to the present, the data starts at 00:00 UT and ends at 24:00 UT, with a time resolution of 1 h for a total of 25 global ionosphere maps per day. The latitude of the data ranges from 87.5°S to 87.5°N with 2.5° intervals, and its longitude ranges from 180°W to 180°E with 5° intervals, for a total of 5183 grid points. This study used the GIMs data of the 24th solar activity cycle (2008–2018) provided by CODE to investigate winter anomaly, annual anomaly, and the contribution of winter anomaly on annual anomaly. First, data with a temporal resolution of 2 h were linearly interpolated to uniform the amount of data for different years. Second, we transformed these data from the UT to the local time (LT) for further research (see Equation (1)). In the formula, *lon* is the geographic longitude of each grid point; *ut* is the universal time; *lt* is the local time; *dh* and *dmin* are the hours and minutes of difference between *lt* and *ut*, respectively. For ease of description, this study abbreviates CODE GIMs as CODG.

$$\begin{cases} dh = round(lon/15) \\ dmin = round(lon/15 - dh) \times 60 \\ lt = ut + (dh + dmin/60) \end{cases} \tag{1}$$

Solar extreme ultraviolet (EUV) radiation is the best parameter to study solar radiation temporal variation characteristics and solar-ionospheric effects. However, the space-based EUV observation records lack continuity and have a short observation history. Therefore, some solar activity indices are often used as the proxy for EUV when measuring solar activity levels. The daily $F_{10.7p}$ has a higher correlation coefficient with solar extreme ultraviolet than other indices, which better describes the level of solar activity [38]. Equation (2) defines the daily $F_{10.7p}$, where $F_{10.7}$ is 10.7 cm solar radiation flux, and $F_{10.7A}$ is the 81-day moving average of $F_{10.7}$. The unit of $F_{10.7p}$ is *sfu* (1 *sfu* = 10^{-22} Wm^{-2}Hz^{-1}).

$$F_{10.7p} = (F_{10.7} + F_{10.7A})/2 \tag{2}$$

The Kp index is usually used to describe the overall level of global geomagnetic activity. According to the scales (https://www.swpc.noaa.gov/noaa-scales-explanation, accessed on 10 October 2023) provided by the National Oceanic and Atmospheric Administration, the Kp value smaller than 5 implies minor geomagnetic activity. In this study, we studied winter anomaly, annual anomaly, and the contributions of winter anomaly to annual anomaly during the geomagnetic activity quiet period. Therefore, we excluded TEC data with Kp values greater than 5.

To characterize the relationship between winter anomaly and geomagnetic field configurations, we calculated the geomagnetic inclination contour by using the 13th-generation International Geomagnetic Reference Field (IGRF-13) model [39]. To verify the physical mechanism of the winter anomaly, we obtained the O/N_2 of solar cycle 24 from the Global Ultraviolet Imager (GUVI).

2.2. Approach

In this study, the correlation coefficient was used to describe the correlation between TEC and the daily $F_{10.7p}$ to verify the reliability of the regression approach (see Equation (3)). In the formula, R is the correlation coefficient; A denotes TEC; B denotes the daily $F_{10.7p}$; N is the number of data; μ_A and σ_A are the mean and standard deviation of TEC; μ_B and σ_B are the mean and standard deviation of the daily $F_{10.7p}$. The correlation is low linear when R is less than 0.4, significant when R is between 0.4 and 0.7, and high linear when R is greater than 0.7 [40].

$$R = \frac{1}{N-1} \sum_{i=1}^{N} \left(\frac{A_i - \mu_A}{\sigma_A} \right) \cdot \left(\frac{B_i - \mu_B}{\sigma_B} \right) \tag{3}$$

In general, winter anomaly is reflected by comparing the local time variation of TEC in summer and winter. However, ignoring the difference in solar activity levels between summer and winter introduces bias in the analysis of winter anomaly. Therefore, Yasyukevich et al. [9] investigated the winter anomaly by using a linear regression approach, which can calculate the winter anomaly index (WAI) at the same level of solar activity. The approach is extended as shown in Equations (4) and (5), where TECS_{LT} and TECW_{LT} are the summer and winter TEC values calculated based on the regression equation in different local times; a, b, c, d are coefficients to be estimated.

$$\begin{aligned} \text{TECS}_{LT} &= a(lat, lon) \times F_{10.7p}(doy) + b(lat, lon) \\ \text{TECW}_{LT} &= c(lat, lon) \times F_{10.7p}(doy) + d(lat, lon) \end{aligned} \tag{4}$$

$$\text{WAI} = \frac{\text{TECW}_{LT}}{\text{TECS}_{LT}} \tag{5}$$

In this study, June was used to represent the Northern Hemisphere summer, and December was used to represent the Northern Hemisphere winter. At each grid point and local time, we performed linear regressions of TEC and $F_{10.7p}$ for summer and winter based on Equation (4), respectively. The coefficients were estimated by using least squares to obtain the corresponding regression equations. Thus, the WAI can be calculated for different local times after specifying $F_{10.7p}$ on the basis of Equation (5). We counted the local time corresponding to the maximum of these winter anomaly indices during 08:00–16:00 LT (ignoring low latitudes). Its global distribution was shown in Figure 1. We found that it mainly ranges between 11:00–13:00 LT and shifts backward with increasing solar activity levels. Therefore, we take the maximum value of the WAI for 11:00–13:00 LT as the final index. We describe the presence and strength of winter anomaly through the WAI. If WAI is less than 1, there is no winter anomaly; if WAI is greater than 1, there is a winter anomaly, and the larger the WAI is, the more significant the winter anomaly is.

For ease of description, the minimum solar activity level corresponding to the occurrence of winter anomaly was represented by $F_{10.7p}$ triggering values (FTVs). We limited the range of FTVs to 66–181 *sfu* since the values of $F_{10.7p}$ for 2018–2018 range from 66 to 181 *sfu*. In this study, the global map of FTVs was calculated based on the regression equation at 12:00 LT, which is as follows: (1) At each grid point, we specify $F_{10.7p}$ in the range of

66–181 *sfu* with an interval of 1 *sfu* and substitute it into the corresponding summer and winter regression equations for 12:00 LT. Thus, the summer and winter TEC sequences calculated by using the regression equation are obtained at different levels of solar activity. (2) If TECW_{LT} is always greater than TECS_{LT} at the same level of solar activity, then FTVs are set to 66 *sfu*, in which case winter anomaly is always considered to occur. (3) If TECW_{LT} is always smaller than TECS_{LT} at the same level of solar activity, then FTVs are set to NAN, in which case winter anomaly is always considered to be absent. (4) If there is an intersection of TECW_{LT} and TECS_{LT}, then we calculate the FTVs based on Equation (6), where a, b, c, d are coefficients of regression equations in Equation (4). Winter anomaly is considered to occur when the level of solar activity is greater than FTVs.

$$\text{FTVs} = (d - b)/(a - c) \tag{6}$$

In this study, we described the magnitude of the annual anomaly by annual anomaly index (AI), the relative size of TEC in December to that in June, as shown in Equation (7). In the formula, TEC^{12} is the global TEC average for December, and TEC^{06} is the global TEC average for June.

$$\text{AI} = \left(\text{TEC}^{12} - \text{TEC}^{06}\right)/\text{TEC}^{06} \tag{7}$$

By decomposing the annual anomaly index to the hemispheric scale, we derived the proportion of the winter hemisphere contributing to the annual anomaly (WHCP) and that of the winter anomaly contributing to the annual anomaly (WACP). See Appendix A for a detailed derivation.

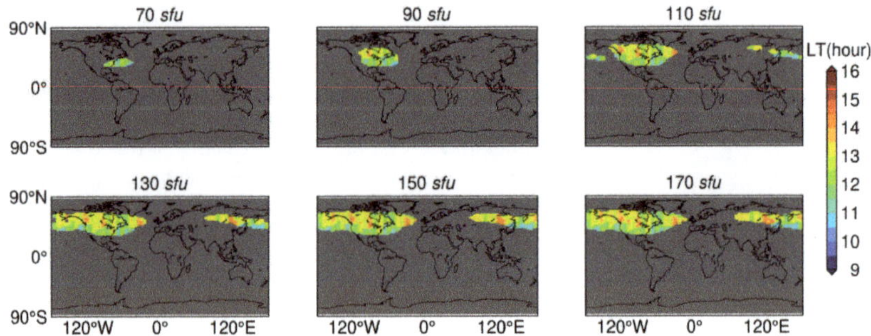

Figure 1. The global distribution of local time corresponds to the maximum of winter anomaly indices during 08:00–16:00 LT.

3. Results

3.1. Viability Analysis

We showed the global distribution of the correlation coefficients between TEC and the daily $F_{10.7p}$ at different local times (06:00–20:00 LT) in June and December (see Figures 2 and 3). In June, the correlation coefficients are lower near the Antarctic Peninsula. Additionally, the correlation coefficients are also relatively lower before 08:00 LT and after 18:00 LT in the middle latitudes of the Southern Hemisphere. The TEC calculated using the regression equation in this region may not describe the true variation in TEC. In December, the correlation coefficients are lower near the North Pole. Additionally, the correlation coefficients are also relatively lower before 08:00 LT and after 20:00 LT in the middle latitudes of the Northern Hemisphere. The TEC calculated using the regression equation in this region may not describe the true variation in TEC. Outside of these regions, correlation coefficients are over 0.9. Therefore, one can reliably use the regression approach to study anomalies when choosing the appropriate geographic location and local time.

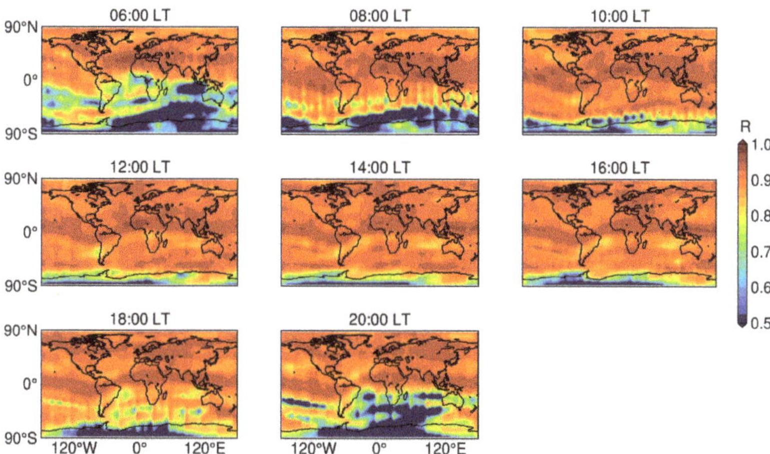

Figure 2. The global distribution of correlation coefficient between TEC and $F_{10.7p}$ during 06:00–20:00 LT in June.

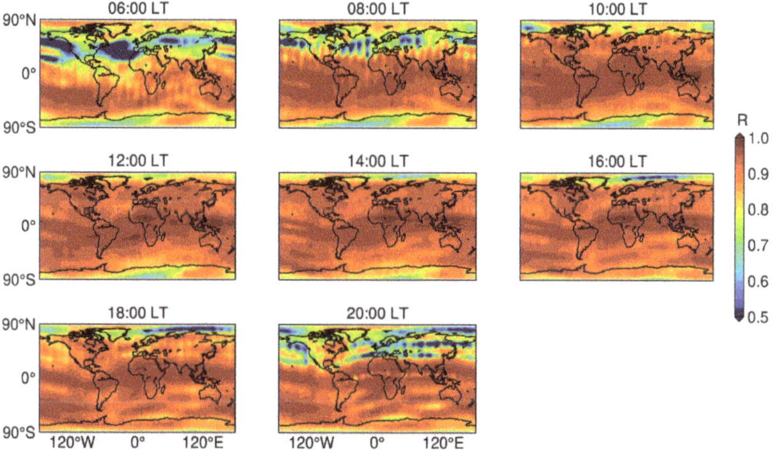

Figure 3. The global distribution of correlation coefficient between TEC and $F_{10.7p}$ during 06:00–20:00 LT in December.

In this study, points A (25.5°N, 5°W) and B (52.5°N, 95°W) were selected to verify the conformity of CODG with the TEC calculated by regression equations at different levels of solar activity, as shown in Figures 4 and 5. In these figures, TECS and TECW represent the summer TEC and winter TEC values calculated based on the regression equation, respectively. We selected four levels of solar activity: 90 *sfu*, 105 *sfu*, 120 *sfu*, and 135 *sfu*. The difference between the solar activity level corresponding to CODG and the labeled solar activity level is ±3 *sfu*. Point A is located at low latitudes and exhibits minimal seasonal variations in TEC, whereas point B is located at middle latitudes and displays more significant seasonal variations in TEC. Both at point A and point B, CODG conformed better with TECS and less well with TECW, but both conformed to the requirements of this study. Overall, the study of winter anomaly and annual anomaly based on the regression approach is reliable.

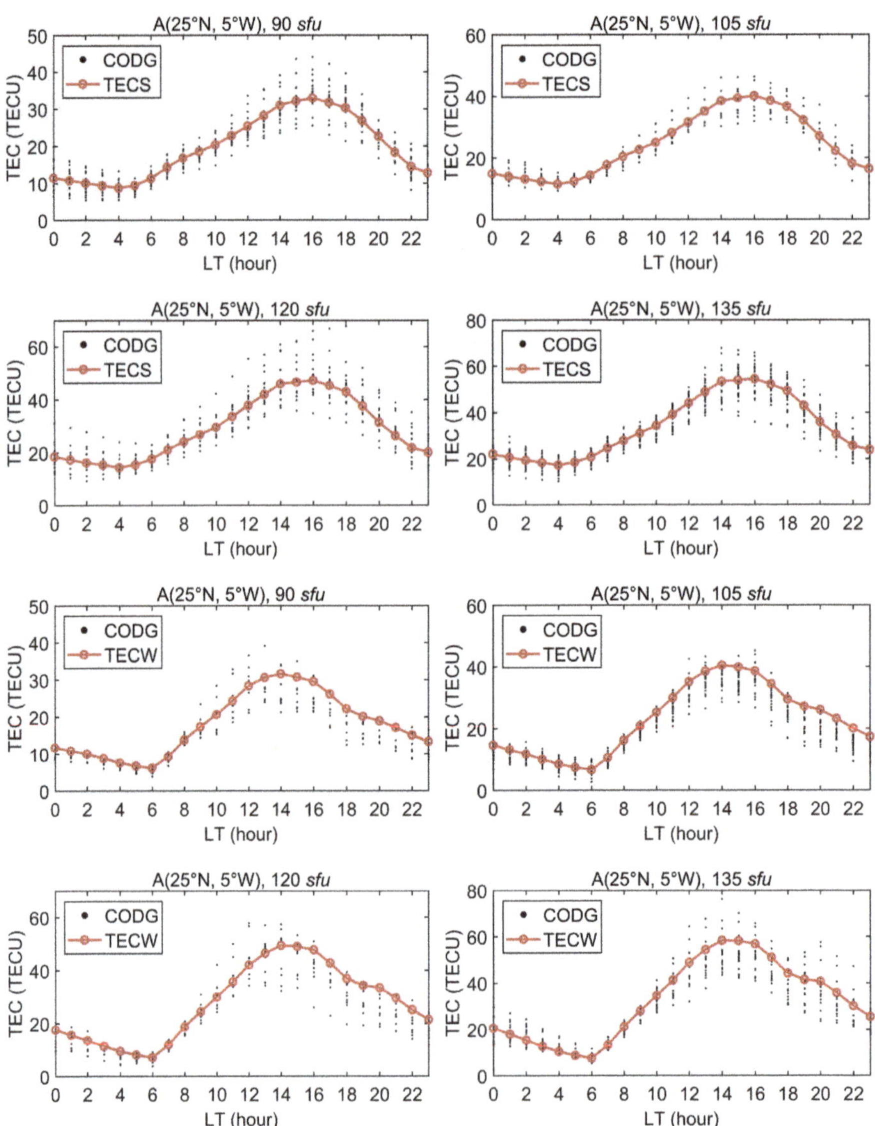

Figure 4. The conformity of CODG with TECS and TECW at point A (25°N, 5°W) for different levels of solar activity and local time.

A spurious winter anomaly index can be obtained by dividing the mean winter and summer TEC values at each grid point. The method incorporates the difference in solar activity levels between winter and summer. For ease of description, this method is called the averaging method. Figure 6 depicts the global distribution of the winter anomaly index from 2008–2018 based on the averaging method, with the $F_{10.7p}$ difference between December and June in parentheses in the subplot. Figure 6 does not reflect a positive correlation between winter anomaly and solar activity due to differences in winter and summer solar activity levels. The winter anomaly index calculated by this method is overestimated when the level of solar activity in winter is higher than that in summer.

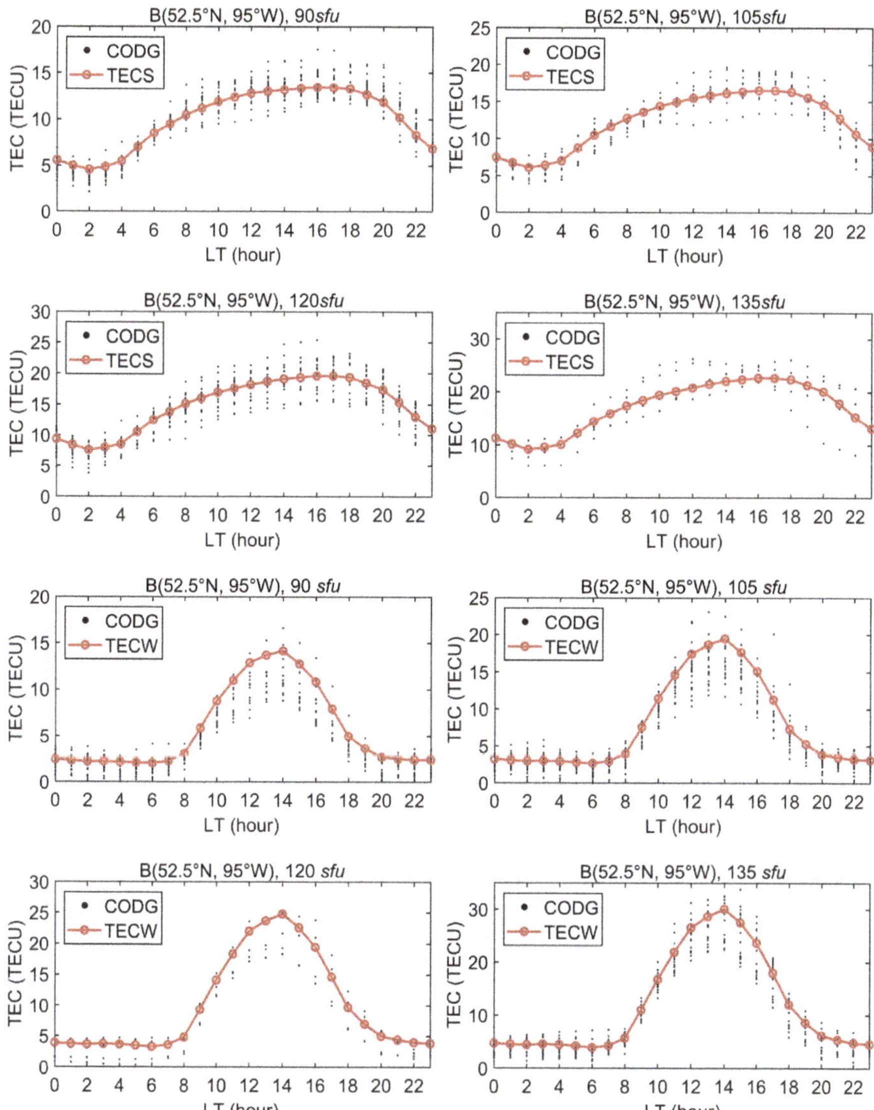

Figure 5. The conformity of CODG with TECS and TECW at point B (52.5°N, 95°W) for different levels of solar activity and local time.

In Equations (4) and (5), assigning different $F_{10.7p}$ to $TECS_{LT}$ and $TECW_{LT}$ can calculate the winter anomaly index with differences in solar activity levels. Based on this property, we can use the regression approach to calculate the winter anomaly index that includes the difference between winter and summer solar activity levels. Thus, we can compare the maps of the winter anomaly index calculated based on averaging and regression methods. Table 1 provides the average values of $F_{10.7p}$ from 2008–2018 for June and December. This study calculated the corresponding winter anomaly index by substituting $F_{10.7p}$ for June into $TECS_{LT}$ and $F_{10.7p}$ for December into $TECW_{LT}$. Figure 7 depicts the global distribution of the winter anomaly index from 2008–2018 based on the regression method. The similarity of the global distribution of the winter anomaly index in Figures 6 and 7 validates the superiority of the regression approach.

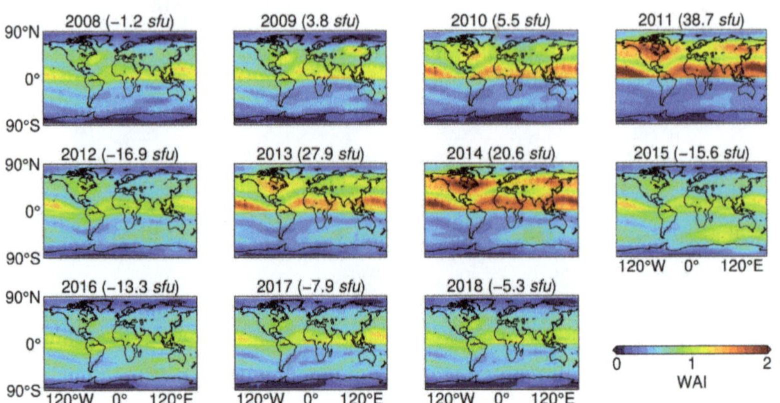

Figure 6. Global distribution of winter anomaly index based on the averaging method, with $F_{10.7p}$ difference between December and June in parentheses in the subplot.

Table 1. The $F_{10.7p}$ of June and December for each year (*sfu*).

Year	2008	2009	2010	2011	2012	2013	2014	2015	2016	2017	2018
June	68.3	70.9	75.9	98.2	127.8	117.5	130.5	123.0	87.1	77.3	73.2
December	67.1	74.7	81.4	136.9	110.9	145.4	151.1	107.4	73.8	69.4	67.9

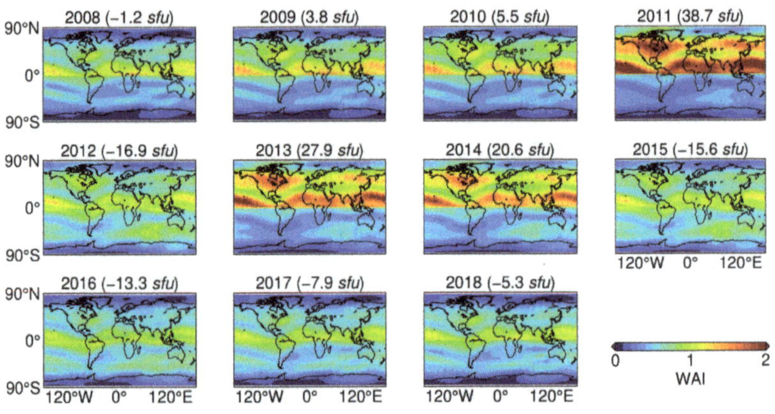

Figure 7. Global distribution of winter anomaly index based on the regression approach, with $F_{10.7p}$ difference between December and June in parentheses in the subplot.

3.2. Winter Anomaly

We depicted the global distribution of the winter anomaly index at different levels of solar activity (see Figure 8). When $F_{10.7p} < 90$ *sfu*, the winter anomaly is weak and occurs only near the southeastern region of North America. When solar activity levels are enhanced, winter anomaly increases in scope and magnitude. When $F_{10.7p} > 110$ *sfu*, the winter anomaly gradually extends to the Far East. The black dotted line in Figure 8 is the 60° geomagnetic inclination contour, which matches well with the winter anomaly region in North America, suggesting that the distribution of winter anomaly may be related to the geomagnetic inclination. The winter anomaly index is higher compared to its surroundings in the marine region in the southwest of Australia. However, the winter anomaly index of this region is less than one, which may be related to the fact that the regression equation was established in this study during the geomagnetic activity quiet period. To verify this, we also calculated the global distribution of the winter anomaly index by using the TEC data with Kp > 3 (see Figure 9). It can be seen that the winter anomaly in Australia is captured during periods of stronger geomagnetic activity. Therefore, winter anomaly does not necessarily occur in the Southern Hemisphere.

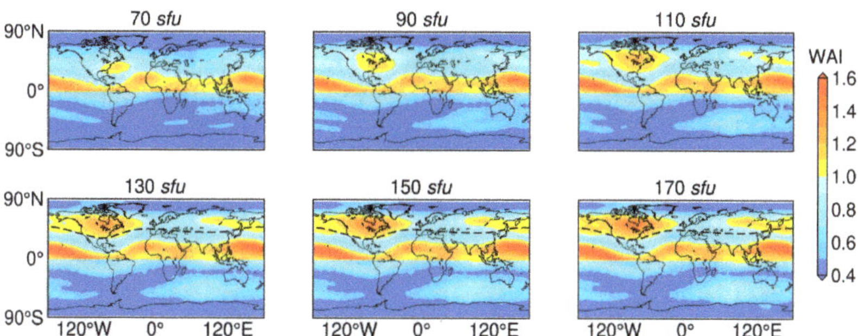

Figure 8. Global distribution of winter anomaly index at different levels of solar activity (Kp ≤ 5), the dotted black line is the 60° geomagnetic inclination contour.

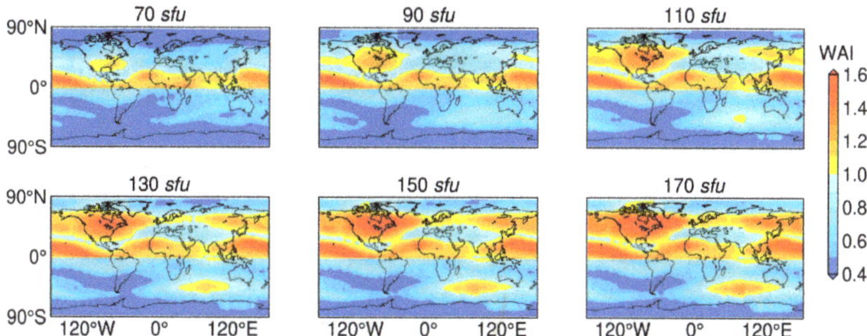

Figure 9. Global distribution of winter anomaly index at different levels of solar activity (Kp > 3).

Figure 10 depicts the interannual variations in the global distribution of winter and summer ratio of O/N_2 (ON2I). In this figure, the part of ON2I less than 0 is not shown. ON2I is the largest in the Far East, while the winter anomaly is strongest in North America. Yasyukevich et al. [9] concluded that it may be related to longitudinal variations of the vertical plasma transport induced by the thermosphere wind in local winter. In general, seasonal variations in O/N_2 not only explain winter anomaly near North America and Australia but also the positive correlation between winter anomaly and solar activity.

In this study, the global distribution of FTVs was calculated (see Figure 11). FTVs are the smallest in the southeastern region of North America, which means that this region is the most prone to winter anomaly, and it corresponds to the 70–90 *sfu* situation in Figure 8. The FTVs are relatively larger in the western regions of North America and the Far East, which means that the winter anomaly in these regions occurs at higher levels of solar activity, corresponding to the 130–170 *sfu* situation in Figure 8. The map can provide a reference for single-point ionospheric modeling and illustrate the solar activity level at which winter anomaly is considered for each grid point.

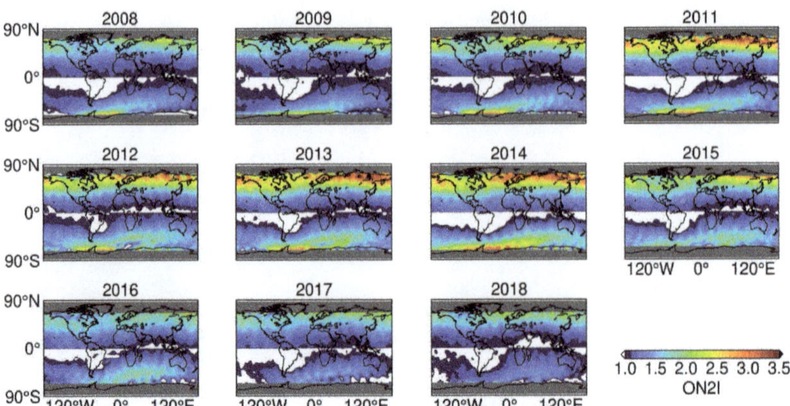

Figure 10. Interannual variations in the global distribution of winter and summer ratio of O/N$_2$.

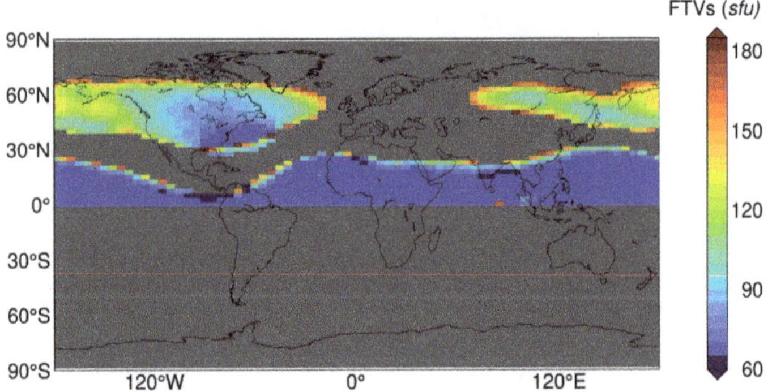

Figure 11. The global distribution of the minimum solar activity level corresponding to the occurrence of winter anomaly (Kp ≤ 5).

3.3. Annual Anomaly

Monthly differences in solar activity levels introduce a bias in the analysis of monthly variations in TEC. In this study, the monthly TEC was calculated under different solar activity levels and local time by using regression equations. This methods can ensure that each month corresponds to the same level of solar activity (see Figure 12). With the enhanced level of solar activity, the TEC values of each month increased, and the contrast in TEC between months became more pronounced. The TEC value in December is greater than in June, reflecting the annual anomaly. The TEC values in March and October are greater than the other months, reflecting the semiannual anomaly.

Based on the results in Figures 2 and 3, we selected the main grid points in the Southern Hemisphere (2.5°S–65°S) and the Northern Hemisphere (2.5°N–65°N) to accurately calculate the annual anomaly index as well as the contribution of the winter hemisphere and winter anomaly to the annual anomaly. We calculated the difference between the global TEC averages for December and June as well as the annual anomaly index. We then provided its variation with local time and solar activity (see Figure 13). In Figure 13a, 12:00 LT and 14:00 LT corresponds to the maximum value of the difference, followed by 10:00 LT, and 16:00 LT is the minimum. In Figure 13b, 12:00 LT corresponds to the maximum value of the annual anomaly index, followed by 10:00 LT and 14:00 LT, and 16:00 LT is the minimum. The annual anomaly index increases with increasing levels of solar activity, with greater growth at low levels of solar activity and smaller growth at high levels of solar activity. The annual anomaly index can be up to 0.344 at 12:00 LT (at this time, $F_{10.7p} = 170$ *sfu*), which indicates that the TEC in December was approximately 34.4% larger than the TEC in June.

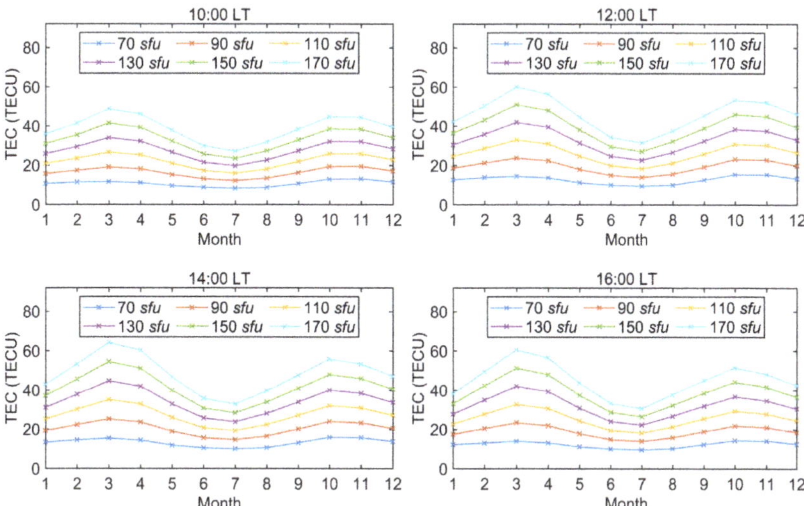

Figure 12. Variation in monthly TEC calculated by using regression equations with local time and solar activity.

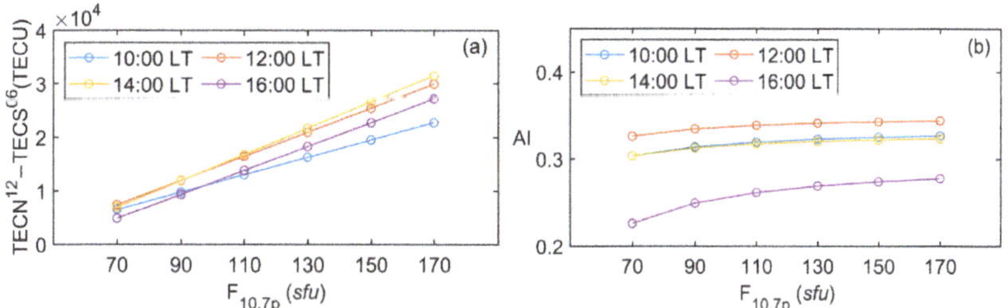

Figure 13. (**a**) Variation of the global TEC difference between December and June with local time and solar activity; (**b**) Variation of annual anomaly index with local time and solar activity.

3.4. Contribution of the Winter Anomaly to Annual Anomaly

We calculated the mean TEC of the southern and northern hemispheres in June and December. We then provided its variation with local time and solar activity (see Figure 14). The TEC in the summer hemisphere ($TECN^{06} + TECS^{12}$) is consistently larger than the TEC in the winter hemisphere ($TECN^{12} + TECS^{06}$), which is caused by the smaller solar zenith angle in the summer hemisphere according to the Chapman model. At 12:00 LT and 14:00 LT, the TEC in the Northern Hemisphere winter ($TECN^{12}$) is larger than that in the Northern Hemisphere summer ($TECN^{06}$) when $F_{10.7p} > 130$ *sfu*, which is related to the winter anomaly enhanced with the levels of solar activity. The TEC in the Northern Hemisphere winter ($TECN^{12}$) is larger than the TEC in the Southern Hemisphere winter ($TECS^{06}$) due to a stronger winter anomaly in the Northern Hemisphere, and their difference is a partial source of the annual anomaly. The TEC in the Southern Hemisphere summer ($TECS^{12}$) is consistently larger than the TEC in the Northern Hemisphere summer ($TECN^{06}$), which may be related to variations in the Sun–Earth distance, and their difference is another source of annual anomaly.

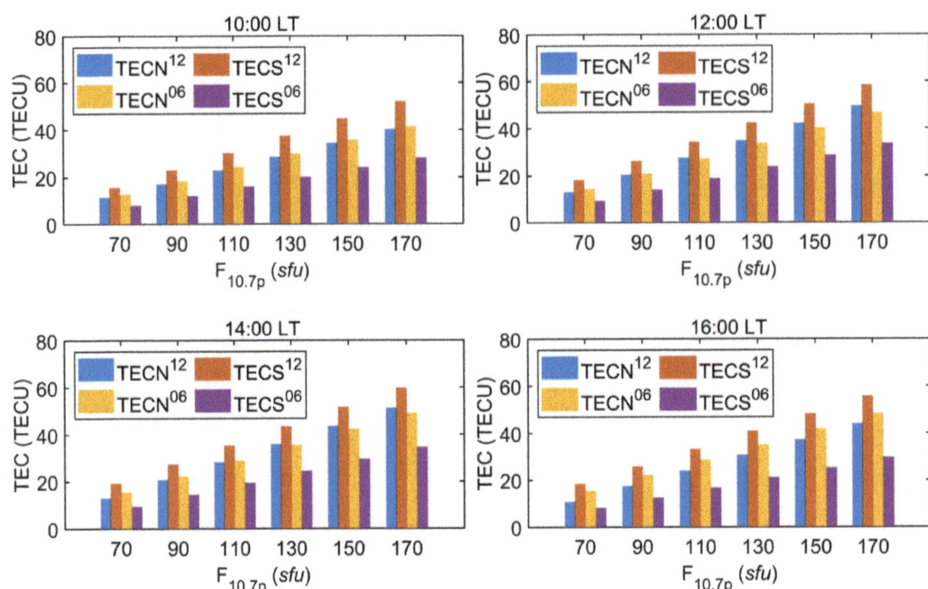

Figure 14. Variation of the mean TEC of the southern and northern hemispheres in June and December with local time and solar activity.

Figure 15c depicts the percentage contribution of the winter hemisphere to the annual anomaly with local time and solar activity. The winter hemisphere contributes more than 50% to the annual anomaly when the solar activity level exceeds 90 *sfu*. The percentage contribution of the winter hemisphere to the annual anomaly can be up to 65% when the solar activity level exceeds 130 *sfu*. At moderate-to-high solar activity levels, the percentage contribution of the winter hemisphere to the annual anomaly is largest at 16:00 LT, followed by 14:00 LT and 12:00 LT, and smallest at 10:00 LT. We summed the TEC at the selected grid points in each hemisphere in June and December, then calculated the contribution to the annual anomaly for the winter hemisphere and the summer hemisphere, respectively. Figure 15a depicts the total contribution of the winter hemisphere to the annual anomaly with local time and solar activity. At 10:00 LT, the growth of the total contribution of the winter hemisphere is the smallest, leading to a decreased WHCP. Figure 15b depicts the total contribution of the summer hemisphere to the annual anomaly with local time and solar activity. At 16:00 LT, the growth of the total contribution of the winter hemisphere is the largest, leading to an increased WACP. Figure 15d depicts the proportion of the winter anomaly contributing to the annual anomaly with local time and solar activity. There is a significant growth in this proportion at 12:00 LT and 14:00 LT, which is related to the solar activity dependence of winter anomaly. At 12:00 LT, winter anomaly can contribute up to 32% to the annual anomaly (at this time, $F_{10.7p}$ = 170 *sfu*). In contrast, there is a slight growth in this proportion at 10:00 LT and 16:00 LT, which is related to the local time dependence of winter anomaly. By comparing WHCP and WACP, we can conclude that the lower latitudes of the Northern Hemisphere also contribute significantly to the annual anomaly.

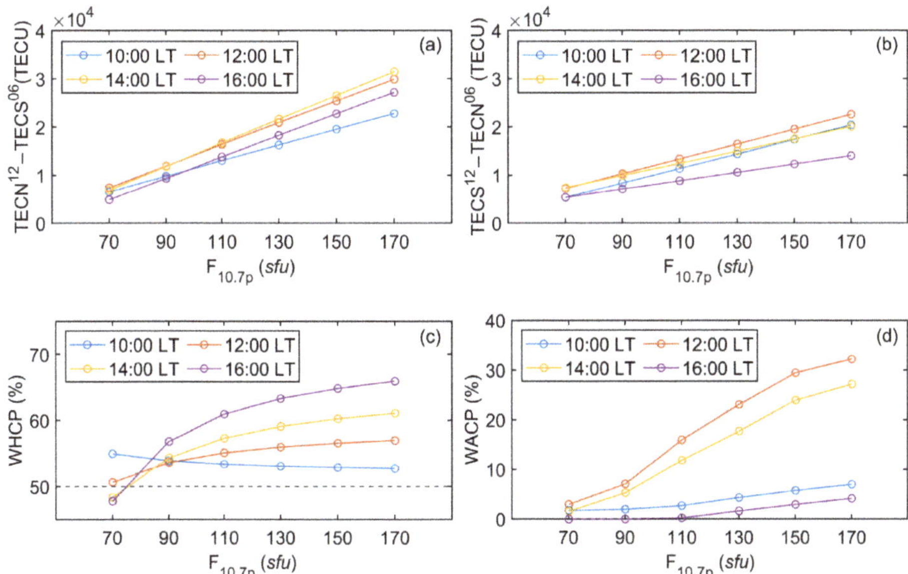

Figure 15. (**a**) Variation in the total winter hemispheric contribution to the annual anomaly with local time and solar activity; (**b**) Variation in the total summer hemispheric contribution to the annual anomaly with local time and solar activity; (**c**) Variation in the proportion of the winter hemisphere contributing to the annual anomaly with local time and solar activity; (**d**) Variation in the proportion of the winter anomaly contributing to the annual anomaly with local time and solar activity.

4. Summary and Conclusions

In this study, we investigated the winter anomaly, annual anomaly, and the proportion of the winter anomaly contributing to the annual anomaly based on CODG data of the 24th solar activity cycle (2008–2018) during the geomagnetic activity quiet period (Kp $\leq$ 5) using a regression approach. First, the reliability of the regression approach was verified. The pattern of variation of the winter anomaly index was then analyzed, and the global distribution of the FTVs was given. Next, the annual anomaly was analyzed. Finally, the contribution of the winter anomaly and winter hemisphere to the annual anomaly was quantitatively analyzed.

We analyzed the global distribution of the correlation coefficients between TEC and $F_{10.7p}$, as well as the agreement between CODG and TEC calculated by regression equations. These all indicate that the regression approach has high accuracy. In addition, the winter anomaly index, which includes the difference between winter and summer solar activity levels, was calculated using the regression and averaging methods, and they are in good agreement.

We measured the local time corresponding to the maximum of winter anomaly indices during 08:00–16:00 LT and found that winter anomaly is more significant at 11:00–13:00 LT. By analyzing the global distribution of the winter anomaly index at different solar activity levels, we verified the positive correlation between winter anomaly and solar activity levels and found that the winter anomaly region extends from North America to the Far East with enhanced solar activity levels. We also found that the distribution of winter anomaly in North America may be related to the geomagnetic inclination. In addition, we verified that winter anomaly near Australia is more significant when geomagnetic activity is stronger. By analyzing the interannual variation of ON2I, we verified that seasonal variations in O/N_2 not only explain winter anomaly near North America and Australia but also the positive correlation between winter anomaly and solar activity.

We established the global distribution of FTVs by calculating the intersection of the regression equations, which can be used as a reference for single-point ionospheric modeling.

We found that the annual anomaly was strongest at 12:00 LT (December TEC was 34.4% larger than the June TEC) by analyzing the variation of the annual anomaly index with local time and solar activity level. In addition, we also verified the positive correlation between annual anomaly and solar activity and found that the growth rate of the annual anomaly index is lower at higher levels of solar activity.

By analyzing the variation of WHCP with local time and solar activity level, we found that the TEC in the winter hemisphere will be the main factor determining the size of the annual anomaly when the solar activity level is stronger. For example, the winter hemisphere contributes more than 50% to the annual anomaly when the solar activity level exceeds 90 *sfu*, while the contribution of the winter hemisphere can be up to 65% when the solar activity level exceeds 130 *sfu*. By analyzing the variation of WACP with local time and solar activity level, we found that winter anomaly can contribute up to 32% to the annual anomaly.

We hope that the results of this study will deepen the current knowledge of scholars about winter and annual anomalies, and provide a reference for establishing and improving empirical ionospheric models. Since NmF2 is an important parameter for the study of ionospheric anomalies, the quantitative analysis of winter anomaly, annual anomaly, and the contribution of winter anomaly to annual anomaly through NmF2 should be considered in the next study.

Author Contributions: Conceptualization, writing—original draft preparation, writing—review and editing, and methodology, K.W., J.F. and Z.Z.; validation, K.W.; data curation, Z.Z. and B.H.; formal analysis, K.W. and J.F.; funding acquisition, J.F. All authors have read and agreed to the published version of the manuscript.

Funding: This work was supported by the National Natural Science Foundation of China under Grant Nos. 42274040 and 41804032, by the Specialized Research Fund for State Key Laboratories, by the State Key Laboratory of Geo-Information Engineering under Grant No. SKLGIE2021-M-3-2, and by the Youth Project of Natural Science Foundation of Shandong Province under Grant No. ZR2022QD078.

Data Availability Statement: The GIMs data provided by the CODE in the study are available at http://ftp.aiub.unibe.ch/CODE/ (accessed on 10 October 2023). The $F_{10.7}$ and Kp data provided by the OMNIWEB in the study are available at https://omniweb.gsfc.nasa.gov/form/dx1.html (accessed on 10 October 2023). The geo-magnetic inclination data provided by the IGRF-13 in the study are available at https://wdc.kugi.kyoto-u.ac.jp/igrf/index.html (accessed on 10 October 2023). The O/N_2 data provided by the GUVI in the study are available at http://guvitimed.jhuapl.edu/ (accessed on 10 October 2023).

Acknowledgments: We are grateful for the GIM products provided by the CODE, the $F_{10.7}$ and Kp data provided by the OMNIWEB, the geomagnetic inclination data provided by the IGRF-13, and the O/N_2 data provided by the GUVI. We acknowledge the use of MATrix LABoratory (MATLAB R2023a) software, the Generic Mapping Tools (GMT v6.4.0). We would like to thank the anonymous reviewers for their helpful suggestions.

Conflicts of Interest: The authors declare no conflict of interest.

Appendix A

The following content in Appendix A comprises the supplemental descriptions in Section 2.2 on how to deduce the proportion of the winter hemisphere contributing to the annual anomaly and that of the winter anomaly contributing to the annual anomaly.

To explore the proportion of the winter hemisphere contributing to the annual anomaly, we decomposed the annual anomaly index to the hemispheric scale, as shown in Equation (A1). In the formula, TEC^{12} (TEC^{06}) represents the global TEC average for December (June); $TECN^{12}$ ($TECS^{12}$) represents the total TEC in the Northern Hemisphere (Southern Hemisphere) in December; TECN^{06} (TECS^{06}) represents the total TEC in the Northern Hemisphere (Southern Hemisphere) in June. In this study, June was used to represent Northern Hemisphere summer, and December was used to represent Northern Hemisphere winter. TECN^{12} and TECS^{06} are both winter, TECN^{06} and TECS^{12} are both summer. The summation of $\text{TECN}^{12} - \text{TECS}^{06}$ and $\text{TECS}^{12} - \text{TECN}^{06}$ can describe the annual anomaly. Therefore, we make $\text{TECN}^{12} - \text{TECS}^{06}$ as the total contribution of the winter hemisphere to the annual anomaly, and $\text{TECS}^{12} - \text{TECN}^{06}$ as the total contribution of the summer hemisphere to the annual anomaly.

$$
\begin{aligned}
\text{AI} \ &= \ \left(\text{TEC}^{12} - \text{TEC}^{06}\right) / \text{TEC}^{06} \\
&= \ \left[\left(\text{TECN}^{12} + \text{TECS}^{12}\right) - \left(\text{TECN}^{06} + \text{TECS}^{06}\right)\right] / \left(\text{TECN}^{06} + \text{TECS}^{06}\right) \\
&= \ \left[\left(\text{TECN}^{12} - \text{TECS}^{06}\right) + \left(\text{TECS}^{12} - \text{TECN}^{06}\right)\right] / \left(\text{TECN}^{06} + \text{TECS}^{06}\right)
\end{aligned}
\tag{A1}
$$

The annual anomaly index can be calculated by summing the magnitude of the contribution from the winter hemisphere (WHC, see Equation (A2)) and that of the summer hemisphere (SHC, see Equation (A3)). With this idea of decomposition, we can calculate the proportion of the winter hemisphere contributing to the annual anomaly (WHCR, see Equation (A4)).

$$
\text{WHC} = \left(\text{TECN}^{12} - \text{TECS}^{06}\right) / \left(\text{TECN}^{06} + \text{TECS}^{06}\right)
\tag{A2}
$$

$$
\text{SHC} = \left(\text{TECS}^{12} - \text{TECN}^{06}\right) / \left(\text{TECN}^{06} + \text{TECS}^{06}\right)
\tag{A3}
$$

$$
\text{WHCR} = (\text{WHC} / \text{AI}) \times 100\%
\tag{A4}
$$

To explore the proportion of the winter anomaly contributing to the annual anomaly, we further decompose Equation (A2). The process of decomposition is shown in Equation (A5), where YWA represents grid points where winter anomaly is present, and NWA represents grid points where winter anomaly is not absent. It is worth noting that winter anomaly is not considered to exist at low latitudes, even if the winter anomaly index is greater than one.

$$
\begin{aligned}
\text{WHC} \ &= \ \left(\text{TECN}^{12} - \text{TECS}^{06}\right) / \left(\text{TECN}^{06} + \text{TECS}^{06}\right) \\
&= \ \left[\left(\text{TECN(YWA)}^{12} + \text{TECS(NWA)}^{12}\right) - \left(\text{TECN(YWA)}^{06} + \text{TECS(NWA)}^{06}\right)\right] / \left(\text{TECN}^{06} + \text{TECS}^{06}\right) \\
&= \ \left[\left(\text{TECN(YWA)}^{12} - \text{TECS(YWA)}^{06}\right) + \left(\text{TECS(NWA)}^{12} - \text{TECN(NWA)}^{06}\right)\right] / \left(\text{TECN}^{06} + \text{TECS}^{06}\right)
\end{aligned}
\tag{A5}
$$

The WHC can be calculated by summing the magnitude of the contribution from the winter anomaly (WAC, see Equation (A6)) and the magnitude of the contribution from the non-winter anomaly (NWAC, see Equation (A7)). With this idea of decomposition, we can calculate the proportion of winter anomaly contributing to the annual anomaly (WACR, see Equation (A8)).

$$
\text{WAC} = \left(\text{TECN(YWA)}^{12} - \text{TECS(YWA)}^{06}\right) / \left(\text{TECN}^{06} + \text{TECS}^{06}\right)
\tag{A6}
$$

$$
\text{NWAC} = \left(\text{TECS(NWA)}^{12} - \text{TECN(NWA)}^{06}\right) / \left(\text{TECN}^{06} + \text{TECS}^{06}\right)
\tag{A7}
$$

$$
\text{WACR} = (\text{WAC} / \text{AI}) \times 100\%
\tag{A8}
$$

References

1. Appleton, E.V.; Naismith, R. Some Further Measurements of Upper Atmospheric Ionization. *Proc. R. Soc. Lond. Ser. A Math. Phys. Sci.* **1935**, *150*, 685–708. [CrossRef]
2. Mendillo, M.; Huang, C.-L.; Pi, X.; Rishbeth, H.; Meier, R. The Global Ionospheric Asymmetry in Total Electron Content. *J. Atmos. Sol.-Terr. Phys.* **2005**, *67*, 1377–1387. [CrossRef]
3. Berkner, L.V.; Wells, H.W. Non-Seasonal Change of F2-Region Ion-Density. *Terr. Magn. Atmos. Electr.* **1938**, *43*, 15–36. [CrossRef]
4. Rishbeth, H. How the Thermospheric Circulation Affects the Ionospheric F2-Layer. *J. Atmos. Sol.-Terr. Phys.* **1998**, *60*, 1385–1402. [CrossRef]
5. Berkner, L.V.; Wells, H.W.; Seaton, S.L. Characteristics of the Upper Region of the Ionosphere. *Terr. Magn. Atmos. Electr.* **1936**, *41*, 173–184. [CrossRef]
6. Appleton, E.V. Two Anomalies in the Ionosphere. *Nature* **1946**, *157*, 691. [CrossRef]
7. Thampi, S.V.; Lin, C.; Liu, H.; Yamamoto, M. First Tomographic Observations of the Midlatitude Summer Nighttime Anomaly over Japan. *J. Geophys. Res. Space Phys.* **2009**, *114*, 1–10. [CrossRef]
8. Torr, M.R.; Torr, D.G. The Seasonal Behaviour of the F2-Layer of the Ionosphere. *J. Atmos. Terr. Phys.* **1973**, *35*, 2237–2251. [CrossRef]
9. Yasyukevich, Y.; Yasyukevich, A.; Ratovsky, K.; Klimenko, M.; Klimenko, V.; Chirik, N. Winter Anomaly in NmF2 and TEC: When and Where It Can Occur. *J. Space Weather Space Clim.* **2018**, *8*, A45. [CrossRef]
10. Azpilicueta, F.; Nava, B. A Different View of the Ionospheric Winter Anomaly. *Adv. Space Res.* **2021**, *67*, 150–162. [CrossRef]
11. Pavlov, A.V.; Pavlova, N.M. Variations in Statistical Parameters of the NmF2 Winter Anomaly with Latitude and Solar Activity. *Geomagn. Aeron.* **2012**, *52*, 335–343. [CrossRef]
12. Zhao, B.; Wan, W.; Liu, L.; Mao, T.; Ren, Z.; Wang, M.; Christensen, A.B. Features of Annual and Semiannual Variations Derived from the Global Ionospheric Maps of Total Electron Content. *Ann. Geophys.* **2007**, *25*, 2513–2527. [CrossRef]
13. Lee, W.K.; Kil, H.; Kwak, Y.-S.; Wu, Q.; Cho, S.; Park, J.U. The Winter Anomaly in the Middle-Latitude F Region during the Solar Minimum Period Observed by the Constellation Observing System for Meteorology, Ionosphere, and Climate. *J. Geophys. Res. Space Phys.* **2011**, *116*, 1–10. [CrossRef]
14. Meza, A.; Natali, M.; Fernandez, L. Analysis of the Winter and Semiannual Ionospheric Anomalies in 1999–2009 Based on GPS Global International GNSS Service Maps. *J. Geophys. Res. Space Phys.* **2012**, *117*, 1319. [CrossRef]
15. Huo, X.L.; Yuan, Y.B.; Ou, J.K.; Zhang, K.F.; Bailey, G.J. Monitoring the Global-Scale Winter Anomaly of Total Electron Contents Using GPS Data. *Earth Planets Space* **2009**, *61*, 1019–1024. [CrossRef]
16. Mikhailov, A.V.; Perrone, L. Comment on "The Winter Anomaly in the Middle-Latitude F Region during the Solar Minimum Period Observed by the Constellation Observing System for Meteorology, Ionosphere, and Climate" by W. K. Lee, H. Kil, Y.-S. Kwak, Q. Wu, S. Cho, and J. U. Park. *J. Geophys. Res. Space Phys.* **2014**, *119*, 7972–7978. [CrossRef]
17. Klimenko, M.V.; Klimenko, V.V.; Zakharenkova, I.E.; Ratovsky, K.G.; Yasyukevich, A.S.; Yasyukevich, Y.V. Altitudinal Extent of Winter Anomaly and Its Manifestation in the Total Electron Content. *Russ. J. Phys. Chem. B* **2019**, *13*, 884–891. [CrossRef]
18. Rishbeth, H.; Setty, C.S.G.K. The F-Layer at Sunrise. *J. Atmos. Terr. Phys.* **1961**, *20*, 263–276. [CrossRef]
19. King, G.A.M. The Dissociation of Oxygen and High Level Circulation in the Atmosphere. *J. Atmos. Sci.* **1964**, *21*, 231–237. [CrossRef]
20. Burns, A.G.; Wang, W.; Qian, L.; Solomon, S.C.; Zhang, Y.; Paxton, L.J.; Yue, X. On the Solar Cycle Variation of the Winter Anomaly. *J. Geophys. Res. Space Phys.* **2014**, *119*, 4938–4949. [CrossRef]
21. Zou, L.; Rishbeth, H.; Müller-Wodarg, I.C.F.; Aylward, A.D.; Millward, G.H.; Fuller-Rowell, T.J.; Idenden, D.W.; Moffett, R.J. Annual and Semiannual Variations in the Ionospheric F2-Layer. I. Modelling. *Ann. Geophys.* **2000**, *18*, 927–944. [CrossRef]
22. Qian, L.; Burns, A.G.; Wang, W.; Solomon, S.C.; Zhang, Y.; Hsu, V. Effects of the Equatorial Ionosphere Anomaly on the Interhemispheric Circulation in the Thermosphere. *J. Geophys. Res. Space Phys.* **2016**, *121*, 2522–2530. [CrossRef]
23. Torr, D.G.; Torr, M.R.; Richards, P.G. Causes of the F Region Winter Anomaly. *Geophys. Res. Lett.* **1980**, *7*, 301–304. [CrossRef]
24. Yonezawa, T. The Solar-Activity and Latitudinal Characteristics of the Seasonal, Non-Seasonal and Semi-Annual Variations in the Peak Electron Densities of the F2-Layer at Noon and at Midnight in Middle and Low Latitudes. *J. Atmos. Terr. Phys.* **1971**, *33*, 889–907. [CrossRef]
25. Rishbeth, H.; Müller-Wodarg, I.C.F.; Zou, L.; Fuller-Rowell, T.J.; Millward, G.H.; Moffett, R.J.; Idenden, D.W.; Aylward, A.D. Annual and Semiannual Variations in the Ionospheric F2-Layer: II. Physical Discussion. *Ann. Geophys.* **2000**, *18*, 945–956. [CrossRef]
26. Rishbeth, H.; Müller-Wodarg, I.C.F. Why Is There More Ionosphere in January than in July? The Annual Asymmetry in the F2-Layer. *Ann. Geophys.* **2006**, *24*, 3293–3311. [CrossRef]
27. Su, Y.Z.; Bailey, G.J.; Oyama, K.-I. Annual and Seasonal Variations in the Low-Latitude Topside Ionosphere. *Ann. Geophys.* **1998**, *16*, 974–985. [CrossRef]
28. Gowtam, V.S.; Ram, S.T. Ionospheric Winter Anomaly and Annual Anomaly Observed from Formosat-3/COSMIC Radio Occultation Observations during the Ascending Phase of Solar Cycle 24. *Adv. Space Res.* **2017**, *60*, 1585–1593. [CrossRef]
29. Rishbeth, H. Thermospheric Targets. *Eos Trans. Am. Geophys. Union* **2007**, *88*, 189–193. [CrossRef]

30. Roma-Dollase, D.; Hernández-Pajares, M.; Krankowski, A.; Kotulak, K.; Ghoddousi-Fard, R.; Yuan, Y.; Li, Z.; Zhang, H.; Shi, C.; Wang, C.; et al. Consistency of Seven Different GNSS Global Ionospheric Mapping Techniques during One Solar Cycle. *J. Geod.* **2018**, *92*, 691–706. [CrossRef]
31. Feng, J.; Wang, K.; Li, W.; Han, B.; Zhao, Z.; Zhang, T.; Meng, D. Analysis of Temporal and Spatial Variation Characteristics of Midlatitude Summer Nighttime Anomaly in Low and Middle Solar Activity Period. *Adv. Space Res.* **2023**, *71*, 4351–4360. [CrossRef]
32. Feng, J.; Han, B.; Zhao, Z.; Wang, Z. A New Global Total Electron Content Empirical Model. *Remote Sens.* **2019**, *11*, 706. [CrossRef]
33. Feng, J.; Yuan, Y.; Zhang, T.; Zhang, Z.; Meng, D. Analysis of Ionospheric Anomalies before the Tonga Volcanic Eruption on 15 January 2022. *Remote Sens.* **2023**, *15*, 4879. [CrossRef]
34. Feng, J.; Zhang, T.; Li, W.; Zhao, Z.; Han, B.; Wang, K. A New Global TEC Empirical Model Based on Fusing Multi-Source Data. *GPS Solut.* **2022**, *27*, 20. [CrossRef]
35. Jakowski, N.; Hoque, M.M.; Mayer, C. A New Global TEC Model for Estimating Transionospheric Radio Wave Propagation Errors. *J. Geod.* **2011**, *85*, 965–974. [CrossRef]
36. Hoque, M.M.; Jakowski, N.; Orús-Pérez, R. Fast Ionospheric Correction Using Galileo Az Coefficients and the NTCM Model. *GPS Solut.* **2019**, *23*, 41. [CrossRef]
37. Schaer, S. *Mapping and Predicting the Earth's Ionosphere Using the Global Positioning System*; Institutfür Geodäsie und Photogrammetrie, Eidg, Technische Hochschule Zürich: Zürich, Switzerland, 1999.
38. Lei, J.; Liu, L.; Wan, W.; Zhang, S.-R. Variations of Electron Density Based on Long-Term Incoherent Scatter Radar and Ionosonde Measurements over Millstone Hill. *Radio Sci.* **2005**, *40*, 1–10. [CrossRef]
39. Alken, P.; Thébault, E.; Beggan, C.D.; Amit, H.; Aubert, J.; Baerenzung, J.; Bondar, T.N.; Brown, W.J.; Califf, S.; Chambodut, A.; et al. International Geomagnetic Reference Field: The Thirteenth Generation. *Earth Planets Space* **2021**, *73*, 49. [CrossRef]
40. Oinats, A.V.; Ratovsky, K.G.; Kotovich, G.V. Influence of the 27-Day Solar Flux Variations on the Ionosphere Parameters Measured at Irkutsk in 2003–2005. *Adv. Space Res.* **2008**, *42*, 639–644. [CrossRef]

Article

Analysis of Ionospheric Anomalies before the Tonga Volcanic Eruption on 15 January 2022

Jiandi Feng [1,2], Yunbin Yuan [3], Ting Zhang [3,*], Zhihao Zhang [1] and Di Meng [1]

[1] School of Civil Engineering and Geomatics, Shandong University of Technology, Zibo 255000, China; jdfeng@whu.edu.cn (J.F.); zzh1049822978@163.com (Z.Z.); diidmeng@163.com (D.M.)

[2] State Key Laboratory of Space Weather, Chinese Academy of Sciences, Beijing 100190, China

[3] Innovation Academy for Precision Measurement Science and Technology, Chinese Academy of Sciences, Wuhan 430071, China

* Correspondence: zhangting@apm.ac.cn

Abstract: In this paper, GNSS stations' observational data, global ionospheric maps (GIM) and the electron density of FORMOSAT-7/COSMIC-2 occultation are used to study ionospheric anomalies before the submarine volcanic eruption of Hunga Tonga–Hunga Ha'apai on 15 January 2022. (i) We detect the negative total electron content (TEC) anomalies by three GNSS stations on 5 January before the volcanic eruption after excluding the influence of solar and geomagnetic disturbances and lower atmospheric forcing. The GIMs also detect the negative anomaly in the global ionospheric TEC only near the epicenter of the eruption on 5 January, with a maximum outlier exceeding 6 TECU. (ii) From 1 to 3 January (local time), the equatorial ionization anomaly (EIA) peak shifts significantly towards the Antarctic from afternoon to night. The equatorial ionization anomaly double peak decreases from 4 January, and the EIA double peak disappears and merges into a single peak on 7 January. Meanwhile, the diurnal maxima of TEC at TONG station decrease by nearly 10 TECU and only one diurnal maximum occurred on 4 January (i.e., 5 January of UT), but the significant ionospheric diurnal double-maxima (DDM) are observed on other dates. (iii) We find a maximum value exceeding NmF2 at an altitude of 100~130 km above the volcanic eruption on 5 January (i.e., a sporadic E layer), with an electron density of 7.5×10^5 el/cm^3.

Keywords: Tonga volcanic eruption; ionospheric anomaly; anomaly detection method; EIA

Citation: Feng, J.; Yuan, Y.; Zhang, T.; Zhang, Z.; Meng, D. Analysis of Ionospheric Anomalies before the Tonga Volcanic Eruption on 15 January 2022. *Remote Sens.* **2023**, *15*, 4879. https://doi.org/10.3390/rs15194879

Academic Editor: Fabio Giannattasio

Received: 5 September 2023
Revised: 6 October 2023
Accepted: 7 October 2023
Published: 9 October 2023

1. Introduction

Natural disasters such as earthquakes and volcanic eruptions pose a serious threat to the safety of human life and property. Since the mechanisms of such geophysical activities are not yet clear, their forecasting problems remain a difficult area of research at present. Leonard and Barnes [1] first detected ionospheric disturbances associated with earthquakes using ionosonde and Doppler sounders after the 1964 Alaska earthquake. Numerous studies have shown that geophysical activities such as earthquakes, volcanic eruptions and nuclear explosions can cause anomalous ionospheric perturbations [2–6]. Based on the Japanese GPS tracking network, Heki [7] detected a significant positive anomalous precursor of total electron content (TEC) in the ionosphere around the source area of the 11 March 2011 Mw9.0 earthquake in Japan at 40 min before the earthquake, and its amplitude was close to 10% of the background TEC. Le et al. [8] used the global ionospheric maps (GIM) to statistically analyze the pre-earthquake ionospheric anomalies for 736 earthquakes (Mw $\geq$ 6.0) worldwide from 2002 to 2010. They found that the incidence of anomalies in the days before the earthquakes is generally greater than the background days, especially for large-magnitude and low-depth earthquakes. These findings are consistent with the results of Heki [7].

Similar to earthquakes and tsunamis, volcanic eruptions generate space-atmosphere disturbances. The impact of volcanic eruptions can trigger mesospheric gravity waves and mesospheric airglow waves, which result in co-volcanic ionospheric disturbances (CVID). CVIDs occur 10~45 min after volcanic eruptions and usually have a quasi-periodic shape, and spread between 0.5 and 1.1 km/s [9–11]. The manner and evolution of volcanic eruptions are determined by the hydrodynamics that control the rise of magma [12]. The intensity of volcanic eruptions is usually estimated by a feature similar to seismic magnitude, the volcanic explosivity index (VEI), which ranges from 0 to 8. It has been shown that only eruptions with VEI between 2 and 6 have a record of ionospheric response [13].

Many scholars have conducted a series of studies on the anomalous ionospheric disturbances caused by volcanic eruptions. Heki [14] used data from the GNSS earth observation network system (GEONET) to study the ionospheric response to the 1 September 2004 eruption of the Asama volcano in Japan and detected CVIDs in the ionospheric TEC for the first time. TEC is also often used to estimate the energy of volcanic eruptions [15,16]. Shults et al. [17] introduced the term "Ionospheric Volcanology", i.e., the use of ionospheric physical observables in volcanological studies. For example, Shults, Astafyeva and Adourian [17] used a method similar to that proposed by Afraimovich et al. [18] for detecting seismic ionospheres to detect volcanic eruptions. This ionosphere-based method not only locates the source, but also estimates the onset of the source and can be used in areas where seismometers are not installed. Liu, Zhang, Shah and Hong [11] also studied the atmospheric–ionospheric disturbances caused by the April 2015 Calbuco volcanic eruption. They observed the amplitude of TEC perturbations of 0.1~0.4 TECU about one hour after the eruption with data from 50 GPS stations. To explore TEC anomalies prior to two geophysical events, volcanic eruptions and earthquakes, Li et al. [19] analyzed the TEC time series prior to the April 2015 Calbuco volcanic eruption and the April 2015 Nepal earthquake using GIM. The experimental results show that the intensity of TEC anomalies before volcanic eruptions is larger, while the duration of TEC anomalies before earthquakes is longer, which may be related to its specific physical mechanism. Then, Li, et al. [20] analyzed the statistical global TEC variations indicated by VEI4+ prior to volcanic eruptions from 2002 to 2015 using GIM data from the Center for Orbit Determination in Europe (CODE). They found that the incidence of TEC anomalies before large volcanic eruptions is related to volcano type and geographical location.

The submarine volcano of Hunga Tonga–Hunga Ha'apai (HTHH) in Tonga has erupted at 04:05:54 UT on 15 January 2022, and its intensity caused widespread international concern [21–24]. Ionospheric disturbances caused by volcanic eruptions were detected by GNSS receivers in their region and in parts of the world during and after the eruption [25–29]. Themens et al. [30] tracked the traveling ionospheric disturbances (TIDs) associated with volcanic eruptions using measurements from 4735 globally distributed GNSS receivers, and identified two large-scale traveling ionospheric disturbances (LSTIDs) and several medium-scale traveling ionospheric disturbances (MSTIDs). Saito [31] observed two types of TIDs with different characteristics 3 h and 7 h after the eruption using the data of the Japanese regional GNSS receiver network, and the disturbance amplitudes were ± 0.5 TECU and ± 1.0 TECU, respectively. The CVID and TIDs can change the propagation environment of radio waves to some extent, thus affecting the performance of GNSS navigation and positioning [31–33]. In addition, Carter et al. [34] found that small-scale ionospheric disturbances triggered by volcanic eruptions in Tonga increased the convergence time of the precise point position.

However, most of the existing studies on Tonga volcanoes have been conducted using ground-based GNSS data to investigate the characteristics of ionospheric TEC variations after volcanic eruptions. The ionospheric anomalies before the volcanic eruption should also be considered, and it is more reliable to investigate the ionospheric anomalies based on multiple data sources. Therefore, in this paper, we utilize TEC data from GNSS stations, CODE GIM and FORMOSAT-7/COSMIC-2 occultation electron density to investigate the anomalous perturbations of ionospheric TEC prior to the eruption of the Tonga volcano on 15 January 2022 by applying two anomaly detection methods.

2. Data and Methods

2.1. Ionospheric Data

In this paper, three GNSS stations (TONG, LAUT and SAMO) within 10° (about 1110 km) from the volcanic eruption location were selected to investigate the anomalous variations of ionospheric TEC before the volcanic eruption. Two GNSS stations (KOKB and CHTI) were selected to study the ionospheric diurnal double-maxima (DDM) that may be associated with volcanic eruption. The positions of the GNSS stations are shown in Figure 1. The main error of ionospheric TEC based on carrier phase smoothing pseudorange calculation is differential code bias (DCB) of satellite and receiver. We apply the Reg-Est algorithm provided by the Ionospheric Research Laboratory (IONOLAB), Hacettepe University, Turkey, to calculate the ionospheric TEC. The algorithm combines pseudorange observations and carrier phase observations to invert GPS-TEC, and removes the influence of DCB at the same time [35]. It is applicable to all stations of International GNSS Service (IGS) and can calculate TEC in near-real time [36].

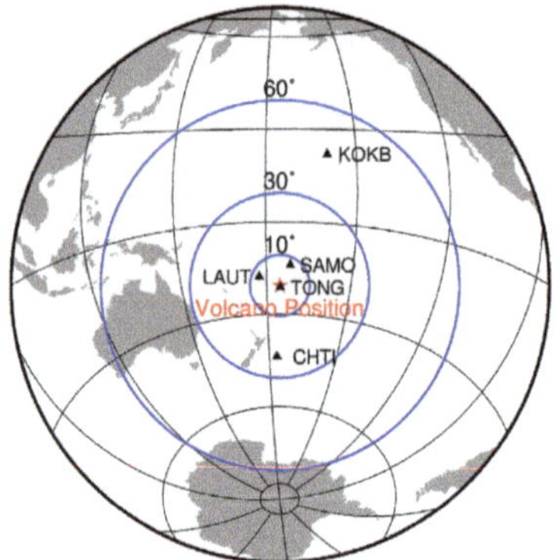

Figure 1. Location of volcanic eruption in Tonga (red pentagram) and distribution of GNSS stations (black triangle).

Since the establishment of the Ionosphere Working Group by IGS in 1998, the number of GNSS stations worldwide has been increasing and GIM products have been fully developed. Among them, the final product provided by CODE has a high accuracy and is widely used in ionosphere-related research. The product is calculated from a spherical harmonic function of 15 orders and 15 degrees, taking 5° × 2.5° along the latitude and longitude, with a total of 5183 grid points.

FORMOSAT-7/COSMIC-2, a follow-on GNSS Radio Occultation (GNSS-RO) mission jointly developed by the United States and Taiwan, China, was successfully launched and placed into a low-inclination orbit on 25 June 2019. The FORMOSAT-7/COSMIC-2 mission daily provides about 4000 occultation events, significantly increasing the number of atmospheric and ionospheric observations, including providing TEC data arcs and maps of the vertical distribution of electron density. These contribute to a better understanding of the structure and electrodynamics of the equatorial and low-latitude ionosphere.

2.2. Space Weather and Geophysical Activity Index

The anomalous variations in the ionosphere are influenced by solar activity and the solar–terrestrial space environment, and the perturbations caused by the space environment are usually widespread [37]. This means that anomalies in ionosphere-related parameters may not only appear in volcanic eruption regions. Therefore, local anomalies near the eruption region may be potentially related to the eruption. When analyzing the abnormal variations in the ionosphere before the volcanic eruption, it is necessary to comprehensively analyze the variations in the space environment during this period to exclude the interference of space environment factors. In this paper, the Bz component of the interplanetary magnetic field (IMF), as well as the Kp index, Dst index, Ap index, solar wind plasma speed and 10.7 cm solar radio flux (F10.7) for a total of 16 days before and after the volcanic eruption are selected as criteria to judge the degree of disturbance in the solar–terrestrial space environment.

2.3. Sliding Interquartile Range

The sliding interquartile range method is widely used in pre-seismic ionospheric TEC anomaly detection, and is similar to the moving average method [6,38]. Compared with the conventional method, it can calculate the background value more accurately, and then calculate the upper and lower limits of ionospheric TEC based on the background value to detect the presence of anomalous values. Using the 20-day data as an example, the data are arranged from smallest to largest as $x_1, x_2, \cdots, x_{20}$, then

$$Q_1 = \frac{x_5 + x_6}{2} \tag{1}$$

$$Q_2 = \frac{x_{10} + x_{11}}{2} \tag{2}$$

$$Q_3 = \frac{x_{15} + x_{16}}{2} \tag{3}$$

$$IQR = Q_3 - Q_1 \tag{4}$$

where Q_1–Q_3 are the upper interquartile, median and lower interquartile, respectively. First, a suitable sliding time window is selected and the TEC values under the window are arranged from smallest to largest and divided into four equal parts. The parity line is represented as Equations (1)–(3) in order, and then the interquartile range value is represented as Equation (4).

Since the solar activity cycle is 27 days, this paper uses 27 days as the sliding window to detect the ionospheric TEC perturbation condition before the volcanic eruption. The upper bound of TEC anomaly is represented as $UB = Q_2 + 1.5IQR$ and the lower bound as $LB = Q_2 - 1.5IQR$.

2.4. NeuralProphet

Facebook offers an interpretable model called Prophet that can be extended to many predictive applications. The NeuralProphet adds autoregressive and covariate modules on the basis of Prophet. These modules can be configured as classical linear regression or neural networks, and perform better than Prophet on various data sets. The prediction accuracy of NeuralProphet is 55~92% higher than that of Prophet in short-term and medium-term prediction [39].

A core concept of the NeuralProphet model is its modular composability. The model consists of modules, each of which contributes an additional component to the prediction, where h defines the number of steps from one prediction to the future and all modules must produce h outputs. The following equation describes the one-step ahead prediction for $h = 1$.

$$TEC(t) = T(t) + S(t) + E(t) + F(t) + A(t) + L(t) \tag{5}$$

where $T(t)$ represents the trend term at time t, $S(t)$ denotes the seasonal effect at time t, $E(t)$ denotes the event and holiday effect at time t, $F(t)$ denotes the regression effect of future known exogenous variables at time t, $A(t)$ represents the autoregressive effect at time t, and $L(t)$ denotes the regression effect of lagged observations of exogenous variables at time t.

In this paper, the value of h is set to 16 days, and the amount of training and testing data is consistent with that of the sliding interquartile range method. The upper and lower bounds of TEC anomalies are expressed in terms of the error in predicting TEC, as follows.

$$Error = TEC_{actual} - TEC_{predicted} \tag{6}$$

$$UB = Mean(Error) + 2 \times Std(Error) \tag{7}$$

$$LB = Mean(Error) - 2 \times Std(Error) \tag{8}$$

2.5. Wavelet Transform

As a time–frequency localization analysis tool for data, wavelet transform has achieved many achievements in meteorology, astronomy and geophysics [40]. Among them, cross-wavelet transform (XWT) and wavelet coherence (WTC) analysis can better explain the correlation between two time series x_n and y_n in the time–frequency space. It is defined as follows.

$$W^{XY} = W^X W^{Y*} \tag{9}$$

$$D\left(\frac{|W_n^X(s)W_n^{Y*}(s)|}{\sigma_X \sigma_Y} < p \right) = \frac{Z_v(p)}{v} \sqrt{P_k^X P_k^Y} \tag{10}$$

where $*$ represents the complex conjugate. $|W^{XY}|$ is defined as the cross-wavelet power, and P_k^X and P_k^Y are the background power spectra of x_n and y_n, respectively. The Equation (10) shows the formula for the theoretical distribution of cross-wavelet power, where σ_X and σ_Y are the standard deviations. When the background power spectrum is real wavelet $v = 1$ and $v = 2$ when it is complex wavelet, $Z_v(p)$ is the confidence level of the probability p.

$$R_n^2(s) = \frac{|S(s^{-1}W_n^{XY}(s))|^2}{S(s^{-1}|W_n^X(s)|^2) \cdot S(s^{-1}|W_n^Y(s)|^2)} \tag{11}$$

Wavelet coherence spectrum analysis can reflect the covariance strength of two time series in time–frequency space, covering the correlation in their low-energy regions. It is defined as in Equation (11), where S is the smoothing operator. The significance test of the wavelet coherence spectrum is performed using the Monte Carlo method wavelet coherence value R^2 between 0 and 1, and its larger value indicates a stronger correlation.

3. Results

The ionospheric TEC is impacted by solar forcing, geomagnetic forcing and lower atmospheric forcing, and the prerequisite for exploring whether the ionosphere is anomalous before the volcanic eruption presupposes a quiet solar–terrestrial space environment. For this purpose, we obtained the solar wind plasma velocity and F10.7 index, which characterize the intensity of solar activity; the Bz component of the IMF, which characterizes the space environment; and the Kp, Dst and Ap index, which characterize the geomagnetic storm condition, for analysis and exploration. The variations of the indices are shown in Figure 2. We can clearly see that the Bz component exceeds 10 nT, the Kp index is close to 5, the Dst index is close to -30 nT, and the Ap index also exceeds 40 nT on the 8th and 9th from Figure 2, while the space environmental events broadcasted by the Space Environment Prediction Center (SPEC) of the Chinese Academy of Sciences show that geomagnetic storms occurred on these two days. Similarly, a large disturbance occurred in the space environment a few hours before the volcanic eruption. The Bz component was lower than -10 nT, and the Dst index was close to -100 nT, which was a geomagnetic storm.

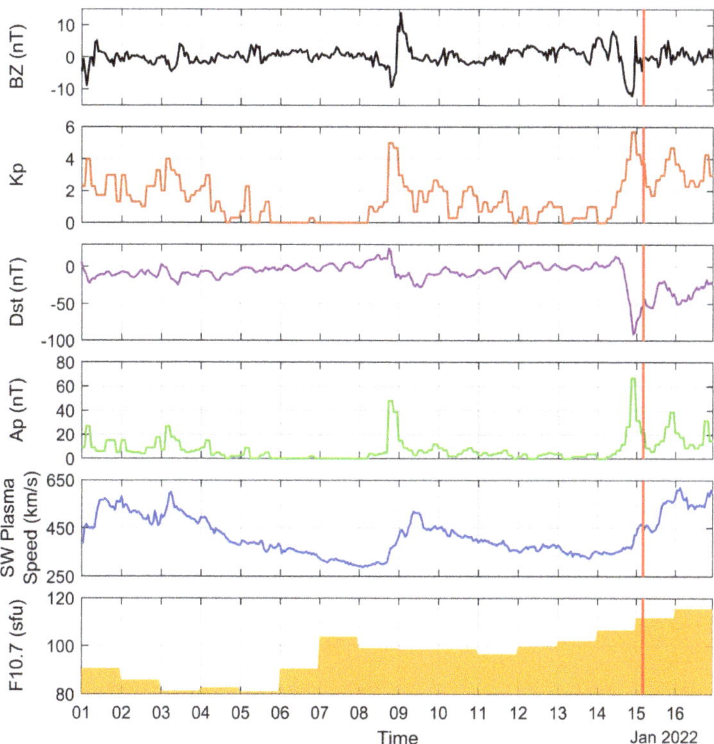

Figure 2. Geomagnetic and solar activity before and after the volcanic eruption from 1 to 16 January 2022. The red vertical line indicates the time of volcanic eruption (same below).

We applied the sliding interquartile range method to detect ionospheric TEC anomalies at TONG, LAUT and SAMO stations near the volcanic eruption, as shown in Figures 3 and 4. The Figure 3 shows the ionospheric TEC variations over the three GNSS stations from 1 to 16 January 2022, where the black line indicates the upper bound of the detected TEC, the blue line indicates the lower bound of the detected TEC, the red line is the variations of the measured TEC, and the red vertical line indicates the moment of volcanic eruption. From the ionospheric TEC values, the values at TONG and LAUT stations are closer, and the TEC over SAMO station is overall higher than those at TONG and LAUT stations. However, the TEC changes at the three stations are consistent, and all show a decrease in TEC peak from 5 January, and this phenomenon continues until 11 January. The ionospheric TEC anomalies detected over the three stations are shown in Figure 4. As shown in Figure 4, the TEC over the three stations exhibited negative anomalies on 5, 6 and 8 January during the 15 days before the eruption. Among them, the negative anomaly on 5 January had the longest duration at TONG station; the maximum anomaly value exceeded 5 TECU at both TONG and SAMO stations, and the peak value of this anomaly was also higher than 2 TECU at LAUT station. On 6 January, negative anomalies of about 5 TECU were also observed at TONG and SAMO stations, but the anomaly at LAUT station was only about 1 TECU and of shorter duration. On 8 January, all three stations showed negative anomalies of about 2 TECU. Positive TEC anomalies were observed over TONG and LAUT stations on the 14th and 15th, with anomalies of about 2 TECU on the 14th and peak values near 5 TECU on the 15th. One day after the eruption, only TONG station showed negative anomalies of about 5 TECU and positive anomalies of more than 10 TECU, and no anomalies were observed at the other two stations. This indicates that the volcanic eruption only caused TEC anomalies at TONG station within a short period of time and failed to affect the distant areas. The ionospheric anomalies during and after the eruption have been studied more and will not be discussed here.

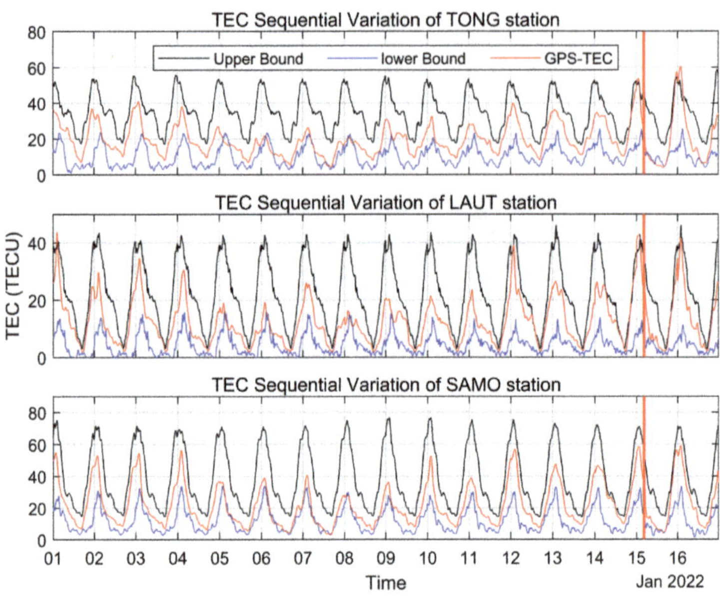

Figure 3. Variations in ionospheric TEC over TONG, LAUT and SAMO stations from 1 to 16 January 2022.

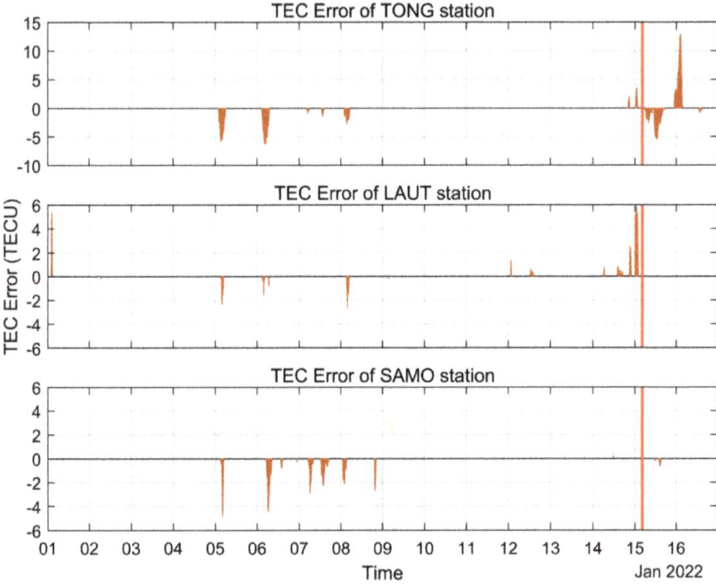

Figure 4. Ionospheric TEC anomalies detected using the sliding interquartile range method over TONG, LAUT and SAMO stations from 1 to 16 January 2022.

Considering the problem of insufficient support of using only one anomaly detection method, we used NeuralProphet as an additional anomaly detection method and applied the same data to detect the ionospheric TEC anomaly before the volcanic eruption, as shown in Figures 5 and 6. Figure 5 shows the time series of GPS-TEC and that predicted by NeuralProphet over the three stations from 1 to 16 January. It can be seen from Figure 5 that the predicted TEC by NeuralProphet reproduces the trend of GPS-TEC better. The ionospheric TEC anomalies detected using NeuralProphet are given in Figure 6, where the black line indicates the upper bound of the detected TEC anomaly, the blue line indicates the lower bound of the detected TEC anomaly, the orange line shows the change in the predicted TEC error, and the red vertical line indicates the moment of volcanic eruption. As can be seen from Figure 6, the TEC anomalies detected using NeuralProphet are similar to those detected using the sliding interquartile range, with negative TEC anomalies also occurring at three stations on the 5th, 6th and 8th, and positive TEC anomalies on the 14th and 15th. This also indicates that the TEC anomalies detected using both detection methods are more reliable.

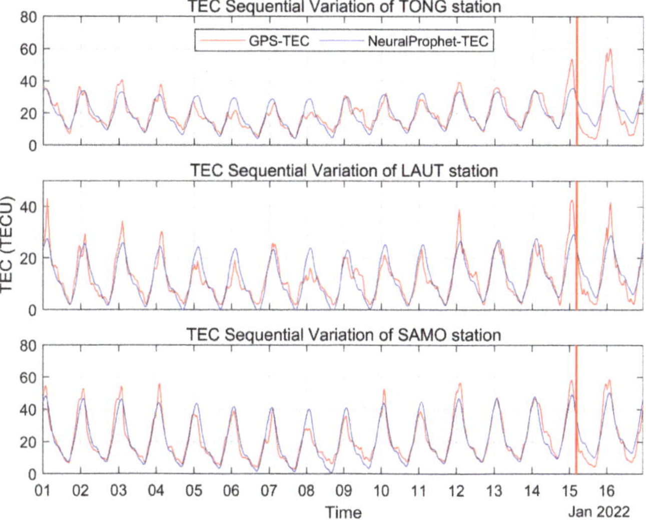

Figure 5. Ionospheric TEC variations (GPS-TEC and NeuralProphet-TEC) over TONG, LAUT and SAMO stations from 1 to 16 January 2022.

Combined with Figures 2–6, there was a large harassment to the ionosphere due to the geomagnetic storms on the 8, 9, 14 and 15 January. Although the various geomagnetic indices did not show the occurrence of geomagnetic storms on the 6th, the F10.7 index on the 6th showed a sudden increase of nearly 10 sfu compared to that on the 5th. The Space Environment Prediction Center also showed the occurrence of a C1.1 solar flare on that day.

In order to further eliminate the interference of the solar–terrestrial space environment on the detection of ionospheric anomalies, we applied cross-wavelet transform and the wavelet coherence spectrum to analyze the correlation between TEC over TONG station and six kinds of solar–terrestrial space environment parameters from 1 to 16 January, as shown in Figures 7 and 8, The right color bar in Figure 7 indicates the cross-wavelet power spectral density, and the arrow direction indicates the phase relationship between the two: to the right indicates that the two sequences are in phase, to the left indicates the opposite phase, vertically down indicates that the former sequence is 1/4 cycle change ahead of the latter sequence, and vertically up indicates that the latter sequence is 1/4 cycle change ahead of the former sequence. Figure 7 shows that the ionospheric TEC has the same resonance period as the six solar–terrestrial space environment parameters, and the TEC has the same phase relationship with Kp and Ap. The wavelet coherence spectrum of TEC with other parameters is given in Figure 8, and the right color bar is the wavelet coherence value, characterizing the strength of coherence. Combining Figures 7 and 8, it can be found that TEC on 8, 9, 14 and 15 January are clearly affected by Bz, Kp, Dst, Ap and SW plasma speed, and TEC on 6 January is affected by F10.7. Therefore, in order to ensure the accuracy of the detected anomalies, the ionospheric anomalies on 6, 8, 9, 14 and 15 January are not explored in this paper, and only the ionospheric anomaly on 5 January is studied.

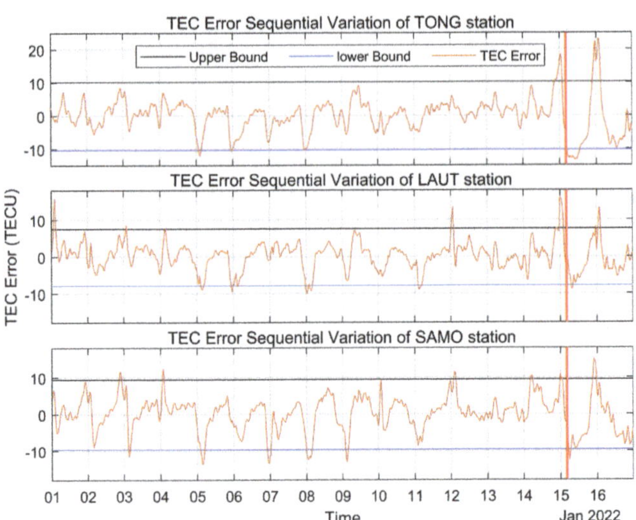

Figure 6. Ionospheric TEC anomalies detected using NeuralProphet over TONG, LAUT and SAMO stations on 1 to 16 January 2022.

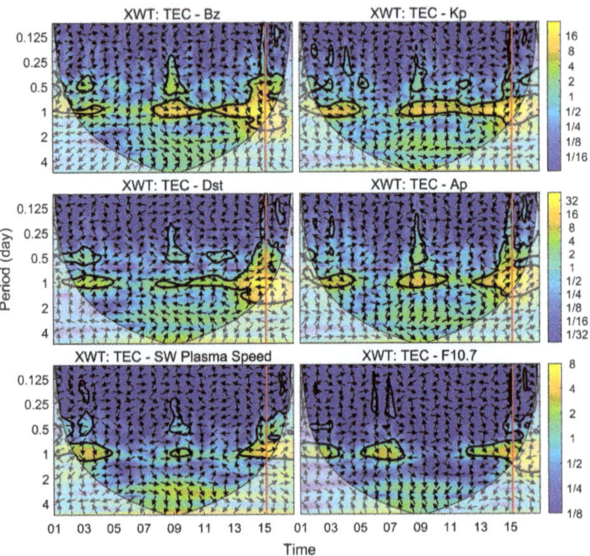

Figure 7. Cross-wavelet transform of TEC time series and spatial weather parameters from 1 to 16 January 2022 at TONG station. The closed area of the thick black line passes the standard red noise test at 95% confidence level, indicating the significance of the period; the cone of influence (COI) area below the thin black solid line is the area of wavelet transform data with large edge effects, and the thick red line indicates the moment of volcanic eruption.

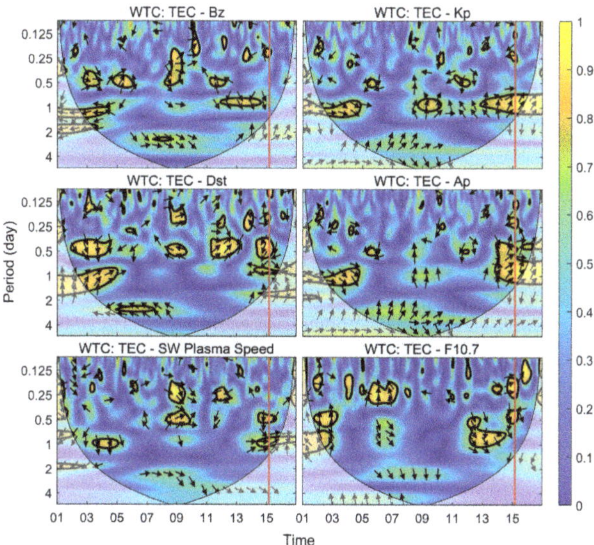

Figure 8. Wavelet coherence spectrum of TEC time series with space weather parameters at TONG station from 1 to 16 January 2022.

To investigate the global distribution of the ionospheric anomalies on 5 January, we used GIM data with a resolution of 1 h provided by CODE for the analysis. We also used the sliding interquartile range method to detect the ionospheric TEC anomaly for each grid point. The sliding window was also set to 27 days, and the upper and lower TEC bounds were calculated for each grid point on that day to find the anomaly value, and the results are shown in Figure 9. The global distribution of ionospheric anomalies on 5 January can be clearly seen in Figure 9. The negative anomaly of 4 TECU in the ionospheric TEC starts at 02:00 UT, and the anomaly area is located southeast of the volcanic eruption. The value of the Ionospheric TEC anomaly increases from 02:00 to 04:00 UT, and the occurrence area gradually approaches the volcanic eruption location. The negative anomaly value exceeds 6 TECU at 04:00 UT, and the center of the anomaly is close to the volcanic eruption location. Starting from 05:00 UT, the ionospheric TEC anomaly value gradually decreases, and the anomaly area also keeps shrinking centered on the volcanic eruption location, and the anomaly basically disappears by 07:00 UT.

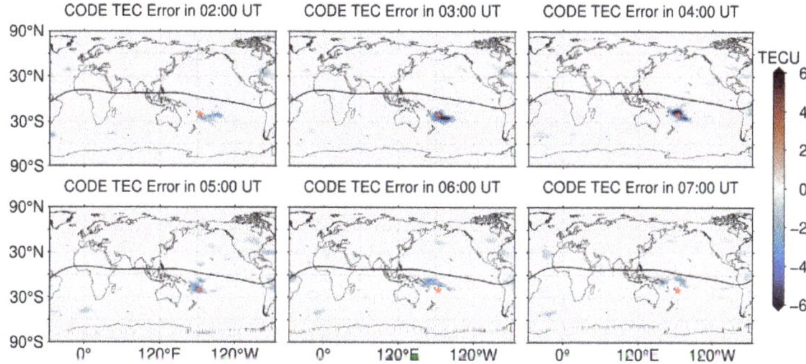

Figure 9. Distribution of the global ionospheric TEC anomaly on 5 January, from 02:00 to 07:00 UT. The red pentagram representing the location of the eruption center and the black solid line indicating the magnetic equator (same below).

Figure 10 shows the distribution of the global ionospheric TEC on 5 January. As can be seen from Figure 10, the intensity of the EIA is greatest at 00:00 UT, where the northern peak anomaly is higher than the southern peak anomaly by about 5 TECU in value and smaller than the southern peak anomaly in coverage. The difference between the EIA and the north-south peak gradually decreases from 00:00 UT, the maximum value of the EIA decreases by 16 TECU at 04:00 UT, and the difference between the north–south peak decreases to 1 TECU. The difference between the EIA and the north–south peak gradually increases, and the peak of the EIA is far away from the eruption center from 05:00 UT. This trend is more consistent with the variations of TEC anomaly in Figure 9, indicating that the influence of EIA at the eruption location gradually becomes weaker.

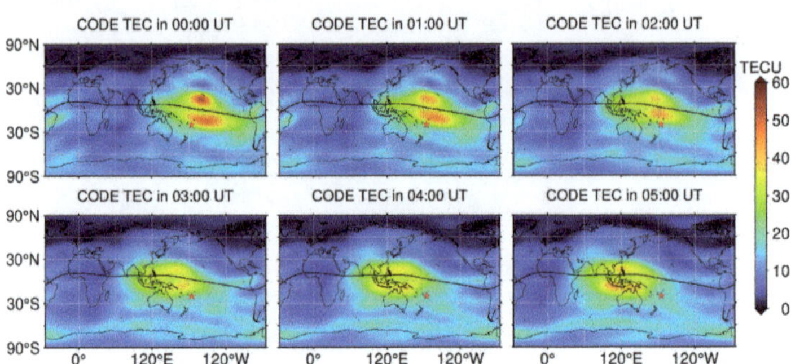

Figure 10. Distribution of global ionospheric TEC on 5 January, from 00:00 to 05:00 UT.

The latitude–time–TEC variation series of the 175°W meridian extracted based on CODE GIM is given in Figure 11 for the period from 1 to 8 January, local time. It can be concluded that the EIA peak shifts significantly toward the Antarctic from 1 to 3 January, appearing from local afternoon to night; the EIA double peak decreases from 4 to 7 January, and the EIA double peak disappears and then merges into a peak on 7 January.

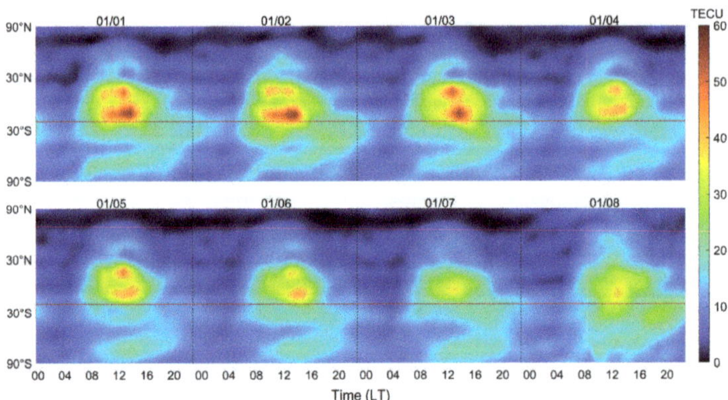

Figure 11. Latitude–time–TEC variations extracted along the 175°W longitude line. The red line is the latitudinal position of the eruption.

To further investigate the ionospheric TEC anomalies at the same longitude, we selected KOKB and CHTI stations, which are at approximately the same longitude as the TONG station, to analyze their anomalous conditions. Among them, KOKB station is located at the northern peak of the EIA, and CHTI station is located at the mid-latitude of the southern hemisphere, where there is no EIA. Figure 12 shows the TEC time series of KOKB, TONG and CHTI stations from 1 to 8 January (local time). It can be seen from Figure 12 that the diurnal ionospheric TEC variation trends are the same at KOKB and

TONG stations, which are located at the EIA peak. However, the daytime TEC variation at CHTI station is more drastic, which may be related to its location. In addition, it is noticeable that the daytime ionospheric peak at TONG station decreases by nearly 10 TECU starting from the 4th day of local time. Interestingly, only one diurnal peak was observed at TONG station on the 4th (5 January UT), while at other times, the ionospheric DDM (also known as the "noontime bite-out") was observed regardless of the ionospheric TEC values.

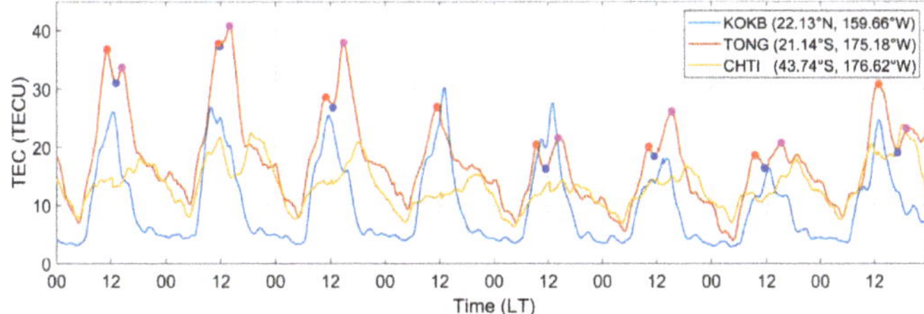

Figure 12. TEC time series of KOKB, TONG and CHTI stations from 1 to 8 January (local time), with the 1st peaks, 2nd peaks and valleys of ionospheric DDM indicated by red, magenta and blue dots, respectively.

Based on the ionospheric electron density data acquired by the FORMOSAT-7/COSMIC-2 occultation, we filtered the data at the location of the time when the ionospheric TEC anomaly occurred in Figure 9 (i.e., 01:30~04:30 UT satellite and GNSS satellite tangent point trajectory close to the volcanic eruption location on 5 January), and obtained a total of eight occultation events for three satellites. Figures 13–15 show the ionospheric electron density profiles and tangent point trajectory locations near the eruption. It is clear from Figure 13 that the electron density appears as an extreme value of 7.5×10^5 el/cm^3 at 100~130 km altitude (about the E layer of the ionosphere), which also far exceeds the peak in the F2 layer of the ionosphere. This phenomenon is called an ionospheric sporadic E (Es) layer, i.e., a thin layer with significantly larger-than-normal electron density at the height of the E layer by chance, ranging from hundreds to thousands of meters thick, and is a significant anomaly on the E layer. However, this phenomenon does not occur in Figures 14c and 15c.

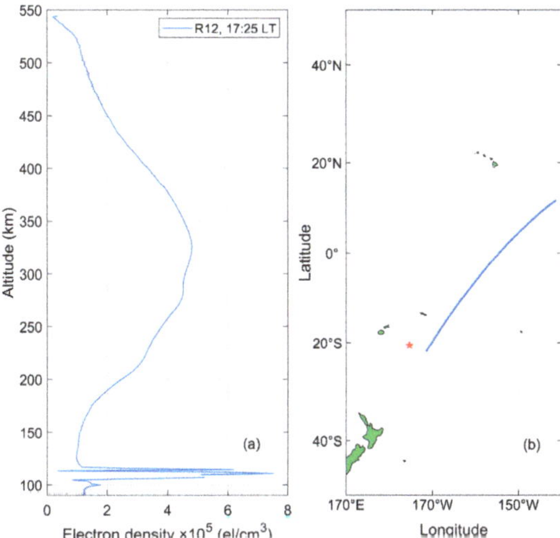

Figure 13. (**a**) Ionospheric electron density profile near the volcanic eruption detected by FOR-MOSAT-7/COSMIC-2 1st satellite on 5 January. The red pentagram is the location of the eruption, and the line segment in (**b**) is the tangent point trajectory of the satellite with GNSS satellite.

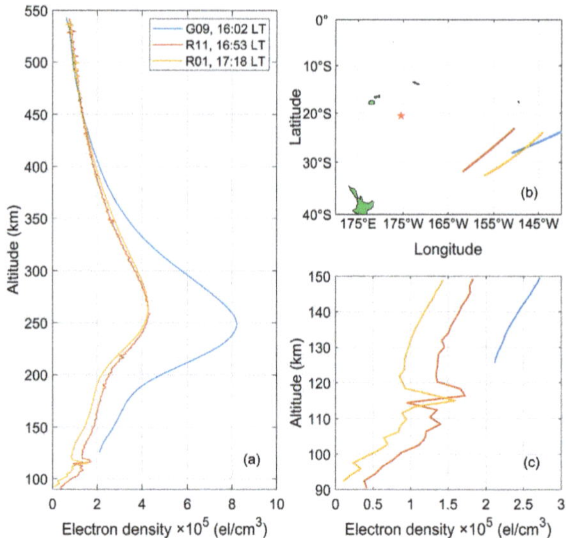

Figure 14. (**a**) Ionospheric electron density profile near the volcanic eruption detected by FOR-MOSAT-7/COSMIC-2 2nd satellite on 5 January. The red pentagram is the location of the eruption, and the line segment in (**b**) is the tangent point trajectory of the satellite with GNSS satellite. Ionospheric electron density profile from 90 to 150 km is shown in (**c**).

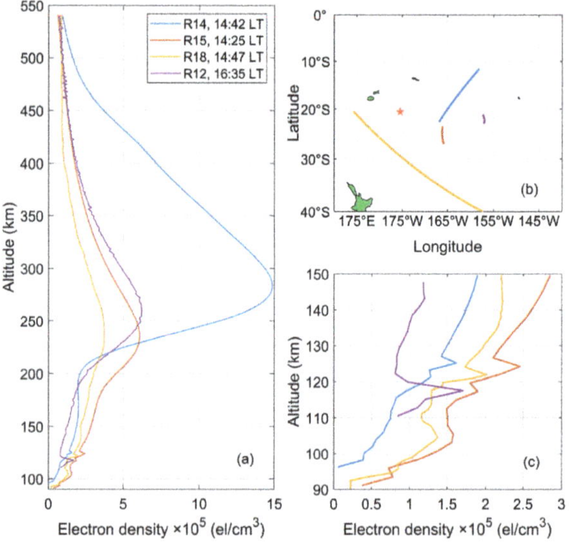

Figure 15. (**a**) Ionospheric electron density profile near the volcanic eruption detected by FOR-MOSAT-7/COSMIC-2 4th satellite on 5 January. The red pentagram is the location of the eruption, and the line segment in (**b**) is the tangent point trajectory of the satellite with GNSS satellite. Ionospheric electron density profile from 90 to 150 km is shown in (**c**).

4. Discussion

In this paper, we detect the TEC anomalies for 16 days before the eruption based on the measured data from three GNSS stations near Tonga volcano, and find inconsistent size anomalies on 5, 6, 8, 9, 14 and 15 January. By analyzing the changes in space weather parameters (Figures 6–8), we exclude the ionospheric TEC anomalies on 6, 8, 9, 14 and 15 January. In addition to solar and geomagnetic activity, low atmospheric forcing has the potential to contribute to variations in the ionosphere [41,42]. In order to study the effect of low atmospheric forcing on the ionosphere, we simulated the meridional and zonal winds in the lower atmosphere on 5, 6, 8 and 9 January based on the level of geomagnetic activity using the HWM14 model, as shown in Figure 16. As analyzed above, there was no significant forcing from solar and geomagnetic activities on 5 January, and from 5 to 9 January, there were no significant changes in the meridional and zonal winds, suggesting that low atmospheric forcing did not have an effect on the ionosphere.

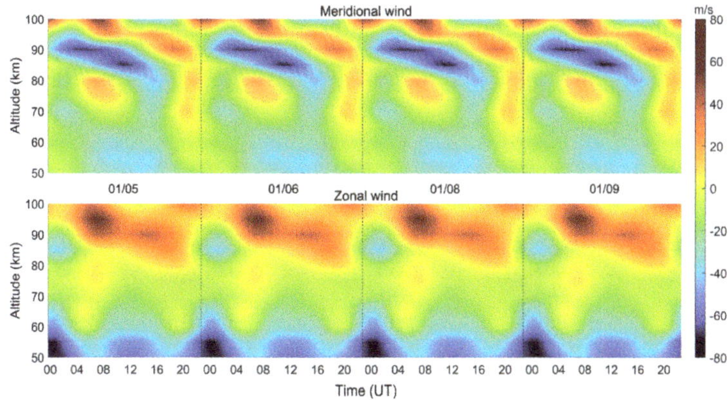

Figure 16. Neutral wind variations from 50 to 100 km above eruption simulated by the HWM14 model during 5–10 January 2022.

The results of the CODE GIM anomaly analysis indicate that the anomaly on 5 January only appeared near the volcanic eruption area and was relatively strong, lasting for about 4 h, while no ionospheric TEC anomalies were observed in other regions of the world during the same period. Liu et al. [43] mentioned in their study of pre-seismic regional ionospheric anomalies that planetary waves and tides usually cause large-scale ionospheric anomalies, and that gravity waves, while they may cause rather small localized features, usually do not last long, and have a single polarity or phase. In this paper, the GIM shows an ionospheric anomaly within a small region of the eruption that lasted 4 h and is unlikely to have been caused by planetary waves and tides or gravity waves.

Since the eruption location is close to the southern peak of the EIA, we studied the EIA of the ionosphere before the eruption (see Figure 11). The analysis results show that the intensity of the EIA double peaks weakened starting on 4 January of local time (i.e., 5 January UT), which is consistent with the GPS-TEC time series results from TONG station in Figure 12. The generator process in the E region during the daytime produces a band electric field in the off-equatorial region, which is mapped to the equatorial F region along the geomagnetic field lines [44], and this electric field pushes the plasma upwards and further diffusion processes and gravity lead to the formation of the EIA [45]. The equatorial E×B electric field and the meridional wind play a decisive role in regulating the strength and position of the EIA [46–49]. We examined the variations in the neutral component using the globally distributed O/N2 data measured by Global Ultraviolet Imager (GUVI) on the TIMED satellite, as shown in Figure 17. From 5 to 10 January, there is no significant change in the ratio of O/N2 in the eruption region, especially in the latitude region where the EIA occurs, which can be explained by the weakening of the intensity of the EIA on 5 January independently of the change in the neutral component. However, on 9 January, a more significant positive value of O/N2 was observed at 30°N, with 135.6 mm O$^+$ occupying a larger proportion, and it was mentioned above that there was a geomagnetic storm event on 9 January, which may be related to the change of the neutral component, and is beyond the scope of this paper.

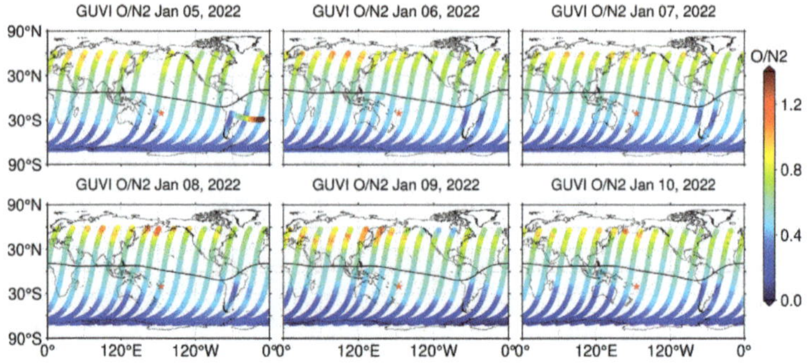

Figure 17. The global distribution of Global Ultraviolet Imager (GUVI)-measured O/N2 ratio during 5–10 January 2022.

The DDM in Figure 12 is caused by the electric field driving the equatorial plasma fountain, moving the plasma to a higher altitude and higher latitude, causing a loss of electron density in the equatorial region around noon. The absence of DDM on 4 January local time is consistent with the cause of the weakening intensity of the EIA wave crest, which may be caused by the electric field drive, and whether it is related to volcanic eruption precursors requires further study.

We find an anomaly phenomenon in the ionospheric E layer on 5 January UT based on electron density data from the FORMOSAT-7/COSMIC-2 occultation, which is particularly notable in Figure 13. It has been shown that the occurrence and spatial and temporal variation of the Es layer are closely related to the wind shear theory, and at low latitudes, especially near the magnetic equator, the wind shear mechanism is not sufficient to form a strong Es layer, and the equatorial Es layer arises from the gradient instability and depends on the electrojet stream [50,51]. This paper alone is not enough to illustrate the relationship between Es layers and volcanic eruption precursors, and we will investigate this phenomenon with more cases in future studies.

5. Conclusions

In this paper, we focus on the anomalous conditions of the ionosphere before the 15 January 2022 eruption of the Tonga volcano. The main results are as follows:

(1)   On 5, 6, 8, 9, 14 and 15 January, ionospheric TEC anomalies were detected at TONG, LAUT and SAMO stations, and most of them were negative anomalies. Combining the space weather parameters and applying the cross-wavelet transform and wavelet coherence spectral analysis, we ruled out the effects of solar activity and geomagnetic disturbances. Using the simulated data of neutral winds, we exclude the effect of lower atmospheric forcing. It is tentatively concluded that the negative TEC anomaly detected by the three GNSS stations on 5 January is related to the volcanic eruption.

(2)   Based on the CODE GIM data, we apply the sliding interquartile range method to detect a negative anomaly in the global ionospheric TEC on 5 January only near the center of the volcanic eruption, with the maximum anomaly exceeding 6 TECUs, which further confirms that the TEC anomaly on 5 January is closely related to the volcanic eruption.

(3)   The sequence of latitude–time–TEC variations along the 175°W meridian shows that the equatorial anomaly wave peaks moved significantly toward the South Pole from the local afternoon to the night from the beginning of the 1st to the 3rd, and the equatorial anomaly double peaks began to decrease from the 4th and disappeared

and merged into a single wave by the 7th. The O/N2 data show that the neutral component did not contribute much to the ionospheric variations on the 5 January.

(4) TONG station shows a decrease in the peak of the diurnal ionosphere by nearly 10 TECU from the 4th local time, while only one diurnal peak occurs on the 4th (i.e., 5 January UT), while all other dates of TONG station show a significant ionospheric DDM. Based on the FORMOSAT-7/COSMIC-2 occultation electron density data, we find an Es phenomenon in the ionosphere near the eruption of the volcano on 5 January (UT), with an extreme value of nearly 7.5×10^5 el/cm^3 at an altitude of 100–130 km well above the peak of the F2 layer of the ionosphere. Whether these two phenomena are related to the volcanic eruption needs to be explored in depth with more cases.

Author Contributions: Conceptualization, writing—original draft, methodology, J.F., Y.Y. and T.Z.; validation, J.F.; data curation, Y.Y., T.Z., Z.Z. and D.M.; visualization, J.F., Y.Y. and T.Z.; formal analysis, J.F., T.Z., Z.Z. and D.M.; writing—review and editing, J.F., Z.Z. and D.M.; funding acquisition, J.F. All authors have read and agreed to the published version of the manuscript.

Funding: This research was funded by the National Natural Science Foundation of China (Grant No. 42274040) and Project Supported by the Specialized Research Fund for State Key Laboratories.

Data Availability Statement: The GNSS observations are from the IGS (https://cddis.nasa.gov/archive/gnss/data/daily/, regis-tration required, accessed on 30 September 2023). The FORMOSAT-7/COSMIC-2 occultation data are from CDAAC (https://data.cosmic.ucar.edu/gnss-ro/cosmic2/provisional/spaceWeather/, accessed on 30 September 2023). The GIM products are from the CODE Analysis Center (http://ftp.aiub.unibe.ch/CODE/, accessed on 30 September 2023). The wind data are derived from CCMC (https://kauai.ccmc.gsfc.nasa.gov/instantrun/hwm/, accessed on 30 September 2023). The solar geomagnetic parameters are from GSFC/SPDF OMNIWeb (https://omniweb.gsfc.nasa.gov/form/dx4.html, accessed on 30 September 2023), and the O/N2 products are from GUVI (http://guvitimed.jhuapl.edu/, accessed on 30 September 2023).

Acknowledgments: We are acknowledge the use of MATrix LABoratory (MATLAB R2022b) software, the generic mapping tools, cross-wavelet and wavelet coherence toolbox [40] and IONOLAB-TEC software [36]. We would like to thank the anonymous reviewers for their helpful suggestions.

Conflicts of Interest: The authors declare no conflict of interest.

References

1. Leonard, R.S.; Barnes, R.A. Observation of ionospheric disturbances following the Alaska earthquake. *J. Geophys. Res.* **1965**, *70*, 1250–1253. [CrossRef]
2. Whitcomb, J.H.; Garmany, J.D.; Anderson, D.L. Earthquake Prediction: Variation of Seismic Velocities before the San Francisco Earthquake. *Science* **1973**, *180*, 632–635. [CrossRef] [PubMed]
3. Pulinets, S. Ionospheric Precursors of Earthquakes; Recent Advances in Theory and Practical Applications. *Terr. Atmos. Ocean. Sci.* **2004**, *15*, 413–435. [CrossRef]
4. Iwata, T.; Umeno, K. Preseismic ionospheric anomalies detected before the 2016 Kumamoto earthquake. *J. Geophys. Res. Space Phys.* **2017**, *122*, 3602–3616. [CrossRef]
5. Xie, T.; Chen, B.; Wu, L.; Dai, W.; Kuang, C.; Miao, Z. Detecting Seismo-Ionospheric Anomalies Possibly Associated with the 2019 Ridgecrest (California) Earthquakes by GNSS, CSES, and Swarm Observations. *J. Geophys. Res. Space Phys.* **2021**, *126*, e2020JA028761. [CrossRef]
6. Ke, F.; Wang, Y.; Wang, X.; Qian, H.; Shi, C. Statistical analysis of seismo-ionospheric anomalies related to Ms > 5.0 earthquakes in China by GPS TEC. *J. Seismol.* **2016**, *20*, 137–149. [CrossRef]
7. Heki, K. Ionospheric electron enhancement preceding the 2011 Tohoku-Oki earthquake. *Geophys. Res. Lett.* **2011**, *38*, L17312. [CrossRef]
8. Le, H.; Liu, J.Y.; Liu, L. A statistical analysis of ionospheric anomalies before 736 M6.0+ earthquakes during 2002–2010. *J. Geophys. Res. Space Phys.* **2011**, *116*. [CrossRef]
9. Dautermann, T.; Calais, E.; Mattioli, G.S. Global Positioning System detection and energy estimation of the ionospheric wave caused by the 13 July 2003 explosion of the Soufrière Hills Volcano, Montserrat. *J. Geophys. Res. Solid Earth* **2009**, *114*, B02202. [CrossRef]

10. Nakashima, Y.; Heki, K.; Takeo, A.; Cahyadi, M.N.; Aditiya, A.; Yoshizawa, K. Atmospheric resonant oscillations by the 2014 eruption of the Kelud volcano, Indonesia, observed with the ionospheric total electron contents and seismic signals. *Earth Planet. Sci. Lett.* **2016**, *434*, 112–116. [CrossRef]

11. Liu, X.; Zhang, Q.; Shah, M.; Hong, Z. Atmospheric-ionospheric disturbances following the April 2015 Calbuco volcano from GPS and OMI observations. *Adv. Space Res.* **2017**, *60*, 2836–2846. [CrossRef]

12. Gonnermann, H.M.; Manga, M. The Fluid Mechanics Inside a Volcano. *Annu. Rev. Fluid Mech.* **2006**, *39*, 321–356. [CrossRef]

13. Astafyeva, E. Ionospheric Detection of Natural Hazards. *Rev. Geophys.* **2019**, *57*, 1265–1288. [CrossRef]

14. Heki, K. Explosion energy of the 2004 eruption of the Asama Volcano, central Japan, inferred from ionospheric disturbances. *Geophys. Res. Lett.* **2006**, *33*, L14303. [CrossRef]

15. Kanamori, H.; Mori, J. Harmonic excitation of mantle Rayleigh waves by the 1991 eruption of Mount Pinatubo, Philippines. *Geophys. Res. Lett.* **1992**, *19*, 721–724. [CrossRef]

16. Kakinami, Y.; Kamogawa, M.; Tanioka, Y.; Watanabe, S.; Gusman, A.R.; Liu, J.-Y.; Watanabe, Y.; Mogi, T. Tsunamigenic ionospheric hole. In *Geophysical Research Letters*; Wiely: Hoboken, NJ, USA, 2012; Volume 39. [CrossRef]

17. Shults, K.; Astafyeva, E.; Adourian, S. Ionospheric detection and localization of volcano eruptions on the example of the April 2015 Calbuco events. *J. Geophys. Res. Space Phys.* **2016**, *121*, 10303–10315. [CrossRef]

18. Afraimovich, E.L.; Astafieva, E.I.; Kirushkin, V.V. Localization of the source of ionospheric disturbance generated during an earthquake. *Int. J. Geomagn. Aeron.* **2006**, *6*, GI2002. [CrossRef]

19. Li, J.; Meng, G.; You, X.; Zhang, R.; Shi, H.; Han, Y. Ionospheric total electron content disturbance associated with May 12, 2008, Wenchuan earthquake. *Geod. Geodyn.* **2015**, *6*, 126–134. [CrossRef]

20. Li, W.; Guo, J.; Yue, J.; Shen, Y.; Yang, Y. Total electron content anomalies associated with global VEI4+ volcanic eruptions during 2002–2015. *J. Volcanol. Geotherm. Res.* **2016**, *325*, 98–109. [CrossRef]

21. Zhou, M.; Gao, H.; Yu, D.; Guo, J.; Zhu, L.; Yang, L.; Pan, S. Analysis of the Anomalous Environmental Response to the 2022 Tonga Volcanic Eruption Based on GNSS. *Remote Sens.* **2022**, *14*, 4847. [CrossRef]

22. Harding, B.J.; Wu, Y.-J.J.; Alken, P.; Yamazaki, Y.; Triplett, C.C.; Immel, T.J.; Gasque, L.C.; Mende, S.B.; Xiong, C. Impacts of the January 2022 Tonga Volcanic Eruption on the Ionospheric Dynamo: ICON-MIGHTI and Swarm Observations of Extreme Neutral Winds and Currents. *Geophys. Res. Lett.* **2022**, *49*, e2022GL098577. [CrossRef]

23. Zhang, J.; Xu, J.; Wang, W.; Wang, G.; Ruohoniemi, J.M.; Shinbori, A.; Nishitani, N.; Wang, C.; Deng, X.; Lan, A.; et al. Oscillations of the Ionosphere Caused by the 2022 Tonga Volcanic Eruption Observed With SuperDARN Radars. *Geophys. Res. Lett.* **2022**, *49*, e2022GL100555. [CrossRef]

24. Aa, E.; Zhang, S.-R.; Wang, W.; Erickson, P.J.; Qian, L.; Eastes, R.; Harding, B.J.; Immel, T.J.; Karan, D.K.; Daniell, R.E.; et al. Pronounced Suppression and X-Pattern Merging of Equatorial Ionization Anomalies after the 2022 Tonga Volcano Eruption. *J. Geophys. Res. Space Phys.* **2022**, *127*, e2022JA030527. [CrossRef] [PubMed]

25. Astafyeva, E.; Maletckii, B.; Mikesell, T.D.; Munaibari, E.; Ravanelli, M.; Coisson, P.; Manta, F.; Rolland, L. The 15 January 2022 Hunga Tonga eruption history as inferred from ionospheric observations. *Geophys. Res. Lett.* **2022**, *49*, e2022GL098827. [CrossRef]

26. Lin, J.-T.; Rajesh, P.K.; Lin, C.C.H.; Chou, M.-Y.; Liu, J.-Y.; Yue, J.; Hsiao, T.-Y.; Tsai, H.-F.; Chao, H.-M.; Kung, M.-M. Rapid Conjugate Appearance of the Giant Ionospheric Lamb Wave Signatures in the Northern Hemisphere After Hunga-Tonga Volcano Eruptions. *Geophys. Res. Lett.* **2022**, *49*, e2022GL098222. [CrossRef]

27. Heki, K. Ionospheric signatures of repeated passages of atmospheric waves by the 2022 Jan. 15 Hunga Tonga-Hunga Ha'apai eruption detected by QZSS-TEC observations in Japan. *Earth Planets Space* **2022**, *74*, 112. [CrossRef]

28. Hong, J.; Kil, H.; Lee, W.K.; Kwak, Y.-S.; Choi, B.-K.; Paxton, L.J. Detection of Different Properties of Ionospheric Perturbations in the Vicinity of the Korean Peninsula After the Hunga-Tonga Volcanic Eruption on 15 January 2022. *Geophys. Res. Lett.* **2022**, *49*, e2022GL099163. [CrossRef]

29. Ghent, J.N.; Crowell, B.W. Spectral Characteristics of Ionospheric Disturbances Over the Southwestern Pacific from the 15 January 2022 Tonga Eruption and Tsunami. *Geophys. Res. Lett.* **2022**, *49*, e2022GL100145. [CrossRef]

30. Themens, D.R.; Watson, C.; Žagar, N.; Vasylkevych, S.; Elvidge, S.; McCaffrey, A.; Prikryl, P.; Reid, B.; Wood, A.; Jayachandran, P.T. Global Propagation of Ionospheric Disturbances Associated with the 2022 Tonga Volcanic Eruption. *Geophys. Res. Lett.* **2022**, *49*, e2022GL098158. [CrossRef]

31. Saito, S. Ionospheric disturbances observed over Japan following the eruption of Hunga Tonga-Hunga Ha'apai on 15 January 2022. *Earth Planets Space* **2022**, *74*, 57. [CrossRef]

32. Timoté, C.C.; Juan, J.M.; Sanz, J.; González-Casado, G.; Rovira-García, A.; Escudero, M. Impact of medium-scale traveling ionospheric disturbances on network real-time kinematic services: CATNET study case. *J. Space Weather. Space Clim.* **2020**, *10*, 29. [CrossRef]

33. Yang, Z.; Morton, Y.T.J.; Zakharenkova, I.; Cherniak, I.; Song, S.; Li, W. Global View of Ionospheric Disturbance Impacts on Kinematic GPS Positioning Solutions during the 2015 St. Patrick's Day Storm. *J. Geophys. Res. Space Phys.* **2020**, *125*, e2019JA027681. [CrossRef]

34. Carter, B.A.; Pradipta, R.; Dao, T.; Currie, J.L.; Choy, S.; Wilkinson, P.J.; Maher, P.S.; Marshall, R.A.; Harima, K.; LeHuy, M.; et al. The ionospheric effects of the 2022 Hunga Tonga Volcano eruption and the associated impacts on GPS Precise Point Positioning across the Australian region. *ESS Open Arch.* **2023**, *21*, e2023SW003476. [CrossRef]

35. Nayir, H.; Arikan, F.; Arikan, O.; Erol, C.B. Total Electron Content Estimation with Reg-Est. *J. Geophys. Res. Space Phys.* **2007**, *112*, A11. [CrossRef]
36. Sezen, U.; Arikan, F.; Arikan, O.; Ugurlu, O.; Sadeghimorad, A. Online, automatic, near-real time estimation of GPS-TEC: IONOLAB-TEC. *Space Weather.-Int. J. Res. Appl.* **2013**, *11*, 297–305. [CrossRef]
37. Tsurutani, B.; Mannucci, A.; Iijima, B.; Abdu, M.A.; Sobral, J.H.A.; Gonzalez, W.; Guarnieri, F.; Tsuda, T.; Saito, A.; Yumoto, K.; et al. Global dayside ionospheric uplift and enhancement associated with interplanetary electric fields. *J. Geophys. Res. Space Phys.* **2004**, *109*, A8. [CrossRef]
38. Pundhir, D.; Singh, B.; Singh, O.P.; Gupta, S.K.; Karia, S.P.; Pathak, K.N. Study of ionospheric precursors using GPS and GIM-TEC data related to earthquakes occurred on 16 April and 24 September, 2013 in Pakistan region. *Adv. Space Res.* **2017**, *60*, 1978–1987. [CrossRef]
39. Triebe, O.; Hewamalage, H.; Pilyugina, P.; Laptev, N.P.; Bergmeir, C.; Rajagopal, R. NeuralProphet: Explainable Forecasting at Scale. *arXiv* **2021**, arXiv:2111.15397. [CrossRef]
40. Grinsted, A.; Moore, J.C.; Jevrejeva, S. Application of the cross wavelet transform and wavelet coherence to geophysical time series. *Nonlin. Process. Geophys.* **2004**, *11*, 561–566. [CrossRef]
41. Yang, Z.; Gu, S.-Y.; Qin, Y.; Teng, C.-K.-M.; Wei, Y.; Dou, X. Ionospheric Oscillation with Periods of 6–30 Days at Middle Latitudes: A Response to Solar Radiative, Geomagnetic, and Lower Atmospheric Forcing. *Remote Sens.* **2022**, *14*, 5895. [CrossRef]
42. Lei, J.; Huang, F.; Chen, X.; Zhong, J.; Ren, D.; Wang, W.; Yue, X.; Luan, X.; Jia, M.; Dou, X.; et al. Was Magnetic Storm the Only Driver of the Long-Duration Enhancements of Daytime Total Electron Content in the Asian-Australian Sector between 7 and 12 September 2017? *J. Geophys. Res. Space Phys.* **2018**, *123*, 3217–3232. [CrossRef]
43. Liu, J.Y.; Le, H.; Chen, Y.I.; Chen, C.H.; Liu, L.; Wan, W.; Su, Y.Z.; Sun, Y.Y.; Lin, C.H.; Chen, M.Q. Observations and simulations of seismoionospheric GPS total electron content anomalies before the 12 January 2010 M7 Haiti earthquake. *J. Geophys. Res. Space Phys.* **2011**, *116*, A4. [CrossRef]
44. Rishbeth, H. The F-layer dynamo. *Planet. Space Sci.* **1971**, *19*, 263–267. [CrossRef]
45. Martyn, D. Theory of height and ionization density changes at the maximum of a Chapman-like region, taking account of ion production, decay, diffusion and tidal drift. *Phys. Ionos.* **1955**, 254.
46. Lin, C.H.; Liu, J.Y.; Fang, T.W.; Chang, P.Y.; Tsai, H.F.; Chen, C.H.; Hsiao, C.C. Motions of the equatorial ionization anomaly crests imaged by FORMOSAT-3/COSMIC. *Geophys. Res. Lett.* **2007**, *34*, L19101. [CrossRef]
47. Chen, Y.; Liu, L.; Le, H.; Wan, W.; Zhang, H. Equatorial ionization anomaly in the low-latitude topside ionosphere: Local time evolution and longitudinal difference. *J. Geophys. Res. Space Phys.* **2016**, *121*, 7166–7182. [CrossRef]
48. Aswathy, R.P.; Manju, G.; Sunda, S. The Response Time of Equatorial Ionization Anomaly Crest: A Unique Precursor to the Time of Equatorial Spread F Initiation. *J. Geophys. Res. Space Phys.* **2018**, *123*, 5949–5959. [CrossRef]
49. Cai, X.; Qian, L.; Wang, W.; McInerney, J.M.; Liu, H.-L.; Eastes, R.W. Hemispherically Asymmetric Evolution of Nighttime Ionospheric Equatorial Ionization Anomaly in the American Longitude Sector. *J. Geophys. Res. Space Phys.* **2022**, *127*, e2022JA030706. [CrossRef]
50. Whitehead, J.D. Recent work on mid-latitude and equatorial sporadic-E. *J. Atmos. Terr. Phys.* **1989**, *51*, 401–424. [CrossRef]
51. Tsunoda, R.T. On blanketing sporadic E and polarization effects near the equatorial electrojet. *J. Geophys. Res. Space Phys.* **2008**, *113*, A9. [CrossRef]

Article

Ionospheric Total Electron Content Changes during the 15 February 2018 and 30 April 2022 Solar Eclipses over South America and Antarctica

Juan Carlos Valdés-Abreu [1,2], Marcos Díaz [1,2,*], Manuel Bravo [3] and Yohadne Stable-Sánchez [1]

1   Space and Planetary Exploration Laboratory (SPEL), Faculty of Physical and Mathematical Sciences, University of Chile, Av. Tupper 2007, Santiago 8370451, Chile; juanvaldes@ug.uchile.cl (J.C.V.-A.); spel@ing.uchile.cl (Y.S.-S.)

2   Department of Electrical Engineering, Faculty of Physical and Mathematical Sciences, University of Chile, Av. Tupper 2007, Santiago 8370451, Chile

3   Centro de Instrumentación Científica, Universidad Adventista de Chile, km 12 Camino a Tanilvoro, Chillan 3780000, Chile; manuelbravo@unach.cl

*   Correspondence: mdiazq@ing.uchile.cl

Abstract: This is one of the first papers to study the ionospheric effects of two solar eclipses that occurred in South America and Antarctica under geomagnetic activity in different seasons (summer and autumn) and their impact on the equatorial ionization anomaly (EIA). The changes in total electron content (TEC) during the 15 February 2018 and 30 April 2022 partial solar eclipses will be analyzed. The study is based on more than 390 GPS stations, Swarm-A, and DMSP F18 satellite measurements, such as TEC, electron density, and electron temperature. The ionospheric behaviors over the two-fifth days on both sides of each eclipse were used as a reference for estimating TEC changes. Regional TEC maps were created for the analysis. Background TEC levels were significantly higher during the 2022 eclipse than during the 2018 eclipse because ionospheric levels depend on solar index parameters. On the days of the 2018 and 2022 eclipses, the ionospheric enhancement was noticeable due to levels of geomagnetic activity. Although geomagnetic forcing impacted the ionosphere, both eclipses had evident depletions under the penumbra, wherein differential vertical TEC (DVTEC) reached values $< -40\%$. The duration of the ionospheric effects persisted after 24 UT. Also, while a noticeable TEC depletion (DVTEC $\sim -50\%$) of the southern EIA crest was observed during the 2018 eclipse (hemisphere summer), an evident TEC enhancement (DVTEC $> 30\%$) at the same crest was seen during the eclipse of 2022 (hemisphere autumn). Swarm-A and DMSP F18 satellite measurements and analysis of other solar eclipses in the sector under quiet conditions supported the ionospheric behavior.

Keywords: ionosphere; solar eclipse; global positioning system (GPS); satellite measurements; total electron content (TEC); in situ measurements; equatorial ionization anomaly

1. Introduction

The ionospheric TEC changes can be provoked by some sources, such as solar flares, geomagnetic storms, solar eclipses, solar terminators, tropical cyclones, thunderstorms/lightning, volcanic eruptions, earthquakes, tsunamis, and rocket launches, among others [1,2]. These ionospheric disturbance sources can cause blackouts in radio signals and are an important problem for satellite communication and radionavigation systems [2].

Solar eclipses provide a unique opportunity to investigate atmospheric and ionospheric processes from the Earth's surface to the topside ionosphere during the fast sunlight–umbra transitions and vice versa. During solar eclipses, solar radiation is attenuated because it is partially blocked by the Moon. Consequently, the sudden decreases in photo-ionization and photo-electron heating of solar extreme ultraviolets (EUVs) affect

Citation: Valdés-Abreu, J.C.; Díaz, M.; Bravo, M.; Stable-Sánchez, Y. Ionospheric Total Electron Content Changes during the 15 February 2018 and 30 April 2022 Solar Eclipses over South America and Antarctica. *Remote Sens.* **2023**, *15*, 4810. https://doi.org/10.3390/rs15194810

Academic Editor: Fabio Giannattasio

Received: 26 July 2023
Revised: 30 August 2023
Accepted: 4 September 2023
Published: 3 October 2023

all the ionospheric layers [3]. This results in the loss of plasma and a local drop in electron density (N_e) dominating the output rate on an eclipse day relative to a regular day.

The total electron content (TEC in TECu, where 1 TECu $= 10^{16}$ el/m^2) is a measure of N_e in the ionosphere integrated along the line of sight; therefore, it provides accurate information about ionization behavior. During eclipses, the TEC fluctuates and can be characterized by the amplitude as differential TEC (in TECu and/or %), which usually has a reduction; delay value (τ in min) relative to the maximum obscuration time (MOT) of the eclipse; and duration (ΔT in hours) of the disturbance [1]. TEC can be estimated via a radio link between Earth and space. The Global Navigation Satellite System (GNSS) is the most widely used communication system for estimating TEC in the ionosphere. Therefore, we use the Global Positioning System (GPS) in this work, which is one of the constellations of satellites that make up GNSS.

The solar eclipse-induced effects on the ionosphere are not simply local, but can even influence areas outside the eclipse's shadow. These effects may be caused by the transfer of hemisphere magnetic conjugates, changes in the equatorial vertical E $\times$ B drift, the development of a disturbance dynamo, and atmospheric gravity waves (AGWs) that generate traveling ionospheric disturbances (TIDs) and/or traveling atmospheric disturbances (TADs) [4–16]. The supersonic moving shadow of solar eclipses on Earth's atmosphere can cause AGWs. Furthermore, the existence of eclipse waves (AGWs, TADs, and TIDs) affects a number of systems, including radiocommunication, radionavigation, and geolocalization.

Ionospheric effects during high-latitude eclipses are driven by intense ionosphere–magnetosphere coupling, particularly during high levels of geomagnetic activity [17–19]; while ionospheric effects during low-latitude eclipses are influenced by electrodynamics processes associated with the equatorial ionization anomaly (EIA) [20,21]. The equatorial vertical E $\times$ B drift, which induces plasma transfer from the magnetic equator to the crest area and decreases strongly during solar eclipses, as has been seen, for example, during the 4 October 1995 and the 9 March 1997 solar eclipses [22]. It has been seen that the EIA source effect is frequently suppressed during Sun–Earth–Moon alignments (full moon, new moon, and eclipse), even with evidence of equatorial counter-electrojets during solstices [23–26]. On the other hand, depending on the geometric configuration of the eclipse and season of occurrence, it can increase or decrease the EIA crests separately, generating asymmetry. Additionally, Resende et al. [27] showed a summary of current studies that found EIA crests decreasing during solar eclipse occurrences, where TEC decrease was between -50% and -20% in comparison to the background ionosphere. In contrast, some authors found TEC increases around the EIA (e.g., [28–31]). For instance, Chen et al. [29] proved that eclipse-induced wind and temperature may significantly influence the ionospheric enhancement at EIA crests. Moreover, the dynamic processes that occur during a single eclipse are heavily influenced by the eclipse trajectory, time of day, location, synoptic, seasons, geomagnetic conditions, geophysical conditions, and solar variation [15,32]. The trajectories of an eclipse can differ geographically and temporally depending on the observing altitude chosen, which could be important for researching ionospheric dynamics. Research has shown the importance of estimating the eclipse mask at the height of the ionosphere to be studied and not only working with the mask at the surface level [33–35].

Recently, several authors published TEC variations and observed anomalies related to solar eclipses in various geographic locations. Kundu et al. [36] presented the change in ionospheric plasma density during the annular solar eclipse on 26 December 2019, through multi-directional approaches by using ground-based TEC and space-based Swarm satellite and observed a depletion in TEC and N_e that is directly proportional to the solar obscuration function. Several authors have used different instruments (mainly VLF receivers) and computed N_e variation using LWPC simulation also [37–40]. Including ionospheric responses to the partial and annular eclipse in different regions will help the reader to make comparisons with other cases and results obtained from other instruments.

The partial solar eclipses of 15 February 2018 (PSE2018) and 30 April 2022 (PSE2022) were the first eclipses of 2018 and 2022, respectively. These eclipses were visible from over parts of Antarctica, the Pacific, and the Atlantic Oceans. Moreover, both eclipses were observed over southern South America in the afternoon (http://xjubier.free.fr/en/site_pages/SolarEclipsesGoogleMaps.html, last accessed on 20 August 2023). PSE2018 and PSE2022 happened over days with some levels of geomagnetic activity. Therefore, these two partial solar eclipses offer an excellent opportunity to study their effects on the regional ionosphere, under different geometric configurations of the eclipse and season of occurrence.

Recent studies looked at the ionospheric effects of solar eclipses over South America and Antarctica. The 2 July 2019 total solar eclipse (TSE2019) occurred in the southern hemisphere winter (subsolar latitude $\sim$23°N) from 16.92 UT to 21.83 UT. TEC showed clear decreases (-40% to -25%) compared to the background ionosphere and a 57% rise near the southern EIA crest. TIDs presented wavelengths ranging from approximately 100 km to more than 200 km, with periods between 20 and 50 min. Jonah et al. [10,31] reported a TEC enhancement effect that altered the equatorial electrodynamics. Jonah et al. [31] utilized an incoherent scatter radar in Jicamarca and 2500 GNSS stations localized over South and North America. Maurya et al. [10] studied 24 Chilean GPS station datasets localized north and south to the path of totality. The wavelet analysis of the VTEC time series showed the presence of strong AGWs of duration from 30 to 60 min at the GPS stations located the north of totality. Aryal et al. [41] used NASA's Global-scale Observation of Limb and Disk (GOLD) instrument aboard the SES-14 satellite in a geostationary orbit. They reported an important decrease in brightness around totality in OI 135.6 nm and N2 LBH band emissions when compared to baseline observations taken two days earlier. They also observed a large increase in the ΣO/N2 column density ratio ($\sim$80%) within the totality of the eclipse. Vargas et al. [42] presented the Andes Lidar Observatory (ALO), TIMED/SABER temperatures, and ionosonde electron density measurements. Bravo et al. [43] used an ionosonde within the totality path and $\sim$500 GNSS receivers localized over South America, which were compared with two ionosonde measurements. The authors modeled the ionospheric effects in the Sheffield University Plasmasphere-Ionosphere Model (SUPIM). Also, they corroborated the relationship between diminishing solar radiation with densities below 200 km and the E and F1 layer critical frequencies. Eisenbeis and Occhipinti [6] analyzed TEC depletion and applied the omega-k technique, which was based on a 3D Fast Fourier Transform. Yan et al. [15] simulated in the Thermosphere-Ionosphere-Electrodynamics General Circulation Model (TIE-GCM) the thermospheric perturbations induced by four different eclipses, such as TSE2019. TADs are induced by fast-moving shadows form bow wavefronts during the eclipse and evolve into freely propagating large-scale TADs after the eclipse ends. During TSE2019, TADs propagated in a direction with a smaller solar zenith angle that deviated from the eclipse trajectory.

The 14 December 2020 total solar eclipse (TSE2020) took place in the southern hemisphere summer (subsolar latitude $\sim$23°S) between 13.56 UT and 18.88 UT, and its geographic location was similar to TSE2019. The DVTEC values had -40% to -20% in the penumbra compared to the background TEC during TSE2020. Therefore, TSE2020 had similar ionospheric values to the values obtained for TSE2019. On average, the maximum dimming occurred 20 to 30 min after MOT of the eclipse at the location of the GPS stations. However, the southern EIA crest was characterized by an ionospheric decrease (DVTEC between -50% to -30%) during TSE2020, as opposed to TSE2019. Previously, Martinez et al. [44] used SUPIM, and were the first to anticipate the ionospheric effect due to TSE2020. They modified the SUPIM to simulate the behavior of solar obscuration on the ionosphere at low and middle magnetic latitudes. Gómez [45] compared TEC variations using 46 GNSS stations around the path of the umbra of the eclipse to that predicted by the SAMI3. The results showed that TEC perturbations after totality could be triggered by orographic gravity waves (oGWs) not related to the eclipse. The passage of TSE2020 across the Andes Mountains aided the propagation of orographic gravity waves to ionospheric heights. Meza et al. [11] simultaneously analyzed the ionospheric and geomagnetic responses to TSE2020. They used data from GNSS stations located in the Argentine and Chilean sectors. The authors analyzed the relationship between VTEC and geomagnetic changes. Resende et al. [27] analyzed the effects of TSE2020 over the Brazilian sector using two digisonde data sources, GNSS-TEC, and the MIRE numerical model. Additionally, de Haro Barbás et al. [46] analyzed eclipse effects on critical ionospheric frequency variability, considering two ionosondes with similar geographical latitudes and a 95% obscuration percentage. They also compared the ionospheric response with the IRI-2016 model. Shrivastava et al. [13] used 36 Chilean GPS stations localized across both sides of the totality line. Although they observed strong AGWs at the GPS stations located north of the path of totality, the AGWs had no noticeable impact on the VTEC of these sites. They found the presence of large variability in the background VTEC values during TSE2020. The authors also compared the ionospheric effects of TSE2019 and TSE2020. Zhang and Wang et al. [47] examined the space-based measurements (Swarm-B/C, COSMIC, and ICON/MIGHTI missions) of equatorial electrojet, zonal winds, and N_e for TSE2020. They reported the evident decrease in equatorial electrojet ($< -29\%$) during the solar eclipse, in comparison with non-eclipse. Recently, Bravo et al. [48] used data from multiple instruments (ionosondes, an Incoherent Scatter Radar, GNSS stations, and GOLD and Swarm-A missions) to evaluate the ionospheric prediction for a solar eclipse over South America conducted by [44]. Bravo et al. [48] also compared the predictions with three different GNSS-TEC estimation methods. Furthermore, they contrasted their evaluation's findings with earlier studies of the ionospheric behavior of TSE2019 and TSE2020.

Over Antarctica, there was a total solar eclipse on 4 December 2021 (TSE2021). The zone of the penumbral eclipse covered Antarctica and was observed over small southernmost sections of southern South America, Africa, New Zealand, and Australia between 5.48 UT and 9.62 UT. Idosa and Rikitu [49] studied TEC effects due to TSE2021 at six GPS stations in Antarctica. Over the six stations, the TEC value decreased during the eclipse. DVTEC values during the eclipse were about -70% to -50% around 8 UT (in the daytime), and in the nighttime, TEC enhancements were about 30% to 150%. The maximum TEC enhancement was at all sites after the eclipse day. Coyle et al. [50] compared GPS TEC observations to the TIE-GCM model ionosphere. They used ground magnetometer observations located along the $40°$ magnetic meridian between $69°$S and $79°$S and their magnetically conjugate counterpart. They observed a substorm that took place one hour prior to peak totality in the AE index and in the combined array magnetograms. The results showed that TEC frequency oscillations were similar to the magnetometer data. The authors' analyses suggested that these observations were caused by eclipse effects rather than the substorm.

In a recent study, Idosa and Beshir [51] reported the TEC variations during the PSE2022 using three GPS stations localized in Santiago ($33.44°$S, $70.67°$W), Montevideo ($34.90°$S, $56.16°$W), and Falkland ($51.45°$S, $59.00°$W). On the day of the eclipse, prior to the event, it was possible to observe that DVTEC at the three GPS stations reached values greater than 30%. According to their study, the GPS station they used with the highest percentage of obscuration was localized in Santiago (MPO 28% at sea level). During the eclipse, the DVTEC experienced deviations from the background ionosphere over Santiago (-39%), Montevideo (-29%), and Falkland (-60%). Finally, the authors used the three stations to compare the TEC behavior due to the 14 April 2022 moderate geomagnetic storm and PSE2022. They concluded that the TEC response to the analyzed storm was greater than that of the PSE2022-induced effects on the ionosphere. Then, Idosa and Beshir [51] did not present TEC maps or the ionospheric effects of the 2022 partial eclipse on the EIA.

Additionally, some studied eclipses have occurred on days when the geomagnetic field has been at active levels, such as the total solar eclipses of 23 November 2003 and 20 March 2015. Both eclipses happened over polar regions. The eclipse of 2003 happened over Antarctica, during the recovery phase of the great storm on 20 November 2003. Over the partial zone of this eclipse, TEC levels at the time window of the eclipse were about -30% to -17% with respect to the day before and the day after the eclipse, respectively [52]. The eclipse of 20 March 2015 took place across Arctic regions, during the recovery phase of the 2015 St. Patrick's geomagnetic storm that started a few days earlier (2015 March 17), and it was accompanied by a negative ionospheric storm on 20 March 2015. Despite the geomagnetic activity, numerous studies have documented ionospheric changes (DVTEC between $\sim$−50% and −10%) caused by the eclipse of 20 March 2015 relative to the background (e.g., [14,53–56]).

In this paper, we study and compare the effects of PSE2018 and PSE2022 on TEC behavior over South America and Antarctica regions. We also take this opportunity to analyze the ionospheric effects induced by both solar eclipses on the EIA. The analysis is based on more than 390 GPS stations (see Figure 1), Swarm-A satellite (SwA) of the European Space Agency's Swarm mission, and DMSP 5D-3 F18 spacecraft (F18) of the Defense Meteorological Satellite Program (DMSP). Our study is relevant because sources of ionospheric disturbances are a significant issue for space-based communication and radionavigation systems. Due to the fact that GPS is a radio-link-based technology, it can be affected by ionospheric fluctuations. Furthermore, our results were compared with previous studies about the ionospheric response to TSE2019, TSE2020, and TSE2021. We must emphasize that this is one of the first works where the ionospheric behavior due to two eclipses that occurred in South America and Antarctica under geomagnetic activity and in different seasons of the year are studied and compared, as well as the ionospheric effects that both eclipses had on the behavior of the EIA.

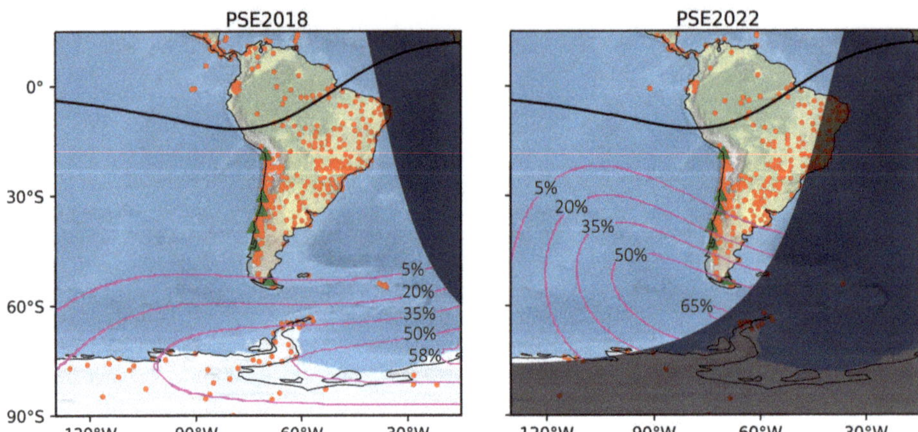

Figure 1. GPS stations (red dots) and six selected GPS stations (green triangles) used in present work. Eclipse obscuration masks at 350 km altitude (magenta lines) at Greatest Eclipse (GE) time, and the magnetic equator (black line) are also shown. The 507 and 398 GPS stations were used in PSE2018 (**left panel**) and PSE2022 (**right panel**), respectively.

2. Materials and Methods

Our study of PSE2018 and PSE2022 is based partially on the methodology described in [2,34,57]. We used the approach proposed in [35] to determine the eclipse masks at 350 km ionospheric height. This work's technique also involves an examination of geophysical and geomagnetic conditions near the dates of two eclipses (15 February 2018, and 30 April 2022). Below, we describe how we work with these data.

2.1. Brief Information about PSE2018 and PSE2022

PSE2018 had its first external contact (P1) at 18.93 UT and its last external contact (P4) at 22.79 UT. Its maximum magnitude was 0.60 at geographic coordinates 71.03°S, and 0.64°W, at 20.86 UT (Greatest Eclipse, GE). The 2018 eclipse happened in the southern hemisphere summer (subsolar latitude $\sim$12°S).

While P1 and P4 time of PSE2022 were at 18.75 UT and 22.63 UT, respectively, their maximum magnitude was 0.64 at geographic coordinates 62.12°S, and 71.48°W, at GE time equal to 20.69 UT. The 2022 eclipse occurred during the southern hemisphere autumn (subsolar latitude ~15°N).

Eclipse obscuration masks at 350 km altitude are shown in Figure 1, in accordance with the method suggested by Verhulst et al. [35]. On 15 February 2018, the Sun was hidden by the Moon at a maximum percentage of obscuration (MPO) of about 60% over Antarctica at 350 km altitude, while on 30 April 2022, at its peak, the Moon covered 66% of the Sun's disk at 350 km altitude.

Moreover, the maximum percentages of obscuration over South America were observed over Puerto Williams due to the trajectories of the two analyzed eclipses. The maximum percentages of obscuration at 350 km altitude caused by PSE2018 and PSE2022 over this city were 35% at 21.62 UT and 63% at 20.98 UT, respectively. Puerto Williams (54.9°S, 67.7°W) is a city in southern Chile, South America. It is called the world's southernmost city.

2.2. Estimation of the Ionospheric TEC

We evaluated the period of days around both eclipses (±5 days compared to the eclipse day). We imposed several conditions for the selection of reference days. First, the geomagnetic conditions for an entire day had to have quiet levels (Kp $\leq 2^+$, Dst > -30 nT), as well as the Bz component of the interplanetary magnetic field (| IMF-Bz | ≤ 10 nT) and the solar wind speed (Vsw < 450 km/s) conditions (see Section 2.4). Second, the days chosen had to have the closest trajectories of the two satellites (SwA and F18) with respect to each eclipse day, both longitudinally and temporally. Third, we wanted to preserve the symmetry of days with regard to the eclipse's day, even though this is less important than the two previous points. Therefore, we worked with the eclipse day, and we took the fifth day before and the fifth day after the eclipse as reference days.

Using the GPS-TEC analysis program version 2.9.5, we estimated the vertical TEC (VTEC) at 350 km altitude from RINEX observation files of the dual-frequency signals (f_1 and f_2) (https://seemala.blogspot.com, last accessed on 25 July 2023) [58] wherein two types of delay are continuously recorded by the dual-frequency GPS receiver: the carrier phases and the pseudoranges of the f_1 and f_2. The acquired data were then utilized to determine the slant TEC (STEC) and VTEC. To limit potential errors, we calculated the VTEC values at a 350 km ionospheric height with a 30° satellite elevation angle as the cut-off (ionospheric pierce point, IPP). TEC values are delivered by software every 30 s and are corrected for receiver and satellite bias using the data obtained from the Center for Orbit Determination in Europe (CODE) (ftp://ftp.aiub.unibe.ch/CODE, last accessed on 25 July 2023).

We evaluated the quality of the files on the chosen days for each GNSS station before making the final selection of its RINEX files. In order to ensure high-quality data, we preprocessed the RINEX observation files using TEQC software (version of 25 February 2019, https://www.unavco.org, last accessed on 20 August 2023) [59]. We made sure that the 24 h RINEX observation files with 30 s intervals were complete and of good quality. For example, we discarded GNSS stations wherein files had general RINEX formatting issues, truncated files, and the like. Also, we verified that there were no errors or data gaps after TEC estimation. This quality checking is critical to making a good comparison of the ionospheric TEC changes between the chosen days. The Chilean network set up by the National Seismological Center (CSN in Spanish); the Argentine network (RAMSAC) [60]; the Brazilian network (RBMC); and UNAVCO provided RINEX files of more than 390 GPS stations that fulfilled our specifications (see Figure 1). This study mainly focused on the South American and Antarctic regions. Then, we used 507 and 398 GPS stations used in PSE2018 (see Figure 1 left panel) and PSE2022 (see Figure 1 right panel), respectively.

Additionally, we use differential VTEC (DVTEC in TECu and %) to analyze ionospheric irregularities, as shown in Equations (1) and (2) [34,57,61].

$$DVTEC_t = VTEC_t - \overline{VTEC_t} \tag{1}$$

$$DVTEC[\%]_t = \frac{DVTEC_t}{\overline{VTEC_t}} \cdot 100 \tag{2}$$

where the mean value of VTEC ($\overline{VTEC_t}$) at the same time of the day, and t represents the epoch. Also, $\overline{VTEC_t}$ is calculated using the reference days, Days of Year (DoYs) 41 and 51, which correspond to the fifth day before and fifth day after the day of the eclipse (DoY 45) the day of the eclipse in 2018. In the same way, if DoY 120 was the day of the eclipse in 2022, we used as reference DoYs 115 and 125.

To represent the ionospheric TEC maps, we take all the VTEC values at IPPs at 350 km altitude within a range of 2.5° latitude and 2.5° longitude (starting from the Geographic Equator) and average them to generate a single VTEC value within a spatial resolution of 2.5° × 2.5° and temporal resolution of 2 min. Furthermore, to correctly carry out the algebra of maps using Equations (1) and (2), we only consider the quadrants (2.5° × 2.5°) that show values during all the days analyzed (reference days and eclipse days) at the same time of the day for each eclipse.

2.3. Ionospheric Observations and Satellite Measurements

We used ionospheric observations obtained from SwA and F18 satellites. We analyzed the eclipse-induced local ionospheric reactions using numerous space-based measurements, such as VTEC, in situ N_e, and electron temperature (T_e) data provided by SwA. This satellite has an inclination of 88° and a circular orbit at about 450 km height over the sea level. We also used in situ N_e measurements provided by F18 in the topside ionosphere at about 850 km height over the sea level with an inclination of 99° and a circular orbit. The measurements from these satellites help us to corroborate the observations on the EIA crests at 350 km, although with different amplitudes since the measurements above 850 km correspond to the plasmasphere.

2.4. Geomagnetic and Geophysical Conditions

The 15 February 2018 (see left panel of Figure 2) and 30 April 2022 partial solar eclipses (see right panels of Figure 2) took place during days with geomagnetically active levels. We downloaded geomagnetic data from OMNIWeb Plus Data (https://omniweb.gsfc.nasa.gov, last accessed on 20 August 2023) and World Data Center (WDC) for Geomagnetism, Kyoto (http://wdc.kugi.kyoto-u.ac.jp, last accessed on 20 August 2023) [62]. Figure 2, from top to bottom, illustrates the 10.7 cm solar radio flux index (F10.7 in 1 sfu = 10^{-22} W/m^2/Hz), Kp index, Dst index (in nT), IMF-Bz (in nT), auroral electrojet index (AE in nT), Vsw (in km/s), and solar wind plasma temperature (Tsw in K) which characterize the geomagnetic conditions during DoYs 41 to 52 of 2018 and DoYs 115 to 126 of 2022.

Figure 2 (left panels) shows F10.7 with 66.3 to 76.6 sfu during DoYs 46 to 51 of 2018. The geomagnetic field presented active levels between DoYs 46 and 51. The maximum 3-h Kp was 5⁻ on DoY 46. Dst peak and maximum AE were −28 nT and 618 nT, respectively. IMF-Bz reached values from −5.6 to 11.7 nT. Vsw had ≥450 km/s (intermediate speeds) between DoYs 48 and 51, with a peak of 619 km/s. Tsw > 100×10^3 K; DoYs 48 to 51 with a maximum equal to 252×10^3 K. Additionally, solar activity showed very low levels during DoYs 46 to 51.

On 15 February 2018, the geomagnetic conditions were at quiet to active levels (see Figure 2 (left panels)). The variations in 3-hourly Kp, Dst, the southward component of Bz, AE, and Vsw reached peak levels of 5⁻, −18 nT, −4.2 nT, 495 nT, and 383 km/s between 15 UT and 18 UT. The maximum Tsw was 66×10^3 K at 13 UT. Also, the AE index was greater than 300 nT from 16 UT to 18 UT, and from 22 UT to 23 UT.

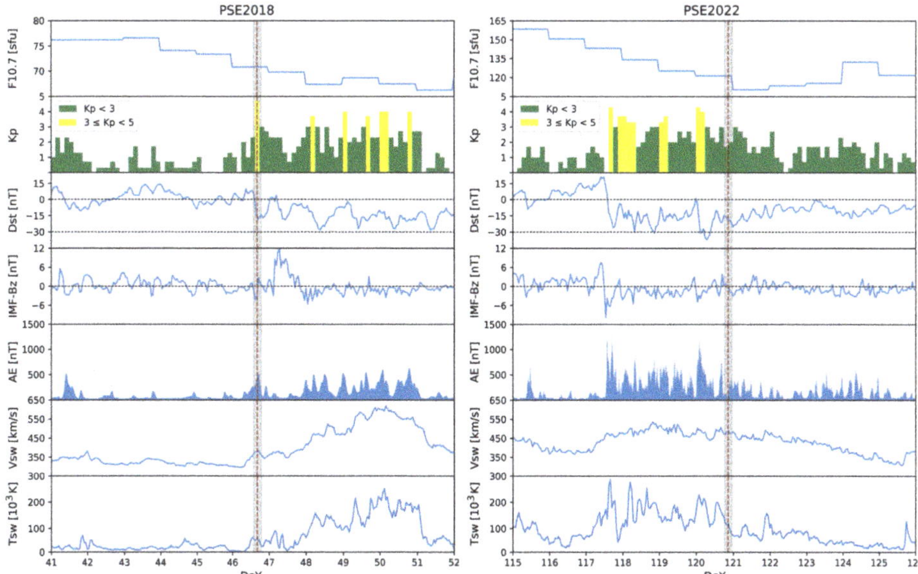

Figure 2. From top to bottom: F10.7, Kp, Dst, IMF-Bz, AE, Vsw, and Tsw used to describe geomagnetic conditions on DoYs 41 to 52 of 2018 (**left panel**) and on DoYs 115 to 126 of 2022 (**right panel**). GE (red dashed line) and P1 to P4 (light grey bar) are shown.

Figure 2 (right panels) illustrates F10.7 with 110.6 to 158.5 sfu during DoYs 115 to 126 of 2022. The geomagnetic field showed active levels between DoYs 117 and 123. The maximum 3-h Kp was 5. Dst showed a weak geomagnetic storm with -37 nT on DoY 120 (day of the 2022 eclipse). IMF-Bz reached values from -9.8 to 7.5 nT. AE presented values greater than 500 nT between DoYs 117 to 121, and it also had some values greater than 1000 nT during those days. Vsw had $\geq$450 km/s (intermediate speeds) between DoYs 117 and 123, with a peak of 535 km/s on DoY 118. Tsw $> 100 \times 10^3$ K; DoYs 48 to 51 with a maximum equal to 252×10^3 K.

On 30 April 2022, the geomagnetic conditions had quiet to minor storm levels (see right panel of Figure 2). There was a weak geomagnetic storm with a Dst peak equal to -37 nT at 8 UT (Dst ≤ -30 nT from 4 UT to 11 UT). The variations in 3-hourly Kp, the southward component of Bz, and Tsw reached peak values of 4^+, -4.8 nT, and 228×10^3 K between 0 UT and 3 UT. Vsw had intermediate speeds with a maximum of 503 km/s at 1 UT, and its peak equal to 507 km/s at 16 UT. Additionally, AE index was greater than 300 nT from 0.5 UT to 11.5 UT (AE > 500 nT, 0.5 to 7 UT), from 15 UT to 17.2 UT (AE > 500 nT, 16 to 17 UT), and from 21.5 to 23.5 UT. Solar activity presented high levels during the 24 h. Also, Active Region 2994 produced an X1.1 solar flare at 13.78 UT.

We should note that minor geomagnetic activities cannot be ignored because they sometimes play crucial roles in ionosphere plasma density, according to Cai et al. [63]. On the other hand, the comparison of the vertical E $\times$ B drift speed measured in the incoherent radar of Jicamarca (https://www.igp.gob.pe/observatorios/radio-observatorio-jicamarca/madrigal, last accessed on 20 August 2023), shows that the 15 February 2018 solar eclipse does not show great differences with the other days (DoYs 45, 47 and 48 of 2018), so there seem to be no significant effects of the geomagnetic activity in the low latitude ionosphere of South America (see Figure 3). There are no measurements for the 30 April 2022 solar eclipse.

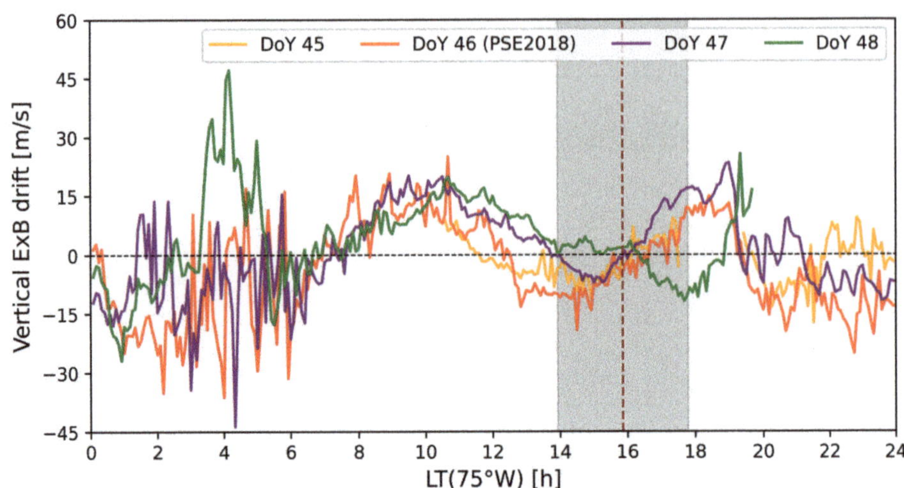

Figure 3. The vertical E × B drift speed measured in the incoherent radar of Jicamarca to compare the eclipse day (DoY 46 of 2018) with respect to DoYs 45, 47, and 48 of 2018. GE (red dashed line) and P1 to P4 (light grey bar) are shown.

2.4.1. Geomagnetic and Solar Activity during Other Eclipses

We also consult the USAF/NOAA Reports of Solar and Geophysical Activity to TSE2019, TSE2020, and TSE2021 (https://www.swpc.noaa.gov, last accessed on 20 August 2023). Solar activity had low levels during these three solar eclipses. The geomagnetic field showed quiet levels during TSE2019 and TSE2020. On 4 December 2021 (TSE2021), the geomagnetic field reached quiet to unsettled levels, where the AE index and Vsw had some activity levels. AE index presented a value of ∼600 nT around 2 UT, decreased until 7 UT, and then it increased (AE index ∼450 nT) until 9 UT. Vsw reached a peak of 534 km/s at ∼3 UT, and it remained above 450 km/s throughout the day.

2.4.2. Earthquake Occurrence

The frequent occurrence of large earthquakes (EQs) on the west coast of South America is due to the fact that this region is on top of the subducting Nazca plate [64]. Therefore, we also investigated whether superficial EQs (depths less than 70 km) with moment magnitudes (M_w) greater than 5 occurred from 10 to 21 February 2018 (by PSE2018), and from 25 April to 6 May 2022 (by PSE2022). This review is relevant because EQs are another important ionospheric disturbance source. We were able to observe that 38 and 30 earthquakes occurred during the analyzed days of 2018 and 2022, respectively (https://earthquake.usgs.gov, last accessed on 25 July 2023). During the days analyzed, however, none of them had a significant influence on the ionospheric behavior across our region of interest.

2.5. Occurrence of Other Ionospheric Disturbance Sources

The occurrence of other ionospheric disturbance sources, such as volcanic eruptions and tsunamis (https://www.ncei.noaa.gov, last accessed on 20 August 2023), was not reported during DoYs 41 to 52 of 2018 and DoYs 115 to 126 of 2022. Also, rocket launches into space from Asia were reported on days 43 and 44 of 2018. On DoYs 117, 119, 120, 122, and 125 of 2022, rocket launches into space from Florida (United States), China, Russia, and New Zealand (https://space.skyrocket.de, last accessed on 20 August 2023) were reported. Then, these rocket launches were far from the region of interest for our study. It is known that thunderstorms/lightning can introduce ionosphere gravity waves

(IGWs) [65,66], but waves in the ionosphere are beyond the scope of our research objective. Therefore, we did not verify the occurrence of thunderstorms/lightning.

3. Results

Here, we provide the main results obtained for PSE2018 and PSE2022. The results found in this paper may be classified as follows: (1) the analysis of ionospheric behavior and TEC maps using GPS stations at a regional scale; and (2) in order to verify our observations with GPS stations, we compare them with Low Earth Orbit (LEO) satellite measurements of the sector involved.

3.1. TEC Changes and Ionospheric Maps Using GPS Stations

Figures 4 and 5 show the summaries of ionospheric maps by comparing the eclipse (VTECe) and the reference days ($\overline{VTEC}$) for PSE2018 and PSE2022, respectively. Eclipse masks are represented from 5% obscuration at 350 km altitude (magenta lines). For both eclipses, we present some particular hours: P1 $-$ 1 h, P1, GE $-$ 0.5 h, GE, GE + 0.5 h, and P4.

Figure 4 illustrates the ionospheric TEC behavior during PSE2018; where 17.95 UT (P1 $-$ 1 h maps), 18.93 UT (P1 maps), 20.36 UT (GE $-$ 0.5 h maps), 20.86 UT with MPO $\sim$60% (GE maps), 21.36 UT (GE + 0.5 h maps), and 22.79 UT (P4 maps). To see TEC maps from other hours, please go to Supplementary Materials Videos S1 and S2.

From 16.85 UT, we could see that TEC is mostly positive relative to the background TEC (DVTEC $\sim$50%) between 30°S and 90°S, and it was $\pm$15% between 30°S and 0° latitude. This TEC enhancement can be clearly seen at 17.95 UT (P1 $-$ 1 h maps). The P1 maps ($\sim$18.93 UT) show that the DVTEC value decreases slightly between 30°S and 90°S, where it was $\sim$20% in this southern South America region. At 19.50 UT, there is a decrease in TEC (DVTEC $\sim$$-$20%) in the Antarctica region between 45°W and 135°W longitude. At 20.15 UT, a notable TEC decrease begins around the southern EIA crest over South America around 30°S, 60°W. This TEC change continues to lower its value (DVTEC $\sim$$-$50%) and expand its region over South America as time passes. This phenomenon can be observed in the GE $-$ 0.5 h (20.36 UT), GE (20.86 UT), GE + 0.5 h (21.36 UT), and P4 (22.79 UT) maps. From 20:50 UT, there is a shift in TEC decrease relative to the background TEC from the southeast (Antarctic sector) that passes through South Georgia and the South Sandwich Islands in the southern Atlantic Ocean and reaches southern South America at about 20.86 UT (GE maps, where MPO $\sim$60%). DVTEC remained positive over the studied sector of the Antarctic Peninsula until $\sim$20.65 UT. In the Antarctic sector, TEC is negative relative to background TEC, and the minimum DVTEC was $\sim$$-$40% at 20.65 UT. At about 20.86 UT (see GE maps), we see the DVTEC depletion move from southwest Antarctica to the northeast, and it reaches southern South America. Additionally, the recovery of TEC and some positive values of DVTEC appear over the Antarctic sector. About 22.80 UT (see P4 maps), we can observe the recovery of TEC in southern Chile and Argentina (up to 45°S latitude), but TEC depletion over South America persisted until after 24 UT. Moreover, after P4, the maximum DVTEC was >40% over the Antarctic region, similar to that before the first contact of the eclipse.

Figure 5 is the same as Figure 4 but shows TEC variations during PSE2022. Figure 5 exhibits some particular hours: 17.75 UT (P1 $-$ 1 h); 18.75 UT (P1); 20.19 UT (GE $-$ 0.5 h); 20.69 UT (GE) with MPO 66%; 21.19 UT (GE + 0.5 h); and 22.63 UT (P4). At least before 16 UT, TEC enhancement was >20% relative to the background over South America (30°S to 60°S). This TEC enhancement is clearly illustrated on the P1 $-$ 1 h and P1 maps. After P1, we could observe a decrease in TEC with <$-$10% along the Antarctic coasts between 60°W and 120°W. TEC decrease (DVTEC $\sim$$-$50%) reaches around Drake Passage ($\sim$60°S, $\sim$60°W) at 20.45 UT. At 20.70 UT, DVTEC had values of around $-$40% over southern South America and Antarctica (see GE + 0.5 h maps). The visible TEC decrease (DVTEC $<$ $-$10%) reaches up to 30°S over South America. The improvement of TEC from the geomagnetic equator to 30°S latitude is reinforced around 21.20 UT, and it is maintained for at least 24 UT (P1 + 0.5 h and P4 maps).

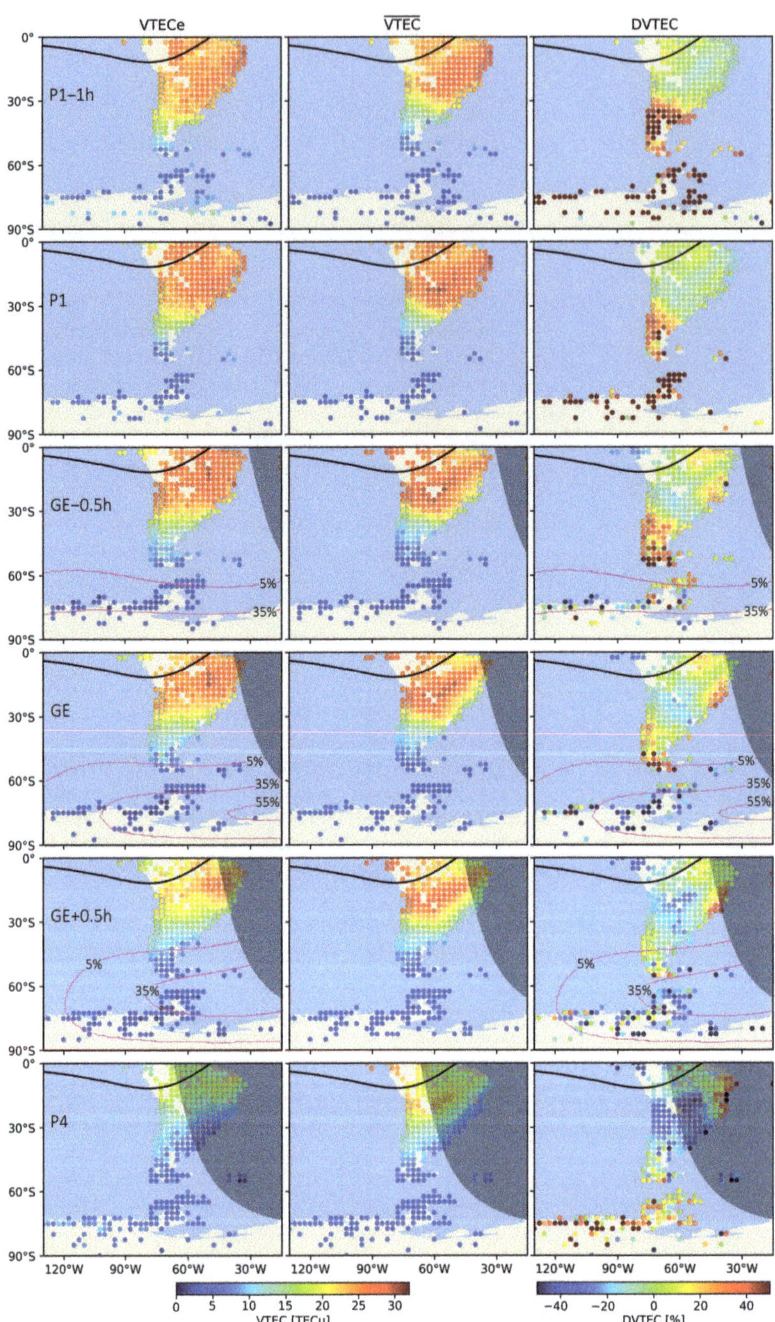

Figure 4. TEC maps during PSE2018 and its eclipse obscuration masks at 350 km ionospheric altitude (magenta lines) are presented. VTECe, $\overline{VTEC}$, and DVTEC (%) are shown from left to right panels. Also, from top to bottom: P1 − 1 h (17.95 UT); P1 (18.93 UT); GE − 0.5 h (20.36 UT); GE (MPO ∼60% at 20.86 UT); GE + 0.5 h (21.36 UT); and P4 (22.79 UT).

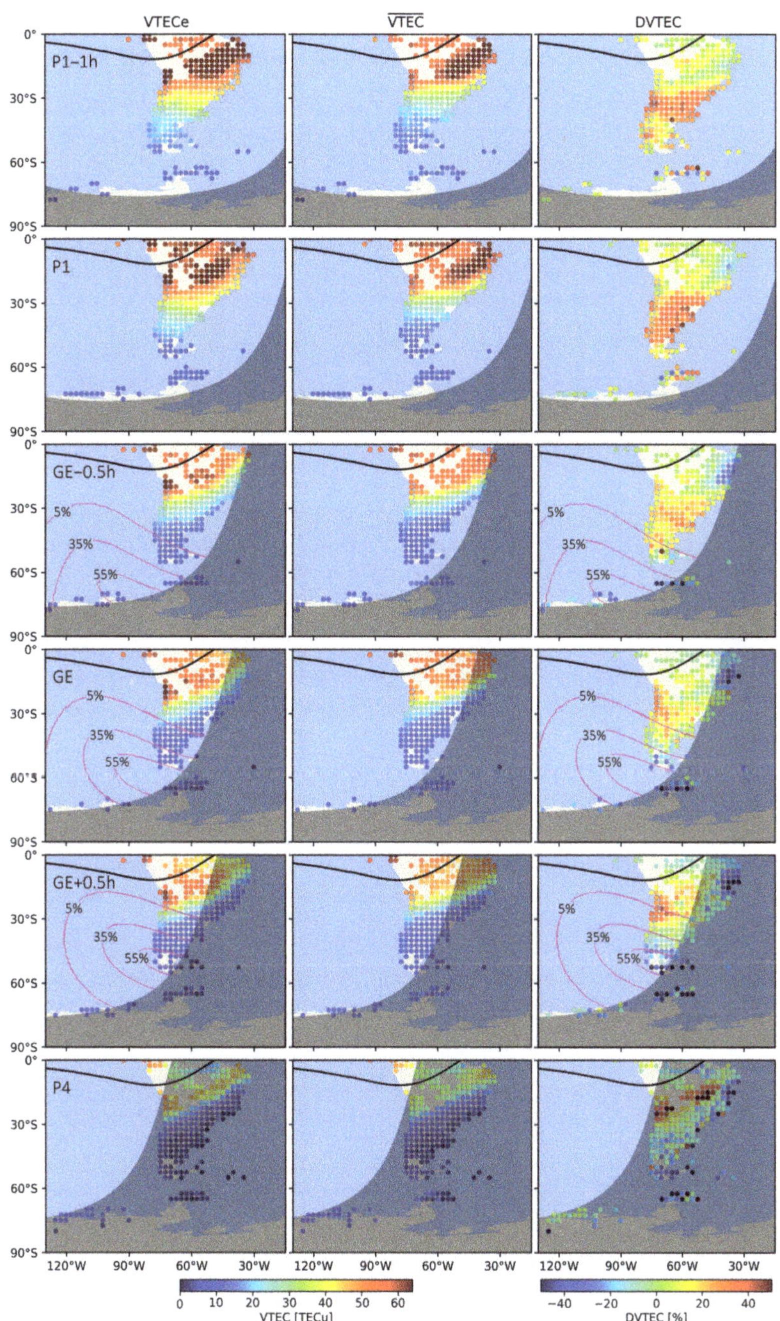

Figure 5. TEC maps during PSE2022 and its eclipse obscuration masks at 350 km ionospheric height (magenta lines) are shown. VTECe, $\overline{VTEC}$, and DVTEC (%) are represented from left to right panels. Additionally, from top to bottom: P1 − 1 h (17.75 UT); P1 (18.75 UT); GE − 0.5 h (20.19 UT); GE (MPO ∼66% at 20.69 UT); GE + 0.5 h (21.19 UT); and P4 (22.63 UT).

To study the effects of the PSEs, we chose six GPS stations located in Chile from among more than 390 GPS stations (see Figure 1), where five stations were under the partial region during both eclipses (LSCH, RCSD, IMCH, QLLN, XPLO, CSOM). Additionally, CYHT station was under the shadow of PSE2022, but it was not under the shadow of PSE2018. Table 1 contains more information about the eclipse conditions and the GPS stations to the ionospheric height of 350 km.

Figure 6 shows the results of the TEC behavior of six selected GPS stations for the eclipse day (VTECe), the reference days ($\overline{VTEC}$), and the results of DVTEC (%). We presented each plot between 12 UT and 24 UT. The brown dotted line presents GE time, and it is inside the light blue bar shading the region between P1 and P4. The black dotted line shows MOT; this line is inside the yellow bar between C1 and C4. Both eclipses occurred during the afternoon at six selected stations. We could clearly see the depletion of TEC during both eclipses. The six selected GPS stations had a recovery between 1.5 UT and 2.5 UT the day after both partial eclipses. Table 1 presents the minimum values of DVTEC for each of the six chosen GPS stations.

On the days of PSE2018 and PSE2022, TEC with respect to the background TEC was -29% ($\tau \sim 7$ to 43 min) and -50% ($\tau \sim 1$ to 45 min), respectively. Although CHYT was not under the shadow of PSE2018, it had a noticeable minimum DVTEC (-28%, -6 TECu) at 22.65 UT, after 5 min that LSCH and after 107 min to GE time. Regarding PSE2022, CHYT did not show a clear TEC reduction near MOT at this station. LSCH had minimum DVTEC values greater than 0% and 0 TECu, but it showed an evident TEC depletion near MOT.

We also estimated DVTEC using the first day before and the first day after the PSE2022 as reference days (DoYs 119 and 121 of 2022) due to the ionospheric behavior of the CHYT and LSCH stations, when we took the fifth day before and the fifth day after this eclipse as reference days (DoYs 115 and 125 of 2022), where DoYs 119 and 121 showed values of $Kp > 2^+$ and Vsw > 450 km (see Figure 2). On this occasion, the eclipse-induced effects were shown clearly at CHYT (DVTEC = 0%, -0.2 TECu, and $\tau = 57$ min) and LSCH (DVTEC = -25%, -4.1 TECu, and $\tau = 32$ min). The minimum DVTEC of CHYT occurred 5 min before the end of the partial eclipse in this station. From the ionospheric response due to the eclipse, using reference days DoYs 115 and 125 of 2022 with respect to employment as reference days DoYs 119 and 121 of 2022, we were able to observe that the GPS stations (IMCH, QLLN, and CSMO) located from $\sim 38°$S geographic latitude to the south showed a difference in the minimum percentage of DVTEC of less than 11% and a difference of $\tau \pm 5$ min.

The six selected GPS stations showed that the ionosphere was impacted by geomagnetic forcing on the days of the 2018 and 2022 eclipses. On the day of the 2018 eclipse, we can analyze the ionospheric changes from 16 UT. Then, we can observe that the DVTEC of the six GPS stations reaches values from $\sim 10\%$ to $\sim 60\%$ (CHYT: 9%, 2 TECu; LSCH: 28%, 5 TECu; RCSD: 46%, 7 TECu; IMCH: 43%, 6 TECu; QLLN: 58%, 6 TECu; CSOM: 58%, 4 TECu) at around 18 UT (see Figure 6). On the day of the 2022 eclipse, the six selected GPS stations had maximum ionospheric values from 42% to 146% with respect to the reference days. Between P1 time and before we could observe the effects of the eclipse on the ionosphere, DVTEC had a maximum value in each of the six stations (CHYT: 24%, 11.7 TECu; LSCH: 34%, 5.9 TECu; RCSD: 34%, 6.6 TECu; IMCH: 34%, 5.2 TECu; QLLN: 21%, 2.9 TECu; and CSOM: 13%, 1.2 TECu) (see Figure 6).

Table 1. Ionospheric TEC changes and eclipse conditions for each of the six chosen GPS stations. According to [35], we estimated the maximum percentage of obscuration (MPO), start of partial (C1), maximum obscuration time (MOT), and end of partial (C4) eclipses at 350 km altitude.

GPS Station			PSE2018							PSE2022						
Code	Lat [°S]	Lon [°W]	MPO [%]	C1 [UT]	MOT [UT]	C4 [UT]	τ [min]	DVTEC [%]	DVTEC [TECu]	MPO [%]	C1 [UT]	MOT [UT]	C4 [UT]	τ [min]	DVTEC [%]	DVTEC [TECu]
CHYT	18.37	70.34	0	0	0	0	12 [a]	−28	−6	8	21.38	22.07	22.70	N.O. [b]	N.O. [b]	N.O. [b]
LSCH	29.91	71.25	3	22.00	22.45	22.87	7	−29	−4.5	30	20.70	21.77	22.73	29	9	1
RCSD	33.65	71.61	6	21.73	22.33	22.90	17	−20	−2.7	38	20.52	21.67	22.68	1	−28	−2.8
IMCH	38.41	73.89	11	21.43	22.18	22.88	35	−12	−1.3	47	20.25	21.48	22.58	20	−37	−2.8
QLLN	43.11	73.66	17	21.15	22.02	22.82	43	−4	−0.4	53	20.07	21.33	22.47	17	−50	−3.6
CSOM	52.73	69.22	32	20.67	21.68	22.63	8	−2	−0.1	62	19.78	21.05	22.18	45	−56	−3.7

[a] In CHYT, τ does not refer to MOT of CHYT, but it refers to MOT of LSCH station. [b] N.O.: We do not observe ionospheric changes near the eclipse time window over the GPS station.

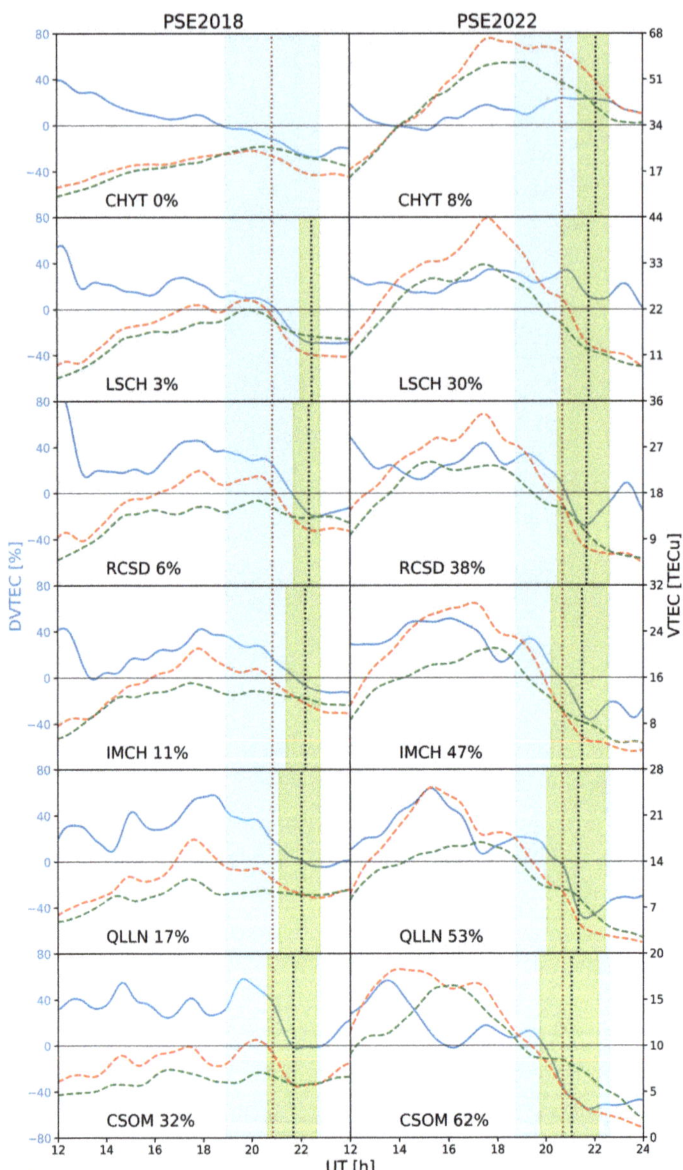

Figure 6. TEC behavior during PSE2018 (**left panel**) and PSE2022 (**right panel**) in selected stations. GPS stations are ordered by latitude. MPO is presented next to each station name. VTECe (red dashed line), $\overline{VTEC}$ (green dashed line), and DVTEC (%) (blue line) are represented. GE (red dotted line), MOT (black dotted line), C1–C4 (yellow bar), and P1–P4 (light blue bar) are also shown.

3.2. Ionospheric Changes Using LEO Satellite Measurements

Figures 7 and 8 show the high ionospheric values in low latitudes and part of the middle latitudes, around the region where the southern EIA crest is located, measured by the SwA and F18 satellites. Figure 7 illustrates the ionospheric behavior for PSE2022 using SwA measurements. We do not show the ionospheric data of PSE2018 because the satellite passes do not cover the region of interest during the eclipse. Then, we present

GPS-VTEC at 850 km altitude (400 km above SwA), and in situ N_e and T_e measurements made by SwA Langmuir probes (at 450 km ionospheric height) on DoY 120 compared to two geomagnetic quiet days (DoYs 115 and 125). We chose the four SwA passes that best fit the eclipse region and eclipse time window over our region of interest (51°W, 74°W, 98°W, and 121°W). Two consecutive descending passes of SwA, the first 74°W (between 20°S and 0°), and the second 98°W (between 30°S and 20°S), show the clear increase in the values of the southern EIA crest (e.g., maximum DVTEC $\sim 65\%$ and >10 TECu).

We were able to observe ionospheric effects under the penumbra centered between 70°S and 30°S. Although Figure 7 cannot be easily seen, we present our findings below where the first SwA pass (51°W) took place more than an hour before P1 (18.75 UT), where TEC changes were from -13% to 20% (from -0.3 to 0.8 TECu); DN_e were from 17% to 64% (from 0.09×10^5 el/cm^3 to 0.68×10^5 el/cm^3); and DT_e had fundamentally around -5%, with respect to the background ionospheric values. The second pass (74°W) was close to P1. The minimum DVTEC was -30% (-1 TECu) at 54.0°S at 18.74 UT. The minimum DN_e was -4% (-0.03×10^5 el/cm^3) at 49.4°S at 18.72 UT; and the minimum DT_e was -34% (from $\sim$3100 K to $\sim$2100 K) at 48°S, in relation to the baseline values. The third satellite pass (98°W) occurred minutes prior to GE (20.69 UT). The minimum DVTEC was -52% (-1.5 TECu) at 64.6°S at 20.34 UT. The minimum DN_e was -45% (-0.34×10^5 el/cm^3) at 57°S at 20.3 UT; and the minimum DT_e was -22% (from $\sim$3200 K to $\sim$2500 K) at $\sim$54°S, with respect to the background values. The fourth selected pass of SwA (121°W) happened before P4 (22.63 UT). The minimum DVTEC was -45% (-2.1 TECu) at 54.2°S at 21.84 UT. The minimum DN_e was -31% (-0.36×10^5 el/cm^3) at 40.5°S at 21.78 UT; and the minimum DT_e was -14% (from 3600 K to 3100 K) at $\sim$53°S, regarding background values. Also, DT_e had values fundamentally around ±100 K.

On the other hand, to study the ionospheric changes during PSE2018, we used in situ N_e measurements from two passes of the F18 satellite over the region of interest (see Figure 8a,b). We compare the ionospheric behavior of the day of the eclipse of 2018 (DoY 46) with respect to the mean N_e of the two geomagnetically quiet days (DoYs 41 and 51). The first satellite pass took place from 13°W to 35°W between 20.63 UT and 20.82 UT, a few minutes before GE (20.86 UT), where the N_e values were greater than 10% in the entire evaluated interval (from 70°S to 30°S). The second F18 pass occurred from 38°W to 61°W between 22.33 UT and 22.52 UT, minutes prior to P4 (22.79 UT). The minimum DN_e was -21% (-0.04×10^5 el/cm^3) around 48°S at 22.44 UT.

Regarding the ionospheric fluctuations during PSE2022, we also used three F18 ascending trajectories (see Figure 8c,d). Similar to F18 measurements of PSE2018, we compare the ionospheric changes of the eclipse day of 2022 (DoY 120) with respect to the mean N_e of two previously selected geomagnetically quiet days (DoYs 115 and 125). Before GE (20.69 UT), the satellite passed from 29°W to 51°W, between 19.73 UT and 19.93 UT. N_e remained positive from $\sim$60°S to $\sim$43°S, its value reached $\sim$20% at $\sim$58°S. The DN_e values were less than 0% from 43°S to the North, with a minimum value equal to -10% (-0.02×10^5 el/cm^3) at $\sim$31°S. F18 traveled from 54°W to 77°W between 21.43 UT and 21.63 UT, about an hour after GE (20.69 UT), where the minimum DN_e was -54% (-0.07×10^5 el/cm^3) at 54°S. The third selected pass of this satellite took place from 79°W to 102°W, (23.13 to 23.33 UT), after P4 (22.63 UT). It had a minimum DN_e equal to -30% (-0.06×10^5 el/cm^3) at 40°S.

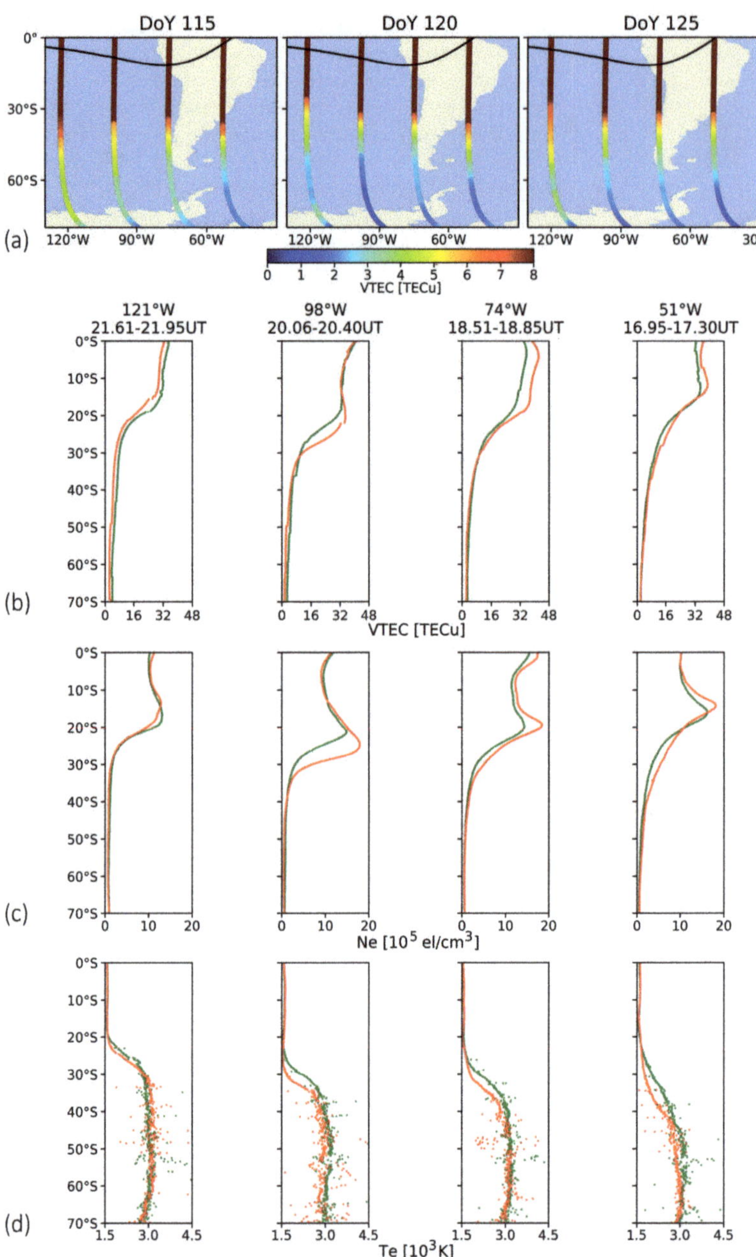

Figure 7. Ionospheric behavior using SwA measurements for PSE2022. (**a**) VTEC data gathered through the satellite orbits taken by SwA at 850 km (400 km above SwA) are presented over the maps during the eclipse day (DoY 120) and the two selected geomagnetically quiet days (DoYs 115 and 125). The black line indicates the magnetic equator. (**b**) Comparison of the latitudinal VTEC profile on the eclipse day (red line) with the mean VTEC of two closest trajectories longitudinally and temporally (DoYs 115 and 125, green line) between 70°S and 0°. (**c**) In situ N_e and (**d**) T_e measurements at 450 km are presented in the same way as VTEC data. The longitudes and time intervals of the satellite passes are also indicated.

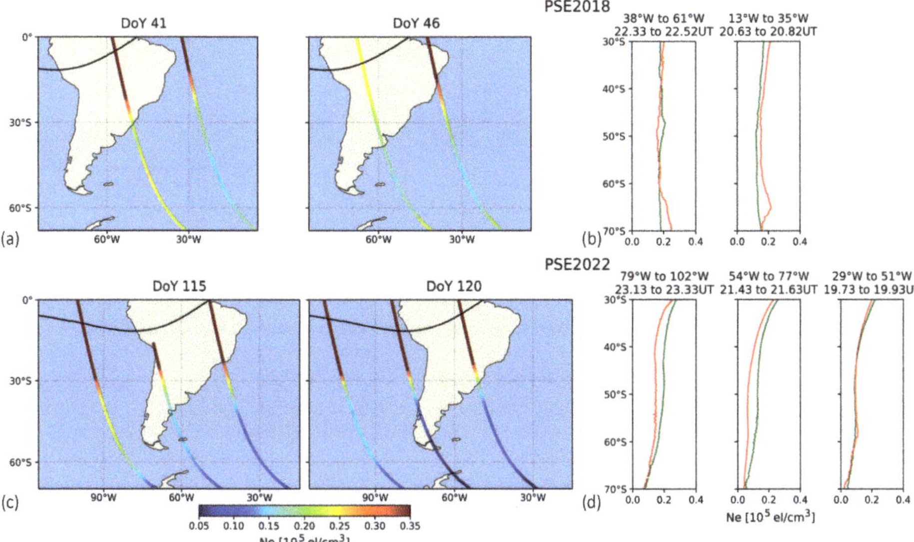

Figure 8. In situ electron density (N_e) data measured by DMSP 5D-3 F18 in the topside ionosphere at about 850 km height for PSE2018 (**a,b**) and PSE2022 (**c,d**). (**a,c**) N_e acquired via satellite orbit the magnetic equator (black line) are displayed on Earth maps. (**b,d**) Profile data on eclipse DoYs (DoY 46 of 2018; and DoY 120 of 2022; in red lines) compared with the mean N_e of the reference DoYs (DoYs 41 and 51 of 2018; and DoYs 115 and 125 of 2022, in green lines). The longitudes and time intervals of the satellite passes are also indicated.

4. Discussion

In this part, we examine our main results for both partial solar eclipses. At 350 km ionospheric altitude, the 15 February 2018 and 30 April 2022 eclipses had MPO of 60% and 66%, respectively. The main goals were to present and compare the TEC behavior of both eclipses over South America and the Antarctic sector. The relevance of these events is that there are few that cross over the south polar region (Antarctica) and culminate in the South American region. Moreover, we compare our ionospheric response to earlier findings for total solar eclipses over South America (TSE2019, and TSE2020) and Antarctica (TSE2021).

4.1. F10.7 and TEC

The background TEC values during the EPS2022 period are notably higher than the background TEC values of the EPS2018 period. A possible explanation for this phenomenon is that the solar flux (F10.7) indices during PSE2022 (F10.7 were from 110.6 to 158.5 sfu) are higher than during PSE2018 (F10.7 were from 66.3 to 76.6 sfu), as shown in Figure 2. Additionally, PSE2022 occurred ∼2.5 years after the start of Solar Cycle 25, while PSE2018 occurred less than 2 years before the end of Solar Cycle 24 (see Figure 9). TEC changes are synchronized with solar cycles. Moreover, the dependency between solar index parameters (e.g., solar flux, extreme ultraviolet radiations, and sunspot number) and TEC has been studied by many researchers (e.g., [67,68]). Therefore, we prefer to conduct the percentage analysis of TEC changes to compare the eclipses instead of using TECu.

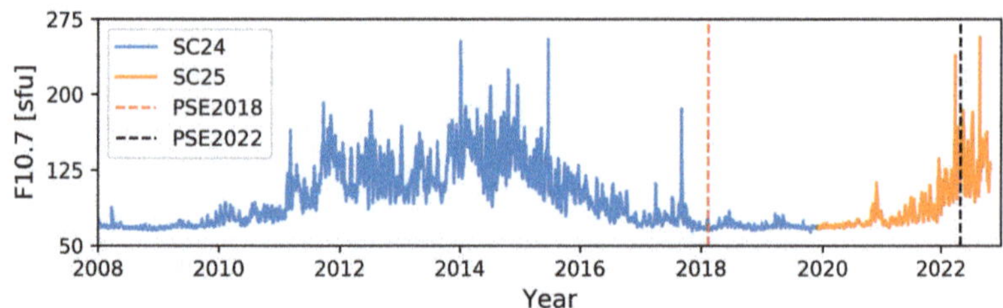

Figure 9. The observed F10.7 Radio Flux during Solar Cycle 24 (SC24, blue line) and Solar Cycle 25 up to day 304 of 2022 (SC25, orange line). 15 February 2018 partial solar eclipse (PSE2018, red dashed line) and 30 April 2022 partial solar eclipse (PSE2022, black dashed line) are also shown.

4.2. Ionospheric TEC Behavior

PSE2018 occurred in a region where we had access to a low density of GPS station measurements. Despite the fact that the PSE2018 and PSE2022 occurred on days with geomagnetic activity (PSE2018: Kp = 5^- and AE = 495 nT; PSE2022: Dst = -37 nT, Kp = 4^+, Vsw = 507 km/s, and AE > 500 nT), we can observe the evident TEC depletions under the Moon's shadow with respect to the background ionosphere (see Figures 4–8 and Table 1). PSE2018 and PSE2022 had similar ionospheric behaviors under the lunar shadow with clear TEC decreases ($<-40\%$) that moved from Antarctica to South America (see Figures 4 and 5). Despite the fact that the maximum obscuration in both eclipses is around 60% at 350 km altitude, the DVTEC values we obtained were consistent with the values presented for the total solar eclipses over South America [6,10,11,13,27,31,43,45,46,48] and Antarctica [49]. Also, τ values were between 1 and 45 min, and the ionospheric effects caused by PSE2018 and PSE2022 persisted after 24 UT because the concentration of electrons does not return to normal levels (see Figure 6 and Table 1). Therefore, the duration (ΔT) of ionospheric changes was over 5 h. This is in agreement with other studies; for example, [34,69] showed that the ionospheric disturbances caused by an eclipse can last more than 7 h. Additionally, the impacts of eclipses can also be seen on a global scale [4,17,34,70] due to the abrupt temperature change and pressure difference brought on by the eclipses-induced rapid cooling of the atmosphere, which may result in AGWs and associated TADs and/or TIDs [4,70]. A thorough examination of these issues, however, is outside the purview of the present paper. Also, the magnetosphere–ionosphere system's electrodynamic interaction caused the polar region under eclipse to cause geospace disturbances that had a larger impact on the ionosphere globally [17].

During PSE2022, we could observe a clear VTEC decrease (-28%, -2.8 TECu) over RCSD (33.65°S, 71.61°W) (see Table 1 and Figure 6). Our results were consistent with the ionospheric behavior (DVTEC = -3 TECu) over Santiago (33.44°S, 70.67°W) reported in [51], for the same eclipse. The ionosphere behavior of the anomaly could be observed at the CHYT and LSCH because these GPS stations are located under the southern EIA crest (see Figures 1, 4 and 5 and Table 1). We could only clearly see the effects induced by the 2022 eclipse at CHYT and LSCH stations when we used non-quiet days as reference days. For example, due to the geographic location of the CHYT station and the geomagnetic activity, this GPS station did not show significant ionospheric variations during the 2022 eclipse when we used selected quiet days as reference days.

We previously showed that the recombination of ionospheric ions and electrons in the absence of sunlight decreased the ionospheric conductivity when the Moon's shadow crossed the ionosphere over South American and Antarctic regions during PSE2018 and PSE2022. However, the enhancement of the ionosphere characterized the geomagnetically active levels on the days of both partial eclipses (see Figures 2 and 4–6). On 15 February 2018, before the P1 time of the PSE2018, we could observe an ionospheric enhancement over the six GPS stations (DVTEC $\sim$ 10% to $\sim$60%) near the time that levels of geomagnetic activity were reached at around 18 UT (Kp = 5^- and AE = 495 nT, see Figure 2).

The influence of geomagnetic activity on changes in the ionosphere TEC was observed most clearly at mid and high latitudes (see Figure 4), but there were no significant effects of the geomagnetic activity in the low-latitude ionosphere over South America (see Section 2.4 and Figure 3). Therefore, GPS stations at higher latitudes (in South America), although they had a higher percentage of obscuration, showed less influence on the effects of the eclipse than GPS stations at lower latitudes with a lower percentage of obscuration (see Table 1 and Figures 4 and 6).

The days prior (since DoY 117) to PSE2022 showed some levels of geomagnetic activity. On 30 April 2022, the ionospheric behavior (DVTEC > 40%) observed in the six selected GPS stations coincided with the levels of geomagnetic activity between 0 UT and 6 UT (Kp = 4^+, Dst = -37 nT, and Vsw > 480 km/s; see Figure 2). Subsequently, the DVTEC values began to decrease until around 12 UT (after sunrise in Chile), and then the ionospheric TEC started to increase until the eclipse impacts on the ionosphere could be observed. After the P1 time of the PSE2022, we could observe positive values of DVTEC over the six GPS stations between 13% and 34% with respect to the background ionosphere. Our findings related to the improvements in the ionospheric TEC at the six stations during the day of the 2022 eclipse were in agreement with the graphs shown by Idosa and Beshir [51]. Therefore, although geomagnetic forcing impacted the ionosphere [63], causing a positive increase in TEC during levels of geomagnetic activity, we could clearly see the ionospheric depletion during both eclipses.

On the other hand, we were able to observe a TEC depletion of the southern EIA crest during PSE2018 but a TEC enhancement of the same crest during PSE2022 (see Figures 4–7). During the 2018 solar eclipse, we were able to observe a clear ionospheric TEC depletion (DVTEC $\sim-50\%$) in the EIA crest over South America. These TEC changes spread over South America as time passed. Our percentage of TEC depletion matches what [27,48] showed in the southern EIA crest for TSE2020 over South America and the EIA depletion summary for different eclipses (DVTEC ~-50 to -20%) by [27]. At least the eclipses of 2018 and 2020 happened in the summer hemisphere. Meanwhile, during the 2022 solar eclipse (hemisphere autumn), we can see a notorious TEC enhancement (DVTEC > 30%) relative to the background in the EIA crest over South America. TEC enhancement in the southern EIA crest obtained by us is comparable to that presented by [31] for TSE2019 (winter hemisphere) over South America, and by [28] for the 2019 December 26 annular solar eclipse (winter hemisphere). Also, we can note that all three eclipses occurred during autumn or winter in their respective hemispheres. Therefore, we demonstrate that our findings of ionospheric behavior in the southern EIA crest for the eclipses of 2018 and 2022 are validated by earlier studies. Moreover, previous research suggests that solar eclipses can cause strong oscillations in the zonal electric field and in the neutral wind circulation pattern, causing a pronounced reduction or marked enhancement in the EIA crests [25,28,29,71].

As a possible explanation for these phenomena, we must consider that, in the summer hemisphere, the Sun's rays arrive perpendicular to the surface, generating a temperature gradient that produces the summer-to-winter neutral wind. This neutral wind transports the plasma by ion-drag effect along the magnetic field lines, in the upward/equatorward direction in the summer hemisphere and in the downward/poleward direction in the winter hemisphere. On the other hand, the vertical E $\times$ B drift at the equator carries plasma to high altitudes, wherein the chemical recombination loss rate is lower and the ambipolar diffusion occurs on both sides along the magnetic field lines (fountain effect) carrying to magnetic latitudes where the chemical recombination loss rate is more significant than the production rate [72]. At the summer EIA crest, an accumulation of plasma occurs due to the opposite directions of the neutral wind and ambipolar diffusion (as modeled by [73]); on the contrary, at the winter EIA crest, the neutral wind and the ambipolar diffusion have the same direction, so there is no accumulation of plasma. Since the solar eclipse is only one day in the entire season, the weakening of the neutral wind due to the shadow of the eclipse should decrease summer EIA crest accumulation and increase winter EIA crest accumulation [74]. This process is observed in the southern EIA crest over South America in PSE2018 (summer), PSE2019 (winter), PSE2020 (summer), and PSE2022 (autumn, close to winter).

SwA and F18 Measurements

In order to verify TEC behavior, we corroborated by SwA and F18 measurements. The minima with respect to the background values of DVTEC, DN_e, and DT_e were -52%, -54% to -21%, and $<-20\%$, respectively, under the penumbra of the eclipses of 2018 and 2022. Our DVTEC and DN_e results are consistent with the data provided in prior publications [34,48,75]. Also, our T_e decrease values are consistent with the findings of earlier studies. For example, previous work reported drops in temperature ($T_e \sim -50$ to -20%) under the shadow of the 21 August 2017 total solar eclipse [76–78] and 26 December 2019 annular solar eclipse [28]. The effects on the ionosphere were observed post-eclipse of 2022 with the satellites. We can observe a minimum $DN_e \sim -30\%$ during the third selected pass of F18 ~ 0.5 h after P4 time. Also, although the pass of SwA for 121°W was over the Pacific Ocean, somewhat away from the shadow of the eclipse of 2022, it had minima DVTEC, DN_e, and DT_e with -45%, -31%, and -14%, respectively. Furthermore, two consecutive passes of SwA showed the notorious increase in the values of the southern EIA crest (DVTEC $\sim 65\%$) between 30°S and 0°.

5. Conclusions

The 15 February 2018 and 30 April 2022 solar eclipses under geomagnetic activity in different seasons (summer and autumn) were good opportunities to study how the eclipse-induced ionization changes affect the EIA and radio link-based systems such as GPS. In this paper, we presented and compared the ionospheric TEC changes during PSE2018 and PSE2022 over South America and Antarctica. Both partial solar eclipses had trajectories over the southern polar region and southern South America. These two eclipses took place on non-quiet geomagnetic days. We used a regional GPS network located around both regions. We create regional TEC maps using data from GPS ground stations. Furthermore, we present in more detail the behavior of the two eclipses at six GPS stations located in Chile, along the west coast of South America. We also selected quiet days as reference days after analyzing the geomagnetic and geophysical conditions.

We suggest performing the percentage analysis of TEC variations to compare eclipses instead of using TECu because TEC changes are synchronized with solar cycles. Therefore, it should also consider solar index parameters (e.g., solar flux, extreme ultraviolet radiation, and sunspot number) and not only take into account elements such as the eclipse trajectory, seasons, times of the day, locations, synoptic, and geomagnetic conditions.

TEC maps of both eclipses showed noticeable depletions under the Moon's shadow with DVTEC $< -40\%$, τ between 1 min and 45 min, and duration of ΔT over 5 h. Additionally, TEC decreased (DVTEC $\sim -50\%$) in the southern EIA crest over South America during the 2018 eclipse (hemisphere summer), but the ionization presented a significant enhancement (DVTEC $> 20\%$) that prevailed for a number of hours after the end of the 2022 eclipse (hemisphere autumn, close to winter).

We show that, despite ionospheric enhancement induced by geomagnetic activity on the days of the two eclipses, the impact of both eclipses could be seen. The ionospheric behavior that we estimated with GPS receivers was corroborated using observations from the SwA satellite (VTEC, N_e, and T_e), and the F18 satellite (N_e). The minima with respect to the background values of TEC, N_e, and T_e were -52%, -54% to -21%, and $<-20\%$, respectively. Therefore, the ionospheric behavior unequivocally showed that electron activity in that layer decreases during both solar eclipses, as expected. Furthermore, our findings about the ionospheric behavior of the 2018 and 2022 eclipses during geomagnetic activity are in line with earlier reports for other solar eclipses in the same sector under quiet conditions, such as those TSE2019 and TSE2020 (over South America). We also corroborate our ionospheric results over the Antarctic sector with TSE2021 that occurred during unstable geomagnetic levels.

As a continuation of this work on the 2018 and 2022 partial solar eclipses, we propose to carry out analyses on a regional and/or global scale of the following aspects: Improve our study with the use of other instruments such as ionosondes and magnetometers, among others, to analyze the alteration of the equatorial vertical $E \times B$ drift and generation of a disturbed dynamo. The effects of thermospheric responses (e.g., temperature and winds) to the eclipse also need to be considered. These partial eclipses offer a fantastic opportunity to investigate whether their waves (AGWs, TADs, and TIDs) travel from the poles to low latitudes. Additionally, research into how solar eclipses affect the precision of GNSS and other radiocommunication systems will aid in understanding and mitigating these effects.

Supplementary Materials: The following supporting information can be downloaded at: https://www.mdpi.com/article/10.3390/rs15194810/s1, Video S1: PSE2018; Video S2: PSE2022.

Author Contributions: Conceptualization, J.C.V.-A., M.D. and M.B.; methodology, J.C.V.-A. and M.D.; software, J.C.V.-A., M.B. and Y.S.-S.; validation, J.C.V.-A., M.D., M.B. and Y.S.-S.; formal analysis, J.C.V.-A., M.B. and Y.S.-S.; investigation, J.C.V.-A., M.D., M.B. and Y.S.-S.; resources, J.C.V.-A., M.D., M.B. and Y.S.-S.; data curation, J.C.V.-A., M.B. and Y.S.-S.; writing—original draft preparation, J.C.V.-A., M.B. and Y.S.-S.; writing—review and editing, M.D.; visualization, J.C.V.-A., M.D. and Y.S.-S.; supervision, M.D.; project administration, M.D.; funding acquisition, M.D. All authors have read and agreed to the published version of the manuscript.

Funding: This research was funded by the Air Force Office of Scientific Research under award numbers FA9550-18-1-0249 and the ANID-PIA grant ATE220057. This work was also partially funded by the ANID-FONDECYT Regular 1211144, the ANID-FONDEF ID23I 10360, and the ANID/Scholarship Program/Doctorado Nacional/2018–21181599.

Data Availability Statement: The RINEX files were downloaded from CSN, RAMSAC, IBGE, and UNAVCO. The satellite and receiver bias data were provided from CODE. The geomagnetic data were available from OMNIWeb Plus Data Documentation, WDC for Geomagnetism, and the Space Weather Prediction Center of NOAA. The United States Geological Survey Comprehensive Catalog of Earthquakes offered earthquake data. The occurrence of volcanic eruptions and tsunamis data were downloaded from the National Centers for Environmental Information of NOAA. The rocket launches were verified from Gunter's Space Page. The satellite data were downloaded from the European Space Agency Swarm mission and the Massachusetts Institute of Technology Haystack Observatory Madrigal database.

Acknowledgments: The authors would like to sincerely thank the librarians from Biblioteca Central of the Faculty of Physical and Mathematical Sciences, University of Chile, for their support on references.

Conflicts of Interest: The authors declare no conflicts of interest. The funders had no role in the design of the study; in the collection, analyses, or interpretation of data; in the writing of the manuscript, or in the decision to publish the results.

References

1. Afraimovich, E.L.; Astafyeva, E.I.; Demyanov, V.V.; Edemskiy, I.K.; Gavrilyuk, N.S.; Ishin, A.B.; Kosogorov, E.A.; Leonovich, L.A.; Lesyuta, O.S.; Palamartchouk, K.S.; et al. A review of GPS/GLONASS studies of the ionospheric response to natural and anthropogenic processes and phenomena. *J. Space Weather Space Clim.* **2013**, *3*, A27. [CrossRef]
2. Valdés-Abreu, J.C. Degradation of the Global Navigation Satellite System Positioning Accuracy Caused by Ionospheric Disturbance Sources. Ph.D. Thesis, Universidad de Chile, Santiago, Chile, 2023.
3. Rishbeth, H. Solar eclipses and ionospheric theory. *Space Sci. Rev.* **1968**, *8*, 543–554. [CrossRef]
4. Aa, E.; Zhang, S.R.; Shen, H.; Liu, S.; Li, J. Local and conjugate ionospheric total electron content variation during the 21 June 2020 solar eclipse. *Adv. Space Res.* **2021**, *68*, 3435–3454. [CrossRef]
5. Coster, A.J.; Goncharenko, L.; Zhang, S.R.; Erickson, P.J.; Rideout, W.; Vierinen, J. GNSS Observations of Ionospheric Variations During the 21 August 2017 Solar Eclipse. *Geophys. Res. Lett.* **2017**, *44*, 12041–12048. [CrossRef]
6. Eisenbeis, J.; Occhipinti, G. TEC Depletion Generated by the Total Solar Eclipse of 2 July 2019. *J. Geophys. Res. Space Phys.* **2021**, *126*, e2021JA029186. [CrossRef]

7.   He, L.; Heki, K.; Wu, L. Three-Dimensional and Trans-Hemispheric Changes in Ionospheric Electron Density Caused by the Great Solar Eclipse in North America on 21 August 2017. *Geophys. Res. Lett.* **2018**, *45*, 10933–10940. [CrossRef]

8.   Le, H.; Liu, L.; Yue, X.; Wan, W. The ionospheric behavior in conjugate hemispheres during the 3 October 2005 solar eclipse. *Ann. Geophys.* **2009**, *27*, 179–184. [CrossRef]

9.   Lei, J.; Dang, T.; Wang, W.; Burns, A.; Zhang, B.; Le, H. Long-Lasting Response of the Global Thermosphere and Ionosphere to the 21 August 2017 Solar Eclipse. *J. Geophys. Res. Space Phys.* **2018**, *123*, 4309–4316. [CrossRef]

10.   Maurya, A.K.; Shrivastava, M.N.; Kumar, K.N. Ionospheric monitoring with the Chilean GPS eyeball during the South American total solar eclipse on 2nd July 2019. *Sci. Rep.* **2020**, *10*, 19380. [CrossRef] [PubMed]

11.   Meza, A.; Eylenstein, B.; Natali, M.P.; Bosch, G.; Moirano, J.; Chalar, E. Analysis of Ionospheric and Geomagnetic Response to the 2020 Patagonian Solar Eclipse. *Front. Astron. Space Sci.* **2021**, *8*, 766327. [CrossRef]

12.   Mrak, S.; Semeter, J.; Nishimura, Y.; Hirsch, M.; Sivadas, N. Coincidental TID Production by Tropospheric Weather During the August 2017 Total Solar Eclipse. *Geophys. Res. Lett.* **2018**, *45*, 10903–10911. [CrossRef]

13.   Shrivastava, M.N.; Maurya, A.K.; Kumar, K.N. Ionospheric perturbation during the South American total solar eclipse on 14th December 2020 revealed with the Chilean GPS eyeball. *Sci. Rep.* **2021**, *11*, 20324. [CrossRef] [PubMed]

14.   Stankov, S.M.; Bergeot, N.; Berghmans, D.; Bolsée, D.; Bruyninx, C.; Chevalier, J.M.; Clette, F.; De Backer, H.; De Keyser, J.; D'Huys, E.; et al. Multi-instrument observations of the solar eclipse on 20 March 2015 and its effects on the ionosphere over Belgium and Europe. *J. Space Weather Space Clim.* **2017**, *7*, A19. [CrossRef]

15.   Yan, M.; Dang, T.; Lei, J.; Wang, W.; Zhang, S.R.; Le, H. From Bow Waves to Traveling Atmospheric Disturbances: Thermospheric Perturbations Along Solar Eclipse Trajectory. *J. Geophys. Res. Space Phys.* **2021**, *126*, e2020JA028523. [CrossRef]

16.   Zhang, S.R.; Erickson, P.J.; Vierinen, J.; Aa, E.; Rideout, W.; Coster, A.J.; Goncharenko, L.P. Conjugate Ionospheric Perturbation During the 2017 Solar Eclipse. *J. Geophys. Res. Space Phys.* **2021**, *126*, e2020JA028531. [CrossRef]

17.   Chen, X.; Dang, T.; Zhang, B.; Lotko, W.; Pham, K.; Wang, W.; Lin, D.; Sorathia, K.; Merkin, V.; Luan, X.; et al. Global Effects of a Polar Solar Eclipse on the Coupled Magnetosphere-Ionosphere System. *Geophys. Res. Lett.* **2021**, *48*, e2021GL096471. [CrossRef]

18.   Davis, C.J.; Lockwood, M.; Bell, S.A.; Smith, J.A.; Clarke, E.M. Ionospheric measurements of relative coronal brightness during the total solar eclipses of 11 August, 1999 and 9 July, 1945. *Ann. Geophys.* **2000**, *18*, 182–190. [CrossRef]

19.   Krankowski, A.; Shagimuratov, I.; Baran, L.; Yakimova, G. The effect of total solar eclipse of October 3, 2005, on the total electron content over Europe. *Adv. Space Res.* **2008**, *41*, 628–638. [CrossRef]

20.   Adekoya, B.; Chukwuma, V. Ionospheric F2 layer responses to total solar eclipses at low and mid-latitude. *J. Atmos. Sol. Terr. Phys.* **2016**, *138–139*, 136–160. [CrossRef]

21.   Cheng, K.; Huang, Y.N.; Chen, S.W. Ionospheric effects of the solar eclipse of September 23, 1987, around the equatorial anomaly crest region. *J. Geophys. Res. Space Phys.* **1992**, *97*, 103–111. [CrossRef]

22.   Tsai, H.; Liu, J. Ionospheric total electron content response to solar eclipses. *J. Geophys. Res. Space Phys.* **1999**, *104*, 12657–12668. [CrossRef]

23.   Mayaud, P. The equatorial counter-electrojet—A review of its geomagnetic aspects. *J. Atmos. Terr. Phys.* **1977**, *39*, 1055–1070. [CrossRef]

24.   Bhaskar, A.; Purohit, A.; Hemalatha, M.; Pai, C.; Raghav, A.; Gurada, C.; Radha, S.; Yadav, V.; Desai, V.; Chitnis, A.; et al. A study of secondary cosmic ray flux variation during the annular eclipse of 15 January 2010 at Rameswaram, India. *Astropart. Phys.* **2011**, *35*, 223–229. [CrossRef]

25.   St.-Maurice, J.P.; Ambili, K.M.; Choudhary, R.K. Local electrodynamics of a solar eclipse at the magnetic equator in the early afternoon hours. *Geophys. Res. Lett.* **2011**, *38*, L04102. [CrossRef]

26.   Panda, S.; Gedam, S.; Rajaram, G.; Sripathi, S.; Bhaskar, A. Impact of the 15 January 2010 annular solar eclipse on the equatorial and low latitude ionosphere over the Indian region. *J. Atmos. Sol. Terr. Phys.* **2015**, *135*, 181–191. [CrossRef]

27.   Resende, L.C.A.; Zhu, Y.; Denardini, C.M.; Chen, S.S.; Chagas, R.A.J.; Da Silva, L.A.; Carmo, C.S.; Moro, J.; Barros, D.; Nogueira, P.A.B.; et al. A multi-instrumental and modeling analysis of the ionospheric responses to the solar eclipse on 14 December 2020 over the Brazilian region. *Ann. Geophys.* **2022**, *40*, 191–203. [CrossRef]

28.   Aa, E.; Zhang, S.R.; Erickson, P.J.; Goncharenko, L.P.; Coster, A.J.; Jonah, O.F.; Lei, J.; Huang, F.; Dang, T.; Liu, L. Coordinated Ground-Based and Space-Borne Observations of Ionospheric Response to the Annular Solar Eclipse on 26 December 2019. *J. Geophys. Res. Space Phys.* **2020**, *125*, e2020JA028296. [CrossRef]

29.   Chen, C.H.; Lin, C.H.C.; Matsuo, T. Ionospheric responses to the 21 August 2017 solar eclipse by using data assimilation approach. *Prog. Earth Planet. Sci.* **2019**, *6*, 13. [CrossRef]

30.   Guo, M.; Xu, N.; Feng, J.; Deng, Z. A prolonged pattern of the ionospheric depletion in the south of the 21 August 2017 solar eclipse path. *J. Atmos. Sol. Terr. Phys.* **2020**, *199*, 105208. [CrossRef]

31.   Jonah, O.F.; Goncharenko, L.; Erickson, P.J.; Zhang, S.; Coster, A.; Chau, J.L.; de Paula, E.R.; Rideout, W. Anomalous Behavior of the Equatorial Ionization Anomaly During the 2 July 2019 Solar Eclipse. *J. Geophys. Res. Space Phys.* **2020**, *125*, e2020JA027909. [CrossRef]

32.   Wang, X.; Berthelier, J.J.; Lebreton, J.P. Ionosphere variations at 700 km altitude observed by the DEMETER satellite during the 29 March 2006 solar eclipse. *J. Geophys. Res. Space Phys.* **2010**, *115*, A11312. [CrossRef]

33.   Chen, G.; Zhao, Z.; Ning, B.; Deng, Z.; Yang, G.; Zhou, C.; Yao, M.; Li, S.; Li, N. Latitudinal dependence of the ionospheric response to solar eclipse of 15 January 2010. *J. Geophys. Res. Space Phys.* **2011**, *116*, A06301. [CrossRef]

34. Valdés-Abreu, J.C.; Díaz, M.A.; Bravo, M.; Báez, J.C.; Stable-Sánchez, Y. Ionospheric Behavior during the 10 June 2021 Annular Solar Eclipse and Its Impact on GNSS Precise Point Positioning. *Remote Sens.* **2022**, *14*, 3119. [CrossRef]
35. Verhulst, T.G.W.; Stankov, S.M. Height Dependency of Solar Eclipse Effects: The Ionospheric Perspective. *J. Geophys. Res. Space Phys.* **2020**, *125*, e2020JA028088. [CrossRef]
36. Kundu, S.; Chowdhury, S.; Palit, S.; Mondal, S.K.; Sasmal, S. Variation of ionospheric plasma density during the annular solar eclipse on December 26, 2019. *Astrophys. Space Sci.* **2022**, *367*, 44. [CrossRef]
37. Clilverd, M.A.; Rodger, C.J.; Thomson, N.R.; Lichtenberger, J.; Steinbach, P.; Cannon, P.; Angling, M.J. Total solar eclipse effects on VLF signals: Observations and modeling. *Radio Sci.* **2001**, *36*, 773–788. [CrossRef]
38. Kumar, S.; Kumar, A.; Maurya, A.K.; Singh, R. Changes in the D region associated with three recent solar eclipses in the South Pacific region. *J. Geophys. Res. Space Phys.* **2016**, *121*, 5930–5943. [CrossRef]
39. Chakrabarti, S.K.; Sasmal, S.; Chakraborty, S.; Basak, T.; Tucker, R.L. Modeling D-region ionospheric response of the Great American TSE of August 21, 2017 from VLF signal perturbation. *Adv. Space Res.* **2018**, *62*, 651–661. [CrossRef]
40. Ghosh, S.; Chowdhury, S.; Kundu, S.; Biswas, S.; Dawn, A.; Ray, S.; Choudhury, A.K.; Bari, M.W.; Bhowmick, D.; Manna, S.; et al. Observations and modeling of D-region ionospheric response of Annular Solar Eclipse on December 26, 2019, using VLF signal amplitude and phase variation. *Astrophys. Space Sci.* **2023**, *368*, 19. [CrossRef]
41. Aryal, S.; Evans, J.S.; Correira, J.; Burns, A.G.; Wang, W.; Solomon, S.C.; Laskar, F.I.; McClintock, W.E.; Eastes, R.W.; Dang, T.; et al. First Global-Scale Synoptic Imaging of Solar Eclipse Effects in the Thermosphere. *J. Geophys. Res. Space Phys.* **2020**, *125*, e2020JA027789. [CrossRef]
42. Vargas, F.; Liu, A.; Swenson, G.; Segura, C.; Vega, P.; Fuentes, J.; Pautet, D.; Taylor, M.; Zhao, Y.; Morton, Y.; et al. Mesosphere and Lower Thermosphere Changes Associated With the 2 July 2019 Total Eclipse in South America Over the Andes Lidar Observatory, Cerro Pachon, Chile. *J. Geophys. Res. Atmos.* **2022**, *127*, e2021JD035064. [CrossRef]
43. Bravo, M.; Martínez-Ledesma, M.; Foppiano, A.; Urra, B.; Ovalle, E.; Villalobos, C.; Souza, J.; Carrasco, E.; Muñoz, P.R.; Tamblay, L.; et al. First Report of an Eclipse From Chilean Ionosonde Observations: Comparison With Total Electron Content Estimations and the Modeled Maximum Electron Concentration and Its Height. *J. Geophys. Res. Space Phys.* **2020**, *125*, e2020JA027923. [CrossRef]
44. Martínez-Ledesma, M.; Bravo, M.; Urra, B.; Souza, J.; Foppiano, A. Prediction of the Ionospheric Response to the 14 December 2020 Total Solar Eclipse Using SUPIM-INPE. *J. Geophys. Res. Space Phys.* **2020**, *125*, e2020JA028625. [CrossRef]
45. Gómez, D.D. Ionospheric Response to the December 14, 2020 Total Solar Eclipse in South America. *J. Geophys. Res. Space Phys.* **2021**, *126*, e2021JA029537. [CrossRef]
46. De Haro Barbás, B.; Bravo, M.; Elias, A.; Martínez-Ledesma, M.; Molina, G.; Urra, B.; Venchiarutti, J.; Villalobos, C.; Namour, J.; Ovalle, E.; et al. Longitudinal variations of ionospheric parameters near totality during the eclipse of December 14, 2020. *Adv. Space Res.* **2022**, *69*, 2158–2167. [CrossRef]
47. Zhang, K.; Wang, H. The Great Reduction of Equatorial Electrojet During the Solar Eclipse on 14 December 2020. *Space Weather* **2022**, *20*, e2022SW003295. [CrossRef]
48. Bravo, M.; Molina, M.; Martínez-Ledesma, M.; de Haro Barbás, B.; Urra, B.; Elías, A.; Souza, J.; Villalobos, C.; Namour, J.; Ovalle, E.; et al. Ionospheric Response Modeling under Eclipse Conditions: Evaluation of December 14, 2020, Total Solar Eclipse Prediction over the South American sector. *Front. Astron. Space Sci.* **2022**, *9*, 1021910. [CrossRef]
49. Idosa, C.U.; Rikitu, K.S. Effects of Total Solar Eclipse on Ionospheric Total Electron Content over Antarctica on 2021 December 4. *Astrophys. J.* **2022**, *932*, 2. [CrossRef]
50. Coyle, S.; Hartinger, M.D.; Clauer, C.R.; Baker, J.B.H.; Cnossen, I.; Freeman, M.P.; Weygand, J.M. The 2021 Antarctic Total Eclipse: Ground Magnetometer and GNSS Wave Observations From the 40 Degree Magnetic Meridian. *J. Geophys. Res. Space Phys.* **2023**, *128*, e2022JA031142. [CrossRef]
51. Idosa Uga, C.; Beshir Seba, E. Ionospheric response to a moderate geomagnetic storm on 14 April 2022 and a partial solar eclipse 30 April 2022. *Indian J. Phys.* **2023**, *47*, 1–11. [CrossRef]
52. Abidin Abdul Rashid, Z.; Awad Momani, M.; Sulaiman, S.; Alauddin Mohd Ali, M.; Yatim, B.; Fraser, G.; Sato, N. GPS ionospheric TEC measurement during the 23rd November 2003 total solar eclipse at Scott Base Antarctica. *J. Atmos. Sol. Terr. Phys.* **2006**, *68*, 1219–1236. [CrossRef]
53. Chernogor, L.; Garmash, K. Magneto-ionospheric effects of the solar eclipse of March 20, 2015, over Kharkov. *Geomagn. Aeron.* **2017**, *57*, 72–83. [CrossRef]
54. Hoque, M.M.; Wenzel, D.; Jakowski, N.; Gerzen, T.; Berdermann, J.; Wilken, V.; Kriegel, M.; Sato, H.; Borries, C.; Minkwitz, D. Ionospheric response over Europe during the solar eclipse of March 20, 2015. *J. Space Weather Space Clim.* **2016**, *6*, A36. [CrossRef]
55. Panasenko, S.V.; Otsuka, Y.; van de Kamp, M.; Chernogor, L.F.; Shinbori, A.; Tsugawa, T.; Nishioka, M. Observation and characterization of traveling ionospheric disturbances induced by solar eclipse of 20 March 2015 using incoherent scatter radars and GPS networks. *J. Atmos. Sol. Terr. Phys.* **2019**, *191*, 105051. [CrossRef]
56. Verhulst, T.G.W.; Sapundjiev, D.; Stankov, S.M. High-resolution ionospheric observations and modeling over Belgium during the solar eclipse of 20 March 2015 including first results of ionospheric tilt and plasma drift measurements. *Adv. Space Res.* **2016**, *57*, 2407–2419. [CrossRef]
57. Valdés-Abreu, J.C.; Díaz, M.A.; Báez, J.C.; Stable-Sánchez, Y. Effects of the 12 May 2021 Geomagnetic Storm on Georeferencing Precision. *Remote Sens.* **2022**, *14*, 38. [CrossRef]

58. Seemala, G.K.; Valladares, C.E. Statistics of total electron content depletions observed over the South American continent for the year 2008. *Radio Sci.* **2011**, *46*, RS5019. [CrossRef]

59. Estey, L.H.; Meertens, C.M. TEQC: The multi-purpose toolkit for GPS/GLONASS data. *GPS Solut.* **1999**, *3*, 42–49. [CrossRef]

60. Piñón, D.; Gomez, D.; Smalley, B.; Cimbaro, S.; Lauría, E.; Bevis, M. The History, State, and Future of the Argentine Continuous Satellite Monitoring Network and Its Contributions to Geodesy in Latin America. *Seismol. Res. Lett.* **2018**, *89*, 475–482. [CrossRef]

61. Tsidu, G.M.; Abraha, G. Moderate geomagnetic storms of January 22–25. 2012 and their influences on the wave components in ionosphere and upper stratosphere-mesosphere regions. *Adv. Space Res.* **2014**, *54*, 1793–1812. [CrossRef]

62. Nose, M.; Iyemori, T.; Sugiura, M.; Kamei, T. Geomagnetic Dst index. *World Data Cent. Geomagn. Kyoto* **2015**, *10*, 15–31 . [CrossRef]

63. Cai, X.; Burns, A.G.; Wang, W.; Qian, L.; Pedatella, N.; Coster, A.; Zhang, S.; Solomon, S.C.; Eastes, R.W.; Daniell, R.E.; et al. Variations in Thermosphere Composition and Ionosphere Total Electron Content Under "Geomagnetically Quiet" Conditions at Solar-Minimum. *Geophys. Res. Lett.* **2021**, *48*, e2021GL093300. [CrossRef]

64. Easton, G.; González-Alfaro, J.; Villalobos, A.; Álvarez, G.; Melgar, D.; Ruiz, S.; Sepúlveda, B.; Escobar, M.; León, T.; Carlos Báez, J.; et al. Complex Rupture of the 2015 Mw 8.3 Illapel Earthquake and Prehistoric Events in the Central Chile Tsunami Gap. *Seismol. Res. Lett.* **2022**, *93*, 1479–1496. [CrossRef]

65. Liu, T.; Yu, Z.; Ding, Z.; Nie, W.; Xu, G. Observation of Ionospheric Gravity Waves Introduced by Thunderstorms in Low Latitudes China by GNSS. *Remote Sens.* **2021**, *13*, 4131. [CrossRef]

66. Osei-Poku, L.; Tang, L.; Chen, W.; Mingli, C. Evaluating Total Electron Content (TEC) Detrending Techniques in Determining Ionospheric Disturbances during Lightning Events in A Low Latitude Region. *Remote Sens.* **2021**, *13*, 4753. [CrossRef]

67. Chakraborty, S.K.; Hajra, R. Solar control of ambient ionization of the ionosphere near the crest of the equatorial anomaly in the Indian zone. *Ann. Geophys.* **2008**, *26*, 47–57. [CrossRef]

68. Mansoori, A.; Khan, P.; Ahmad, R.; Atulkar, R.; Aslam, A.; Bhardwaj, S.; Malvi, B.; Purohit, P.; Gwal, A. Evaluation of long term solar activity effects on GPS derived TEC. *J. Phys. Conf. Ser.* **2016**, *759*, 012069. [CrossRef]

69. Huang, F.; Li, Q.; Shen, X.; Xiong, C.; Yan, R.; Zhang, S.R.; Wang, W.; Aa, E.; Zhong, J.; Dang, T.; et al. Ionospheric Responses at Low Latitudes to the Annular Solar Eclipse on 21 June 2020. *J. Geophys. Res. Space Phys.* **2020**, *125*, e2020JA028483. [CrossRef]

70. Dang, T.; Lei, J.; Wang, W.; Zhang, B.; Burns, A.; Le, H.; Wu, Q.; Ruan, H.; Dou, X.; Wan, W. Global Responses of the Coupled Thermosphere and Ionosphere System to the August 2017 Great American Solar Eclipse. *J. Geophys. Res. Space Phys.* **2018**, *123*, 7040–7050. [CrossRef]

71. Choudhary, R.K.; St.-Maurice, J.P.; Ambili, K.M.; Sunda, S.; Pathan, B.M. The impact of the January 15, 2010, annular solar eclipse on the equatorial and low latitude ionospheric densities. *J. Geophys. Res. Space Phys.* **2011**, *116*, A09309. [CrossRef]

72. Titheridge, J. Winds in the ionosphere—A review. *J. Atmos. Terr. Phys.* **1995**, *57*, 1681–1714 . [CrossRef]

73. Balan, N.; Bailey, G.J. Equatorial plasma fountain and its effects: Possibility of an additional layer. *J. Geophys. Res. Space Phys.* **1995**, *100*, 21421–21432. [CrossRef]

74. Chou, M.Y.; Wu, Q.; Pedatella, N.M.; Cherniak, I.; Schreiner, W.S.; Braun, J. Climatology of the Equatorial Plasma Bubbles Captured by FORMOSAT-3/COSMIC. *J. Geophys. Res. Space Phys.* **2020**, *125*, e2019JA027680. [CrossRef]

75. Cherniak, I.; Zakharenkova, I. Ionospheric Total Electron Content Response to the Great American Solar Eclipse of 21 August 2017. *Geophys. Res. Lett.* **2018**, *45*, 1199–1208. [CrossRef]

76. Goncharenko, L.P.; Erickson, P.J.; Zhang, S.R.; Galkin, I.; Coster, A.J.; Jonah, O.F. Ionospheric Response to the Solar Eclipse of 21 August 2017 in Millstone Hill (42N) Observations. *Geophys. Res. Lett.* **2018**, *45*, 4601–4609. [CrossRef]

77. Hairston, M.R.; Mrak, S.; Coley, W.R.; Burrell, A.; Holt, B.; Perdue, M.; Depew, M.; Power, R. Topside ionospheric electron temperature observations of the 21 August 2017 eclipse by DMSP spacecraft. *Geophys. Res. Lett.* **2018**, *45*, 7242–7247. [CrossRef]

78. Hussien, F.; Ghamry, E.; Fathy, A.; Mahrous, S. Swarm satellite observations of the 21 August 2017 solar eclipse. *J. Astron. Space Sci.* **2020**, *37*, 29–34. [CrossRef]

remote sensing

MDPI

Article

Unveiling the Core Patterns of High-Latitude Electron Density Distribution at Swarm Altitude

Giulia Lovati [1,2], Paola De Michelis [2,*], Tommaso Alberti [2,3] and Giuseppe Consolini [3]

1   Dipartimento di Fisica, Università di Roma Sapienza, 00185 Rome, Italy; giulia.lovati@uniroma1.it
2   Istituto Nazionale di Geofisica e Vulcanologia, Via di Vigna Murata 605, 00143 Rome, Italy; tommaso.alberti@ingv.it
3   INAF—Istituto di Astrofisica e Planetologia Spaziali, Via Fosso del Cavaliere 100, 00133 Rome, Italy; giuseppe.consolini@inaf.it
*   Correspondence: paola.demichelis@ingv.it

Abstract: The ionosphere has distinctive characteristics under different solar and geomagnetic conditions, as well as throughout the seasons, and has a direct impact on our technological life in terms of radio communication and satellite navigation systems. In the pursuit of developing highly accurate ionospheric models and/or improving existing ones, understanding the various physical mechanisms that influence electron density dynamics is critical. In this study, we apply the Multivariate Empirical Mode Decomposition (MEMD) method to the electron density distribution in the mid-to-high latitude (above 50° magnetic latitude) regions in order to identify the dominant scales at which these mechanisms operate. The data were collected by the Swarm mission in the Northern Hemisphere. MEMD allows us to separate the main intrinsic modes and assess their relative contributions to the original one, thereby identifying the most important modes and the spatial scales at which they exert influence. Our study spanned the period from 1 January 2016 to 31 December 2021, which was characterized by low solar activity levels. This choice allowed for a more focused investigation of other variables influencing electron density distribution under similar solar activity conditions. We specifically examined the variations of the resulting modes in relation to different seasons and geomagnetic activity conditions, providing valuable insights into the complex behavior of the ionosphere in response to various external factors.

Keywords: Multivariate Empirical Mode Decomposition; electron density distribution; Swarm satellites; high-latitude ionosphere; seasonal and geomagnetic variability

Citation: Lovati, G.; De Michelis, P.; Alberti, T.; Consolini, G. Unveiling the Core Patterns of High-Latitude Electron Density Distribution at Swarm Altitude. *Remote Sens.* **2023**, *15*, 4550. https://doi.org/10.3390/rs15184550

Academic Editor: Yunbin Yuan

Received: 5 August 2023
Revised: 11 September 2023
Accepted: 11 September 2023
Published: 15 September 2023

1. Introduction

The ionosphere, a layer of Earth's atmosphere, extends from about 60 km above the surface to a height of about 1000 km. It is made up of charged particles such as electrons and ions that form as a result of the ionization of neutral particles caused by solar radiation and energetic particles from the magnetosphere [1]. Aided by technological advances in observation techniques and numerical modeling, researchers have extensively studied the ionosphere over the last century, leading to continuous advances in our understanding of its dynamics. Solar flares, geomagnetic storms, atmospheric tides, and atmospheric waves all influence the behaviour of the ionosphere, resulting in distinct characteristics under varying solar and geomagnetic conditions. These dynamics have immediate consequences for radio communication and satellite navigation systems [2–4]. Despite significant progress in recent decades, certain aspects of the ionosphere, particularly during geomagnetic storms, remain unresolved due to the complex interplay within the interconnected magnetosphere–ionosphere–thermosphere system.

Over the course of the last decade, the Swarm satellite constellation [5] launched by the European Space Agency (ESA) has consistently surveyed the highest-altitude layer of the ionosphere, known as the F layer. This constellation, which consists of three identical satellites orbiting in a polar trajectory at altitudes ranging from 450 and 500 km, collects crucial data on electron density and other physical parameters. This information has created a valuable database, providing new insights into the F layer region. The processes occurring in the ionosphere at middle and high latitudes, particularly in regions directly influenced by interactions with the magnetosphere and solar wind, are of particular interest [6]. At these latitudes, the ionosphere responds to the velocity field transmitted by the solar wind–magnetosphere interaction, exhibiting distinctive planetary-scale patterns [7]. However, the specific pattern observed at any particular time is highly dependent on the prevailing conditions in the interplanetary medium, including the orientation and magnitude of the interplanetary magnetic field (IMF). In addition to large-scale plasma motions that contribute to electron density distribution at high latitudes, the ionosphere is subject to other latitude-independent processes [8]. Photoionization is a notable process that occurs as a result of the ionizing impact of sunlight on the ionospheric plasma. This photoionized plasma can travel long distances, even through dark regions, before significant recombination occurs. The interaction with sunlight shapes the plasma content in the ionosphere, making it an important consideration in understanding the dynamic behavior of the system. Other processes, such as particle precipitation, frictional heating, particle and heat exchange with the plasmasphere, and interaction with thermospheric winds [9], all have significant impact on the polar ionosphere, contributing to the considerable complexity of this region.

In recent years, ionospheric modeling has grown in popularity as a means of reproducing and predicting ionospheric behavior at various temporal and spatial scales. These models are critical in assisting advanced technological systems that rely heavily on applications such as radio signal propagation. Furthermore, they help to improve our understanding of the ionosphere and the near-space environment. By conducting numerical experiments with these models, researchers can study various ionospheric processes without the need to take direct measurements. Ionospheric models are primarily concerned with describing the behavior of electron density in the ionosphere over space and time; however, certain models can account for additional parameters such as ion and electron temperatures, ion densities, and ionospheric drift. The advancement of ionospheric models results from high-quality observations, model improvements, and enhanced computing power. It is critical to identify and differentiate the various physical mechanisms influencing electron density dynamics in order to build more accurate ionospheric models of the middle and high latitude region and to improve existing ones. In this context, it is critical to understand the scales at which these mechanisms operate, their relative importance, and their reliance on external parameters. In this paper, we use data from the Swarm mission to apply the Multivariate Empirical Mode Decomposition (MEMD) method to the electron density distribution in the middle and high latitude ionospheric region. Our goal is to identify and isolate the various contributors and spatial scales that have an impact within this area. The MEMD technique allows the main intrinsic modes in the ionospheric density distribution to be identified; these are associated with different spatial scales, and influence the plasma density dynamics in the selected region. Having extracted these modes, their relative contributions to the original distribution can be assessed, allowing the most significant modes to be identified along with the spatial scales at which they operate. We have concentrated our research on the period from 1 January 2016 to 31 December 2021, which was characterized by a low level of solar activity. This choice allows the effect of other parameters on the electron density distribution to be investigated in greater depth under similar solar activity conditions. Specifically, we looked at how the resulting modes change with seasonal and geomagnetic activity conditions.

2. Data

The Swarm project is a pioneering ESA satellite constellation mission dedicated to Earth Observation (EO) [5]. It consists of three identical satellites (Swarm A, Swarm B, and Swarm C) launched on 22 November 2013. The satellites initially operated in a string-ofpearls configuration before transitioning to the final constellation formation on 17 April 2014. Within the mission, Swarms A and C were designated as the lower pair, operating in close proximity throughout their operation. Positioned at an initial altitude of 462 km and an inclination angle of 87.35 degrees, these satellites conducted detailed observations within this orbit. Swarm B, on the other hand, remained in an higher orbit with an initial altitude of 511 km and an inclination angle of 87.75 degrees. While Swarm B initially followed a trajectory roughly parallel to the lower pair, its orbit has continued to evolve over time due to the inherent dynamics of satellite motion. Each satellite is equipped with various instruments, including the Absolute Scalar Magnetometer (ASM) [10], Vector Field Magnetometer (VFM), Star Tracker (STR), Electric Field Instrument (EFI) [11,12] consisting of Langmuir probes (LPs) and thermal ion imagers (TIIs), Global Positioning System (GPS) Receiver (GPSR) [13], Laser Retro-Reflector (LRR), and Accelerometer (ACC) [14]. These instruments allow for in situ measurements of electric and magnetic fields, plasma density, and temperature, providing valuable insights into the ionosphere.

For this study, we utilized data from the LP instruments; a detailed description of the LPs can be found in Knudsen et al. [11]. Our analysis focused on electron density measurements obtained from the Swarm A satellite at a frequency of 1 Hz during the period from 1 January 2016 to 31 December 2021. These valuable datasets are readily accessible for download through the official website at https://swarm-diss.eo.esa.int/, accessed on 15 January 2022, providing users with the option to utilize either a web browser or an FTP client for retrieval.

The six-year data collection period allowed for a robust analysis of electron density distribution at medium and high latitudes (above 50° magnetic latitude) in the Northern Hemisphere during a period of low solar activity, as indicated by an average value of the solar radio flux at 10.7 cm (2800 MHz) of 80 sfu (solar flux units). Moreover, the collection period enabled the investigation of seasonal variations and their impact on density patterns. To investigate this, the dataset was divided into segments based on the local seasons, with the equinoxes and solstices at the center. Each year was divided into four periods of three months each in order to more accurately capture the seasonal changes. This segmentation approach ensures that the observed spatial distribution of the electron density reflects the corresponding solar illumination conditions at the satellite's position. Furthermore, to examine the effect of geomagnetic activity on the distribution of electron density, we divided the dataset into two distinct periods, one corresponding to a period of low geomagnetic activity and the other to a period of high geomagnetic activity. These periods were identified using the SME index, which supplements the AE index and was obtained from the SuperMAG global magnetometer network [15]. We used the SME index [16] instead of the well-known AE index [17], as the latter was not available for the entire time period under analysis. Previous scientific studies [18] have shown that, similar to the AE index, the SME index represents a reliable indicator of geomagnetic activity at high latitudes. While the SME index uses the same calculation methodology as the AE index, it uses a much larger number of magnetometer stations, typically ten times more than the original index. This increased station coverage makes for improved precision in capturing the dynamics of geomagnetic disturbances by improving the accuracy of timing, intensity, and event location determination. To differentiate between low and high geomagnetic activity within the dataset, we used two SME thresholds, one set at 70 nT and the other at 230 nT, as these thresholds corresponded to the 25th and 75th percentiles of the cumulative distribution during the chosen period. SME values of less than 70 nT indicate periods of low geomagnetic activity at high and middle latitudes, while values greater than 230 nT identify periods of geomagnetic disturbance at these latitudes.

Figure 1 illustrates the electron density (N_e) data that support the analysis presented in this study. Specifically, the figure displays six distinct maps arranged in two columns. The first column corresponds to the period of low geomagnetic activity (SME $\leq$ 70 nT), while the second column represents the period of high geomagnetic activity (SME $\geq$ 230 nT). Each column contains polar views depicting the average spatial distribution of the electron density during the equinox period and during the Northern Hemisphere's summer and winter solstice periods; the spring and autumn equinoxes are grouped together because electron density is subject to the same solar illumination conditions during these periods. To visualize the average polar electron density, data are plotted in a coordinate system that combines the quasi-dipole (QD) latitude and magnetic local time (MLT). The QD coordinate system introduced by Richmond [19] was specifically chosen due to its applicability in studying phenomena associated with horizontally stratified ionospheric currents. In order to align the data with the position of the Sun, MLT was utilized instead of QD longitude. The top of the maps correspond to noon and the bottom to midnight, and a binning grid of $1° \times 1°$ is employed. Each $1°$ bin in MLT corresponds to a time interval of 4 min. The spatial distributions of electron density clearly show a dependence on MLT and latitude, along with variations associated with seasonality and the level of geomagnetic activity. The maps provide useful information about several well-known aspects of electron density in the high-latitude ionosphere. One notable finding is that the electron density values during the day are nearly double those during the night. The solar ionization process, which is the primary mechanism of generating free electrons in the ionosphere, is responsible for this stark contrast. Surprisingly, this characteristic remains consistent regardless of the geomagnetic activity levels, despite a clear seasonal dependence. Another notable feature is the depletion of electron density in the sub-auroral ionospheric region, particularly on the night side. This depletion is a distinguishing feature of the main ionospheric trough (MIT), which acts as a boundary between the auroral and middle latitude regions [20]. The MIT is a highly dynamic structure that serves as a signature of the magnetospheric plasmapause in the nighttime ionosphere. This characteristic remains consistent for different levels of geomagnetic activity and seasons. Finally, a notable feature that is particularly evident during equinoxes is the presence of a region with high electron density values in the mid-latitude and sub-auroral ionospheric region. It is most noticeable during periods of increased geomagnetic activity. This phenomenon is known as the storm-enhanced density (SED), and the mechanisms of its generation and decay have been studied using ground-based and space observations [21]. Panels b and d of Figure 1 show that the SED region forms a plume extending from the day side into the polar cap along the convection streamlines, giving rise to the well-known "tongue of ionization" (TOI). Foster et al. [22] proposed that the TOI represents the ionospheric projection of the equatorial plume into the magnetosphere. The orientation of the IMF, specifically the B_y and B_z components, has a strong influence on the spatial distribution of the TOI within the polar cap region. In the Northern Hemisphere, the TOI moves towards the dawn sector with positive IMF B_y and towards the dusk sector with negative IMF B_y. Furthermore, when negative IMF B_z values are involved the TOI appears narrower and more intense over the polar cap with respect to periods characterized by positive IMF B_z values. This behavior is consistent with the influence of the B_z and B_y IMF components on the polar ion convection pattern; B_z governs the cross polar cap potential and ion convection pattern, while B_y determines the dawn–dusk asymmetry in this convection. While we do not specifically analyze the average distribution of electron density with respect to IMF orientation in this study, it is clear that this description aligns well with the average distribution obtained during periods of high geomagnetic activity, in which the IMF B_z component is negative, as we discuss later.

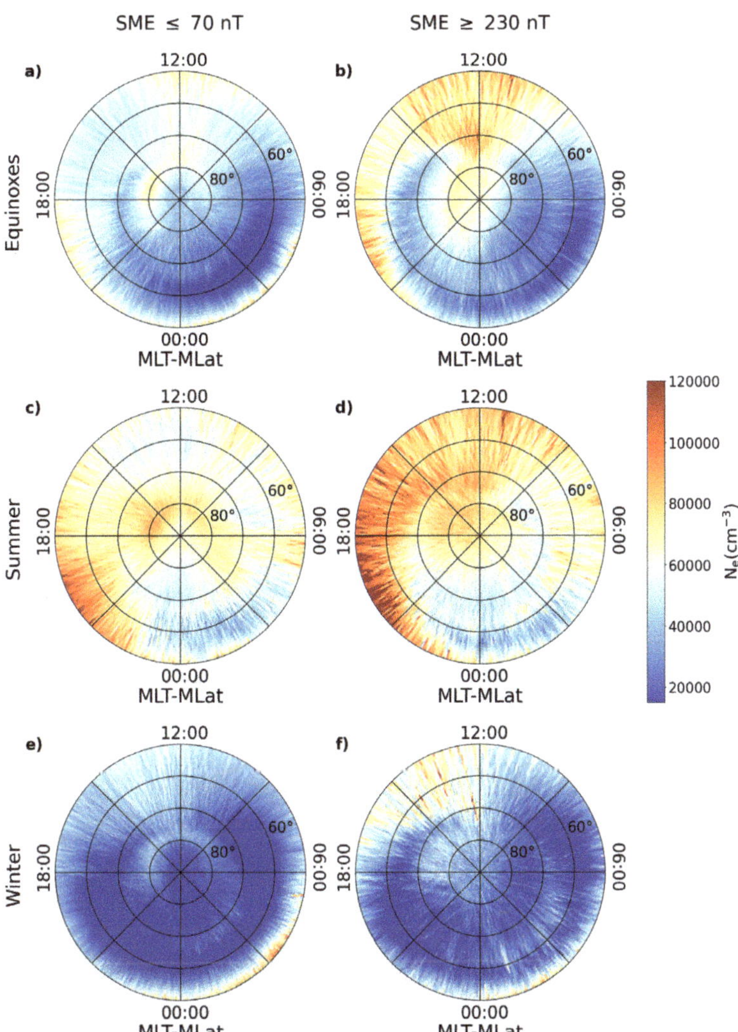

Figure 1. Polar view of the average spatial distribution of electron density (N_e) for low (**a,c,e**) and high (**b,d,f**) geomagnetic activity conditions in the Northern Hemisphere as represented in the QD–magnetic latitude (MLat) and MLT reference system. Each column shows the average spatial distribution of electron density during three different periods of the year: equinoxes, summer, and winter. The maps are based on data collected by the Swarm A satellite from 1 January 2016 to 31 December 2021, and reflect an average altitude of approximately 460 km. The concentric circles are plotted in $10°$ intervals, with the outermost circle corresponding to $50°$.

3. Method: EMD and its Multivariate Extension

Empirical Mode Decomposition (EMD) is a data analysis technique first introduced by Huang et al. [23] with the aim of adaptively representing non-stationary signals. The fundamental concept is that a signal can be decomposed in a finite, often small number of intrinsic oscillatory modes, each one of them reproducing the repeating behaviour of the signal at a specific time scale [24]. The EMD method allows signals that are not compatible with the Fourier Transform restrictions to be analyzed. In fact, Fourier analysis is a powerful method of extracting the energy–frequency distribution of a signal, and is widely used for its simplicity and its general applicability; however, it cannot produce physically meaningful results when dealing with nonstationary data [23]. In such cases, EMD can resolve the role of reducing the signal $s(t)$ into a set of basis signals

$$s(t) = \sum_{n=1}^{N} c_n(t) + res(t) \tag{1}$$

which are derived directly from the data itself [24].

The empirical modes $c_n(t)$ must fulfill two criteria: first, when considering the entire dataset, each of these function must have the same number of extrema (local maxima and/or minima) and zero crossings or differ by one at most; second, when considering any point, the mean value between the envelopes defined by the local maximum and local minimum must be zero [23]. The resulting functions are simple oscillatory modes, which differ from simple harmonic components in that their frequency and amplitude can change along the time frame.

The procedure for extracting the empirical modes, called sifting, involves the following steps:

1. The local extrema of the considered dataset are identified.
2. A cubic spline is used to connect all the local maxima in order to obtain the upper envelope.
3. Another cubic spline is used to connect all the local minima in order to obtain the lower envelope.
4. The mean between the two envelopes (m_1) is computed.
5. The first component h_1 is obtained by subtracting m_1 from the data $h_1 = x(t) - m_1$.
6. If h_1 is a function with zero mean, it is accepted as the first empirical mode, $c_1 = h_1$; otherwise, h_1 undergoes steps 1–5 itself and the procedure is repeated until h_1 fulfills the zero mean criterion.

In addition, it is necessary to apply a stopping criteria to the number of sifting iterations; in this way, the obtained empirical modes preserve the amplitude and frequency modulation as well as their physical sense [23]. One potential stopping criterion is to determine a threshold limit for the size of the standard deviation between two consecutive sifting steps k and $k + 1$:

$$\sigma_k = \sum_j \frac{\left[\delta_k(t_j) - \delta_{k+1}(t_j)\right]^2}{\delta_n(t_j)^2} < \epsilon \tag{2}$$

in which ϵ is usually fixed at 0.3 [23]. To obtain the following empirical modes, the process is successively applied to the residual $r_1 = s(t) - c_1$. The procedure described above is then repeated until the residual becomes a constant, a monotonic function, or a function with only one maximum and minimum from which it is not possible to extract other modes [24]. If necessary, it is then possible to compute the instantaneous frequencies and amplitudes of each obtained mode by applying the Hilbert transform to each of the modes. When adding this final step, the full process is called the Hilbert–Huang Transform [24].

In the presence of multivariate signals, the definition of local extrema becomes more complicated, making the step of local mean estimation non-obvious [25]. Therefore, in these cases it is not possible to apply directly EMD, and becomes necessary to employ its multivariate extension (MEMD). After the first proposals of EMD extension to complex/bivariate [26,27] and trivariate data [28], Rehman and Mandic [25] proposed a way of extending the concept of local extrema to an N-dimensional space. The N-variate signal has to be subdivided into N-dimensional datasets, each of which is then projected along different directions in the N-dimensional space. Each projected signal has its own envelopes for each of the directions, and the local mean can be computed by averaging over the N-dimensional space. The local mean can be computed through two different methods, both involving the selection of a suitable set of direction vectors in the N-dimensional space. In the first method, uniform angular sampling of a unit sphere in an N-dimensional hyperspherical coordinate system is used to obtain a set of direction vectors that covers the whole $(N-1)$ sphere. The second method is based on low-discrepancy point sets, in which discrepancy refers to a quantitative measure of the irregularity or non-uniformity of a distribution. This approach belongs to the class of quasi-Monte Carlo methods [29]. Through either methods, a uniform distribution of direction vectors is obtained that can provide more accurate estimation of the local mean in N-dimensional spaces. Multivariate empirical modes can be successfully obtained following the already mentioned procedure of standard EMD by using multivariate spline interpolation and checking the obtained modes properties [30]. An important difference with EMD is that in MEMD the condition of equality between the numbers of extrema and zero crossings is not imposed, as the definition of extrema is not straightforward for multivariate signals [31].

In this work, we used a Python algorithm to apply the MEMD method, available at https://github.com/mariogrune/MEMD-Python-, accessed on 20 April 2023; it is an adaptation of the procedure described in Rehman and Mandic [25] and is produced in Matlab (freely available at: http://www.commsp.ee.ic.ac.uk/~mandic/research/emd.htm). We chose to use this Python version because of its applicability to input data characterized by any number of channels. In fact, as an input for this MEMD algorithm we utilized the two-dimensional matrix obtained by binning the electron density data into a grid of $1° \times 1°$ in magnetic latitude and MLT, where $1°$ corresponds to 4 min in MLT. Each bin in the map reports the mean electron density value considering all the Swarm measurements that fall within it. Here, we are interested in the region above $50°$ in magnetic latitude; in order to avoid a bad description of the lower latitude border, however, we obtained a map above $45°$. On the other hand, to account for the cylindrical symmetry in MLT, where the values of the first column in MLT (MLT = 00:00) must be the same as those of the last column in MLT (MLT = 24:00), we extended the matrix by replicating the original matrix three times. In this way, the input matrix has a dimension of $45° \times 1080°$. Thus, the MEMD algorithm returns $(N + 1)$ matrices $45° \times 1080°$, representing the N modes plus the residue in which the signal is decomposed. From each of these N matrices, we selected the 360 central columns and the rows representing the latitudes above $50°$. In this way, $(N + 1)$ rectangular maps of $40° \times 360°$ are obtained, represented in polar coordinates, allowing for better visualization of the spatial distributions of the features characterizing each mode.

Each multivariate empirical mode is characterized by a distinctive scale, which can be roughly visualized by considering a fixed magnetic latitude and checking how many times the electron density passes from positive to negative values. This happens with a specific periodicity in MLT depending on the mode. We evaluated this average timescale/period by considering the 40 time series of 360 data points, obtainable from each one of the 40 rows of the aforementioned matrix. It is possible to obtain a periodogram from each series that provides the signal spectral density as a function of the frequency. By averaging the 40 periodograms relative to each fixed magnetic latitude, a single mean periodogram characterized by a peak corresponding to the most significant frequency of the mode can be obtained. The inverse of the frequency corresponding to this Fourier PSD peak provides an indication of the longitudinal variability associated with the considered mode. Because we report our results in MLT coordinates, we are able to represent the average periods in units of time (minutes).

4. Results

We begin our investigation by applying the MEMD method to the electron density distribution during the equinox period when geomagnetic activity is low. Figure 2 presents the results of the decomposition. We obtained a total of eight modes plus the residue. These are represented on polar maps by plotting their average electron density value using QD latitude and MLT. The top of the map corresponds to noon and the bottom to midnight; the MLT equal to 06:00 (on the right) and 18:00 (on the left) is indicated in order to better visualize the morning and evening sectors, and a $1° \times 1°$ binning grid is used.

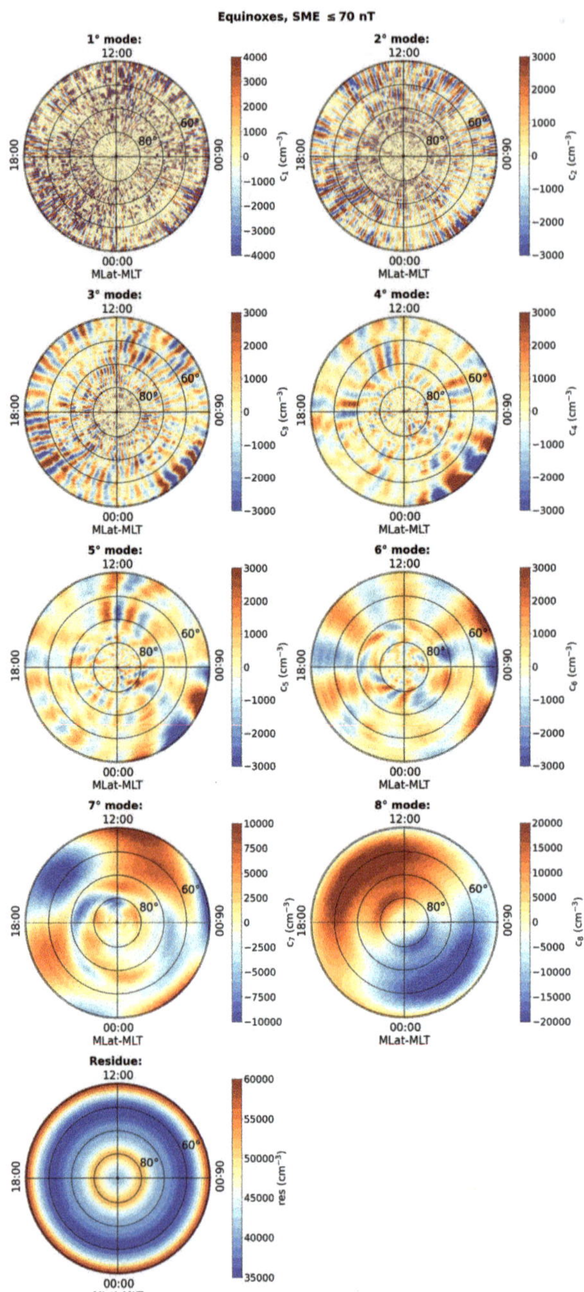

Figure 2. Polar view of the eight modes and the residue obtained from the MEMD of electron density distribution during the equinoctial period, in the Northern Hemisphere, when geomagnetic activity is low. Maps are represented using the QD–magnetic latitude (MLat) and MLT reference system. The concentric circles are plotted in $10°$ intervals, with the outermost circle corresponding to $50°$.

In Figure 3, the left side shows the average period obtained through the latitudinal averaged frequency analysis associated with each mode and right side shows the energy distribution in terms of the normalized variance of each mode with respect to the total variance of the signal. Observing the energy trend of each mode, the first six exhibit extremely low energy levels, with a normalized variance value lower than that of the first mode; as a result, because the EMD acts as a dyadic filter [32] and the first mode is usually associated with the noise content of the signals [33], when the normalized variance is less than that of the first mode it cannot be physically meaningful in the statistical sense.

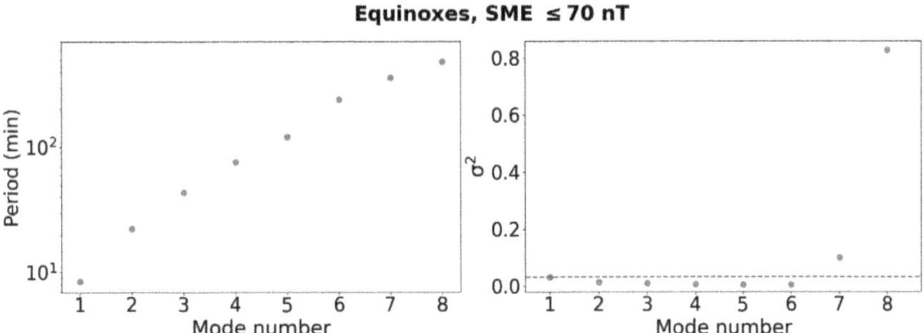

Figure 3. (**Left**) the MLT period of each mode reported in Figure 2 and (**Right**) the energy content of each mode shown in Figure 2 evaluated as the normalized variance. The horizontal dashed line indicates the energy value of the first mode, which is associated with noise.

As a result of analyzing the energy contribution of each mode to the original signal, it is clear that the last two modes and the residue contribute significantly to the description of the initial distribution. By examining the time periods associated with each mode, however, three distinct trends emerge, allowing the obtained modes to be classified into three classes. The first class includes the first two modes, the second class includes the four following modes (from the third to the sixth mode), and the third class includes the seventh and eighth modes.

We examined any variations with the latitude in the average period value for each of these four modes by conducting a comprehensive analysis of the periodicities of the modes in MLT belonging to the second class. The periodogram for a given magnetic latitude value was computed by taking into account the series of density values corresponding to different MLTs. This entailed analyzing 40 distinct periodograms for each mode obtained from a series of 360 data points, with a resolution of $1°$ for both magnetic latitude and MLT. The fundamental frequency value was calculated from each periodogram based on the peak position. The average period was then calculated using the frequency values in MLT, with the period found to be practically constant with the latitude for all four analyzed modes. The average periods obtained for each one of the four modes are (41 ± 4), (81 ± 9), (130 ± 10), and (235 ± 25) min. These periodicities are in good agreement with the harmonics and subharmonics of the satellite's orbital period ($\sim$94 min). In other words, the spatial structure of the electron density distribution associated with these modes closely resembles the satellite traces, implying a possible link to the data acquisition process. However, we note that these modes do not contribute significantly to the overall decomposition, because their energy remains lower than that of the first one. As a result, we combine this second class of modes together and consider a smaller number of modes for further analysis.

Figure 4 illustrates the grouped modes that effectively describe the fundamental modes present in the spatial distribution of electron density in the Northern Hemisphere. The modes (labeled hereinafter by Roman numerical notation) and the residue are among them. The first mode (I mode), which is essentially noise, is the sum of the initial two modes. The second mode (II mode), a combination of the third to sixth initial modes, is associated with the satellite's orbit. The third (III) and fourth (IV) modes are direct results of the initial decomposition, and given their higher energy values represent the primary modes through which the original signal is decomposed.

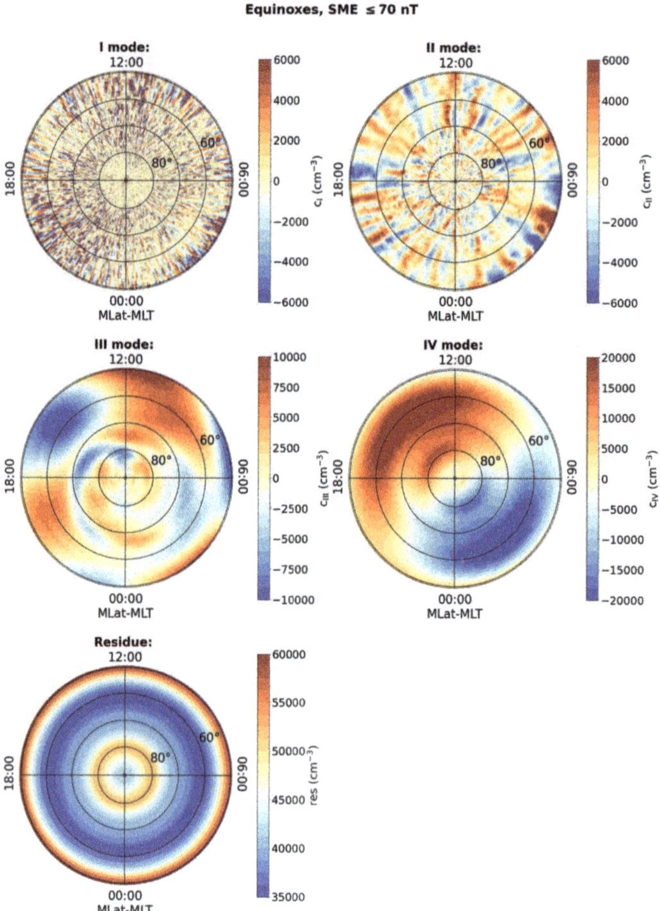

Figure 4. Polar view of the fundamental modes and residue obtained from the Multivariate Empirical Mode Decomposition (MEMD) of electron density distribution during the equinoctial period under low geomagnetic activity in the Northern Hemisphere. Imode corresponds to the sum of the 1° and 2° modes reported in Figure 2, II mode corresponds to the sum of the 3°, 4°, 5°, and 6° modes in Figure 2, and III and IV modes are equal to the 7° and 8° modes in Figure 2, respectively. Maps are represented in the QD–magnetic latitude (MLat) and MLT reference system. The concentric circles are plotted in 10° intervals, with the outermost circle corresponding to 50°.

The same identical procedure was used for the electron density distributions during summer and winter seasons when the geomagnetic activity was low, and was repeated for the geomagnetically disturbed period. In all the analyzed cases, regardless of the season and the geomagnetic activity level, the original electron density distribution is consistently decomposed into eight modes along with the residue. In addition, during all the analyzed configuration, which are not shown here for brevity, the modes' average periods and energy distributions follow the same behaviour as the one obtained in the first study case, i.e., during the equinoctial period and low geomagnetic activity, as reported in Figure 3. This makes it clear that even in the other five configurations under analysis it was possible to distinguish three different classes of modes which contribute different amounts of energy to the original signal. This observation led us to use the same mode grouping approach described in detail previously during the equinoctial period of low geomagnetic activity. In this way, in each case we obtained four decomposed modes accompanied by their residue. The third and fourth modes and the residue obtained for each configuration of season and geomagnetic activity level are the ones that provide the more relevant contributions to the electron density description, and are shown and discussed in the following Section 5.

5. Discussion

Application of the MEMD method to the spatiotemporal distribution of electron density at middle and high latitudes in the Northern Hemisphere under various seasonal and geomagnetic activity conditions reveals the presence of fundamental modes that underlie the initial distributions. Surprisingly, these modes are consistently observed in all six datasets regardless of the specific season or level of geomagnetic activity under consideration. Each electron density dataset can be accurately reconstructed by combining the residue with these four distinct fundamental modes.

To gain a better understanding of the physical processes associated with each mode, we examine its spatiotemporal distribution and how it varies across seasons and levels of geomagnetic activity. We begin with the residue term, which is the basis of the electron density distribution and to which the modes are later added to reconstruct the initial distribution. We find a striking similarity in the spatiotemporal structure of the residues obtained under low geomagnetic activity across the three seasonal periods, as can be seen in the left column of the upper part of Figure 5. The electron density is highest at lower latitudes and gradually decreases towards higher latitudes, reaching a minimum around $60°–65°$. It then begins to rise again, reaching a peak near $80°$ before falling near the magnetic pole. This pattern reveals two distinct bands of maximum intensity, one below $55°$ magnetic latitude and another between $75°$ and $85°$ magnetic latitudes. Lower electron density regions exist primarily between $55°$ and $75°$ latitudes and near the magnetic pole, corresponding to the polar cap. While the spatiotemporal position of these bands remains consistent across seasons, the absolute values of the electron density and intensity ratios between the different bands change.

The upper part of Figure 5 depicts the spatiotemporal distributions of residues, which cannot be effectively represented using a single value scale in all cases. Therefore, to better compare the different configurations using a common scale, the bottom part of Figure 5 displays various electron density profiles as a function of the magnetic latitude for the studied season and level of geomagnetic activity in a single graph. This provides a comprehensive view of the spatial variations in electron density across different latitudes, enabling a detailed examination of the patterns and changes associated with each specific season and level of geomagnetic activity. These profiles correspond to a fixed MLT = 12:00, and clearly show that the electron density is higher during summer than during winter at all latitudes and exceeds the density observed during the equinoxes. This consistent pattern holds regardless of geomagnetic activity. The variation in the baseline electron density, which varies depending on the time period under study, corresponds to the differences observed in the distribution of electron density in the F-layer of the ionosphere between seasons (summer, winter, and equinoxes). When compared to other seasons, the summer has a higher electron density. This is due to an increase in solar radiation, which heats both the atmosphere and the ionosphere. Elevated temperatures promote the ionization of atoms and molecules, resulting in a higher electron density in the F-layer of the ionosphere. This effect is particularly pronounced in the equatorial and lower-latitude regions, where solar radiation is stronger. However, as our research shows, similar phenomena can be observed at higher latitudes as well. Winter, on the other hand, sees a decrease in electron density in the F-layer of the ionosphere. Solar radiation is less intense during this season, and the atmospheric and ionospheric temperatures are generally lower. As a result, there is less ionization of atoms and molecules in the atmosphere, resulting in a lower electron density in the F-layer of the ionosphere. The decrease in electron density is most noticeable in the polar regions, where solar radiation is scarce or non-existent during the winter months. There is an intermediate distribution of electron density in the ionosphere's F-layer during the equinoxes, which mark the transition between summer and winter. When compared to other seasons, the moderate solar radiation and temperatures during these times result in a relatively balanced electron density. Therefore, the spatiotemporal pattern of the residues reflects the fluctuations in electron density within the F-layer of the ionosphere throughout the seasons which arise from the complex interplay between solar radiation, atmospheric temperatures, and ionization processes.

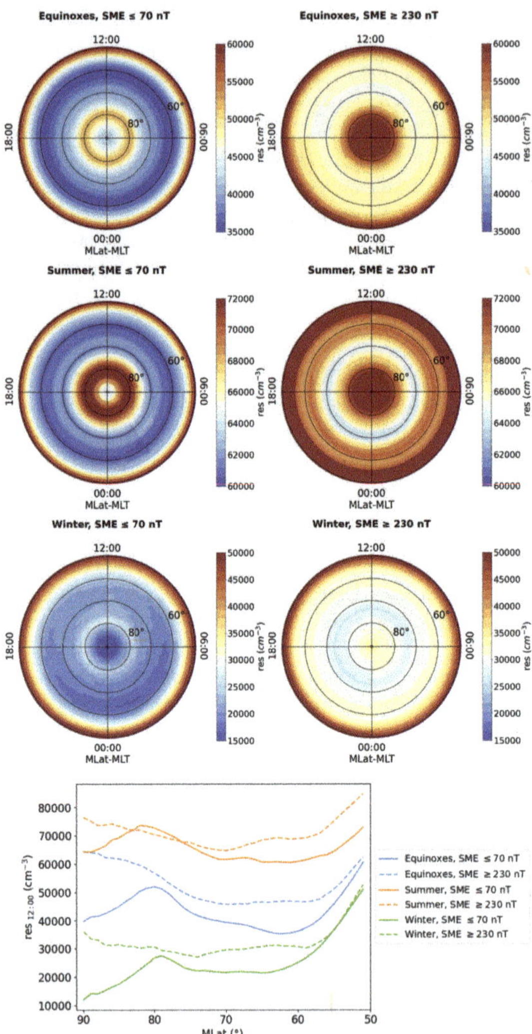

Figure 5. (Upper part): polar view of residue obtained from the MEMD of electron density distribution during the equinoxes (first row), summer (second row), and winter (third row) under low (left column) and high (right column) geomagnetic activity conditions in the Northern Hemisphere. Maps are represented in the QD–magnetic latitude (MLat) and MLT reference system. The concentric circles are plotted in 10° intervals, with the outermost circle corresponding to 50°. **(Bottom part)**: residue trend as a function of the magnetic latitude at a fixed MLT value of 12:00 for the three seasons and two levels of geomagnetic activity.

The seasonal differences in the spatiotemporal distributions of the residues obtained by decomposing the electron density distributions remain visible during periods of high geomagnetic activity. The spatial and temporal distribution of the electron density within the residue is higher during the summer compared to the winter and equinox seasons. Figure 5 reveals that the average level of electron density is affected by both the season and the level of geomagnetic activity, with notable changes occurring during periods of high geomagnetic activity. The minimum in the density distribution within the polar cap region vanishes, replaced by a single region of maximum electron density spanning the entire polar region that begins around 80°. Additionally, the band of the minimum electron density shifts towards higher latitudes. The residue maps exhibit a minimum region around 75°, with a subsequent displacement towards the polar region of approximately 5° compared to the quiet condition. The finding that an increase in geomagnetic activity causes an increase in electron density in the F region of the ionosphere is consistent with previous research. This increase in average electron density can be attributed to a number of well-studied and documented factors. One factor is the enhanced energy input from the magnetosphere during periods of geomagnetic activity. Geomagnetic disturbances such as substorms and magnetic storms can lead to the injection of energetic particles into the ionosphere. These particles ionize the neutral particles in the F region, resulting in an increase in electron density. Another factor is the enhancement of electric fields and plasma convection associated with geomagnetic activity. Electric fields generated by the interaction between the solar wind and the Earth's magnetosphere can accelerate charged particles in the ionosphere, leading to increased electron density. Driven by the motion of plasma within the magnetosphere, plasma convection can transport higher-density plasma in the ionosphere from lower latitudes to higher latitudes. Furthermore, increased geomagnetic activity can lead to changes in thermospheric composition and temperature. These changes can affect the ionization and recombination rates in the F region, influencing the electron density. It is important to note that the relationship between geomagnetic activity and electron density in the F region is complex, and varies depending on the circumstances and location. It is worth noting, however, that the residues obtained from our analysis appear to capture these changes, providing insights into the variations in electron density associated with geomagnetic activity.

We now examine the four modes resulting from each decomposition. The first of these modes, reported only for the equinoctial and low geomagnetic activity case in Figure 4, is analogous to the other five configurations (data not shown), and essentially represents a noise component present in the data. Starting from an analysis of the data distribution resulting from an average, this term could be due to small-scale variations in the original distribution. However, the energy associated with this mode is extremely low, and there are no significant structures in its distribution on the MLat–MLT plane. On the other hand, the second mode (for brevity, data are shown only for the equinoctial and low geomagnetic activity configuration in Figure 4) is quite different. As extensively discussed in the previous section, this mode exhibits a spatial structure that closely resembles the orbit of the satellite from which the data were acquired. The average frequency of this mode is consistent across latitudes, and is comparable to the orbital period of the satellite around the Earth. Finally, the third and fourth modes capture our interest because they are intricately linked to dynamic processes occurring within the ionospheric layer under investigation. These modes exhibit a strong dependence on both the season and the level of geomagnetic activity, making them particularly intriguing and worthy of further exploration.

To gain insight into the dynamic processes that contribute to the distribution of electron density for these two modes regardless of season or activity level, convection cell structures are overlaid onto the maps depicting these modes. At high latitudes, the ionized region of the upper atmosphere undergoes a dynamic circulation known as convection driven by the interaction of the magnetized solar wind and the Earth's magnetosphere. This convection is important in determining the distribution of electron density in the ionosphere. The convection patterns are twin-celled in nature, with the flow moving away from the Sun at the poles and toward the Sun at lower latitudes. The strength of this circulation is influenced by the north–south component of the IMF (represented as B_z), while any asymmetries between the dawn and dusk regions are associated with the east–west component of the IMF (B_y). In general, convection exhibits a highly dynamic nature, displaying sensitivity to intermittent phenomena such as variations in the IMF and magnetospheric processes such as substorms. One effective method for capturing and analyzing these convection patterns is to utilize data from the Super Dual Auroral Radar Network (SuperDARN). SuperDARN is a widespread network of scientific radars designed to monitor the near-Earth space environment. Each radar within the network is capable of measuring the velocity of ionospheric plasma. By combining data from multiple radars, a comprehensive view of plasma motion in the polar ionosphere can be obtained, enabling studies on the electromagnetic interaction between the solar wind and Earth's magnetosphere [34]. Using a long period of SuperDARN data, a dynamic model of convection can be constructed to enable the precise determination of the electrostatic potential distribution in high latitude regions for a wide range of solar wind, IMF, and dipole tilt parameter values [35]. In our case, we obtained representative average convection motions that reflect the interplanetary conditions underlying the originated data by calculating the mean values of the x-component of the solar wind velocity ($v_{sw,x}$), the y- and z-components of the magnetic field, and the dipole tilt for each of the six different datasets analyzed and presented in Figure 1. The obtained values, which refer to the time intervals used to build the different datasets, were then incorporated into the CS10 model [36], enabling us to obtain the average trends of the electrostatic potential distribution. Table 1 reports the values of the solar wind velocity, the two components of the magnetic field, and the dipole tilt used in the different cases. Figure 6 illustrates the spatial and temporal distributions of the electron density for the third and fourth fundamental modes obtained from decomposition during low geomagnetic activity. The figure consists of six maps, with each pair representing the III and IV modes for the analyzed equinoctial, summer, and winter periods. Each map includes the average electrostatic potential level curves overlaid on top of the electron density spatiotemporal distribution. Similarly, Figure 7 follows the same structure as Figure 6 with a focus on the period of high geomagnetic activity.

Table 1. Interplanetary parameters and dipole tilts used as input values for the SuperDARN model relative to each one of the seasonal and geomagnetic activity conditions considered in the present study.

Condition	$v_{sw,x}$ (km/s)	IMF B_y (nT)	IMF B_z (nT)	Dipole Tilt
Equinoxes, SME $\leq$ 70 nT	-367	-0.13	1.36	$0°$
Equinoxes, SME $\geq$ 230 nT	-482	0.10	-2.08	$0°$
Summer, SME $\leq$ 70 nT	-355	0.19	1.17	$10°$
Summer, SME $\geq$ 230 nT	-451	-0.36	-2.06	$10°$
Winter, SME $\leq$ 70 nT	-366	-0.67	0.88	$-10°$
Winter, SME $\geq$ 230 nT	-491	0.24	-1.66	$-10°$

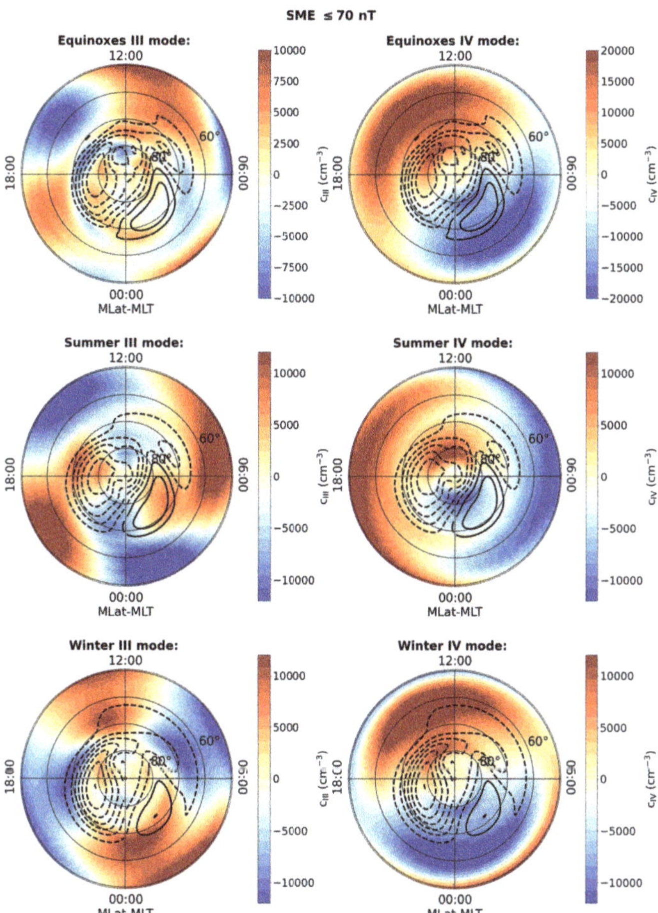

Figure 6. Polar view of the III and IV fundamental modes obtained from the MEMD of electron density distribution during the equinoctial (first row), summer (second row), and winter (third row) periods under low geomagnetic activity in the Northern Hemisphere. SuperDARN polar potential maps obtained using the statistical convection model CS10 are superimposed on each mode as level curves. Maps are represented in the QD–magnetic latitude (MLat) and MLT reference system. The concentric circles are plotted in 10° intervals, with the outermost circle corresponding to 50°.

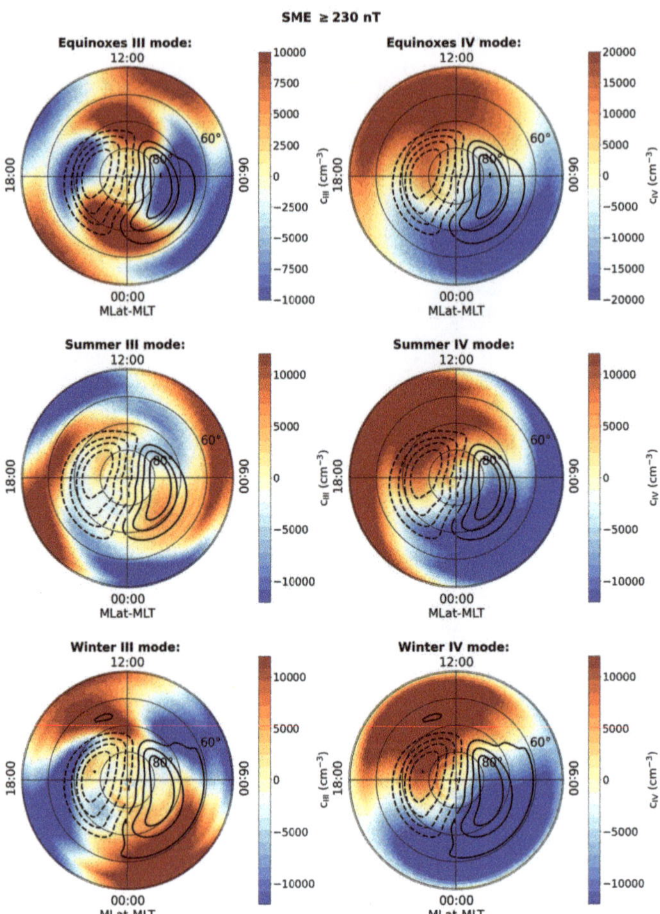

Figure 7. Polar view of the III and IV fundamental modes obtained from the MEMD of electron density distribution during the equinoctial (first row), summer (second row), and winter (third row) periods under high geomagnetic activity in the Northern Hemisphere. SuperDARN polar potential maps obtained using the statistical convection model CS10 are superimposed on each mode as level curves. Maps are represented in the QD–magnetic latitude (MLat) and MLT reference system. The concentric circles are plotted in $10°$ intervals, with the outermost circle corresponding to $50°$.

We begin with the fourth mode and its relationship to the season and geomagnetic activity, as well as the structure of the convection cells superimposed on each map. Here, we are specifically looking at all of the electron density maps shown in the right column of Figures 6 and 7. The electron density spatiotemporal distribution exhibits a clear day–night difference. The spatiotemporal distribution during equinoxes shows higher density values within the MLT interval between 7:00 and 20:00, with a peak occurring between 11:00 and 18:00. This pattern is consistent across all studied latitudes. This day–night asymmetry is primarily attributed to the interaction between solar radiation and the Earth's atmosphere. During the day, solar radiation provides a significant amount of energy that ionizes the upper atmosphere, resulting in higher electron density in the F-layer. Conversely, during the night solar radiation is absent or significantly reduced, leading to a decrease in ionization and electron production. Another contributing factor to the day–night asymmetry is the recombination effect that takes place during the night. Without direct solar radiation, electrons in the ionospheric plasma tend to recombine with ions, leading to a decrease in electron density within the F-layer throughout the nighttime. Furthermore, ionospheric wind movements and diffusion processes can influence electron distribution, adding to the day–night asymmetry. These factors contribute to the creation of plasma flows that shift electron density across different regions of the ionosphere during various phases of the day and night. It is interesting to note that there is a difference in the day–night asymmetry of the electron distribution between the summer and winter seasons. During the summer there is generally higher ionization of the upper atmosphere due to more direct and prolonged solar incidence, resulting in a higher electron density in the F-region during daylight hours. As a result, the amplitude of the day–night asymmetry may be greater in the summer than in the winter. This is precisely what is observed when comparing the results obtained for this fourth mode in summer and winter. However, it is important to note that the day–night asymmetry depends on other factors as well, such as the latitude, solar activity, and presence of geomagnetic events. In this study, we fixed the level of solar activity while analyzing two levels of geomagnetic activity. When examining the characteristics of the fourth mode during high geomagnetic activity across different seasons, as shown in the maps in the right column of Figure 7, it can be seen that the day–night asymmetry follows the same seasonal patterns, with only the intensity level varying. Overall, it can be concluded that the fourth mode, obtained through the decomposition of the different datasets, is the result of a combination of factors including interaction with the solar radiation, the nighttime electron recombination process, and ionospheric wind movements. These processes play a dominant role compared to the other modes, as the fourth mode is the most energetic one.

While the analysis of the convection cell structure makes no significant contribution to a better understanding of the electron density distribution in the fourth mode, it is critical to comprehending the third mode. The third mode is presented in Figures 6 and 7 (left columns) in relation to the season and the level of geomagnetic activity. During periods of high geomagnetic activity, the third mode exhibits a region of high electron density at high latitudes on the day side. This characteristic is consistent throughout the year. This region corresponds to the cusp, which is a distinct area within the day side auroral oval near noon. The cusp is known for enhanced ion and electron precipitation, and is considered the primary location for the transport of magnetosheath plasma into the ionosphere. It typically appears as a single spot that spans a few hours of MLT, and its position is influenced by the B_y and B_z components of the IMF. However, it is important to note that the cusp does not always manifest as a single structure. In certain cases, especially when the IMF B_y component is large or when there are changes in IMF orientation, the cusp can bifurcate into two or more distinct regions. This bifurcation phenomenon adds complexity to the cusp structure, and highlights the dynamic nature of the interaction between the solar wind and Earth's magnetosphere. During periods of high geomagnetic activity, various phenomena contribute to the dynamic nature of the high-latitude mesoscale ionosphere. Precipitating electrons propagate towards the pole, and localized structures of electron density known as polar cap patches can be observed moving into and across the polar cap region; additionally, flux transfer events (FTEs) occur, in which photoionized plasma from closed field lines on the day side is transported into the polar cap across the boundary between open and closed field lines. These processes lead to plasma redistribution and changes in plasma density structures in the surrounding regions. The third mode effectively captures the overall results of electron motion, showing an evident increase in electron density along the two convection cells in the polar cap region and subsequent accumulation on the night side at high latitudes. The outcome of this dynamic is significantly less pronounced during periods of low geomagnetic activity, likely due to the influence of a predominantly positive IMF z-component (see Table 1), which affects the configuration of the convection cells, as well as to lower overall dynamics in this region compared to the geomagnetic disturbance period.

6. Summary and Conclusions

For the first time, the MEMD method was used to examine the spatiotemporal distribution of ionospheric electron density at middle and high latitudes (above 50° magnetic latitude) in the Northern Hemisphere as recorded by the Swarm satellite constellation between 1 January 2016, and 31 December 2021. In this research, we looked at different levels of seasonal and geomagnetic activity and discovered the fundamental modes that underpin the initial distributions. By combining these modes with the residue, we were able to precisely reconstruct each electron density dataset. Interestingly, the same number of modes were consistently observed in all datasets regardless of season or geomagnetic activity. Typically, the number of extracted modes is dependent on factors such as the number of datapoints, the presence of different noise sources, and/or additional short-term contributions. In our case, we expect that the noise is the same for the different seasons, and we have the same number of points; thus, these two aspects should not change the number of modes and only differences in short-term variability between the quiet and disturbed period can change the number of modes. Because this does not occur, ew can conclude that similar scale-dependent processes form the basis of the different spatiotemporal patterns seen across different seasons and geomagnetic conditions.

We were able to identify four distinct modes: the first mode represents noise or small-scale variations; the second is characterized by a constant frequency across latitudes, similar to the satellite's orbit; and seasonality and geomagnetic activity dependence are observed in the third and fourth modes, which are linked to dynamic processes within the ionospheric layer. According to our findings, the third mode describes the dynamic processes in high-latitude ionospheric circulation, which is typical of the polar cap region, while the fourth mode emphasizes day–night differences caused by solar radiation, recombination, and ionospheric wind movements. The four modes do not contribute to the original signal in the same way; the most important are the last two modes, which are the most energetic. All of these modes must be added to a base, which is the residue, in order to reconstruct the initial electron density distribution. The residue is higher in the summer than in the winter at all latitudes, and exceeds the density observed during the equinoxes. This consistent pattern persists regardless of geomagnetic activity, though as geomagnetic activity increases the average value of the electron density increases for all latitudes and MLTs.

It is critical to emphasize that our results accurately depict the ionospheric dynamics at both middle and high latitudes during a quiet solar period. It is known that solar activity has a significant impact on the ionosphere, indicating the need to replicate this analysis using years with increased solar activity to confirm our findings' ongoing relevance or detect any potential changes. In general, we anticipate that the decomposition should remain mostly unchanged, though the relative importance of each mode may shift. With the onset of a period of increased solar activity and the continued operation of the Swarm satellite constellation in orbit, we expect to soon be able to investigate the relationship between the solar cycle and the results obtained in this paper.

In conclusion, our research has revealed important insights into the spatiotemporal distribution of the electron density, thereby advancing our understanding of the ionosphere and its spatiotemporal variations while allowing us to better understand its complex behavior. These findings carry relevance for ionospheric models, offering the ability to pinpoint spatiotemporal electron density distributions tied to distinct dynamic processes within the studied ionospheric area. Additionally, they provide insights into the energy contributions of these processes. By integrating these spatiotemporal distributions and energy factors into models, a more precise and comprehensive depiction of the ionosphere could be achieved. In turn, this advance could lead to more accurate prediction of forthcoming ionospheric conditions. Indeed, these findings could be similarly applied to the recent development of a prediction model of Total Electron Content (TEC) derived from GPS data [37]. In this case, a hybrid approach combining Ensemble Empirical Mode Decomposition (EEMD) and Long Short-Term Memory (LSTM) deep learning was utilized. Similarly, one could consider applying the obtained results by directly employing machine learning techniques, as recently proposed by [38], to construct a global model for TEC using Global Ionospheric Maps data along with features such as geomagnetic and solar activity indexes.

Author Contributions: Conceptualization, P.D.M. and T.A.; methodology, G.L. and T.A.; formal analysis, G.L.; investigation, P.D.M., G.L., T.A. and G.C.; data curation, G.L.; writing—original draft preparation, P.D.M. and G.L.; writing—review and editing, all authors. All authors have read and agreed to the published version of the manuscript.

Funding: T.A. received funding from the "Bando per il finanziamento di progetti di Ricerca Fondamentale 2022" mini-grant of the Italian National Institute for Astrophysics (INAF) for "The predictable chaos of space weather events".

Data Availability Statement: Swarm satellite data can be accessed at http://earth.esa.int/swarm, accessed on 15 January 2022. SuperMAG data are available at https://supermag.jhuapl.edu/, accessed on 18 April 2023. IMF components and solar wind speed data are freely available at https://cdaweb.gsfc.nasa.gov/index.html/, accessed on 2 November 2022.

Acknowledgments: The results presented herein rely on data collected by the ESA's Swarm mission. We thank the European Space Agency that supports the Swarm mission. We gratefully acknowledge the SuperMAG collaborators (https://supermag.jhuapl.edu/info/?page=acknowledgement, accessed on 18 April 2023). The IMF components and solar wind velocity data were obtained from the OMNIWeb Data Explorer website, which is operated by the Goddard Space Flight Center (NASA) Space Physics Data Facility, for which we are grateful. In addition, we acknowledge the national scientific funding agencies of Australia, Canada, China, France, Italy, Japan, South Africa, UK, and USA that funded the SuperDARN radar network. G.L. acknowledges the PhD course in Astronomy, Astrophysics, and Space Science of the University of Rome "Sapienza", University of Rome "Tor Vergata", and Istituto Nazionale di Geofisica e Vulcanologia, Italy.

Conflicts of Interest: The authors declare no conflicts of interest.

Abbreviations

The following abbreviations are used in this manuscript:

ACC	Accelerometer
AE	Auroral Electrojet
ASM	Absolute Scalar Magnetometer
EFI	Electric Field Instrument
EEMD	Ensamble Empirical Mode Decomposition
EMD	Empirical Mode Decomposition
ESA	European Space Agency
EO	Earth Observation
EUV	Extreme Ultra Violet
FTE	Flux Transfer Event
GLONASS	Globalnaya Navigazionnaya Sputnikovaya Sistema
GNSS	Global Navigation Satellite System
GPS	Global Positioning System
GPSR	GPS Receiver
IMF	Interplanetary Magnetic Field
LP	Langmuir Probe
LRR	Laser Retro-Reflector
LSTM	Long Short-Term Memory
MEMD	Multivariate Empirical Mode Decomposition
MIT	Main Ionospheric Trough
MLT	Magnetic Local Time
QD	Quasi-Dipole
SED	Storm-Enhanced Density
STR	Star Tracker
SuperDARN	Super Dual Auroral Radar Network
TEC	Total Electron Content
TII	Thermal Ion Imager
VFM	Vector Field Magnetometer

References

1. Kelly, M.C. *The Earth's Ionosphere: Plasma Physics and Electrodynamics*, 2nd ed.; Elsevier: Amsterdam, The Netherlands, 2009.
2. Klobuchar, J.A. Ionospheric time-delay algorithm for single-frequency GPS users. *IEEE Trans. Aerosp. Electron. Syst.* **1987**, *AES-23*, 325–331. [CrossRef]
3. Prieto-Cerdeira, R.; Orús-Pérez, R.; Breeuwer, E.; Lucas-Rodriguez, R.; Falcone, M. Performance of the Galileo single-frequency ionospheric correction during in-orbit validation. *GPS World* **2014**, *25*, 53–58.
4. Yuan, Y.; Wang, N.; Li, Z.; Huo, X. The BeiDou global broadcast ionospheric delay correction model (BDGIM) and its preliminary performance evaluation results. *Navigation* **2019**, *66*, 55–69. [CrossRef]
5. Friis-Christensen, E.; Lühr, H.; Hulot, G. Swarm: A constellation to study the Earth's magnetic field. *Earth Planets Space* **2006**, *58*, 351–358. [CrossRef]
6. Milan, S.E.; Grocott, A. High Latitude Ionospheric Convection. In *Ionosphere Dynamics and Applications*; Huang, C., Lu, G., Eds.; AGU: Hoboken, NJ, USA , 2021; Volume 3, p. 21. [CrossRef]
7. Cowley, S.W.H. TUTORIAL: Magnetosphere-Ionosphere Interactions: A Tutorial Review. *Geophys. Monogr. Ser.* **2000**, *118*, 91. [CrossRef]
8. Prölss, G.W.; Bird, M.K. *Physics of the Earth's Space Environment: An Introduction*; Springer: Berlin/Heidelberg, Germany, 2004.
9. Tashchilin, A.; Romanova, E. Numerical modeling the high-latitude ionosphere. In Proceedings of the Solar-Terrestrial Magnetic Activity and Space Environment: Proc. COSPAR Colloquium, Beijing, China, 10–12 September 2001; Volume 14.
10. Fratter, I.; Léger, J.M.; Bertrand, F.; Jager, T.; Hulot, G.; Brocco, L.; Vigneron, P. Swarm Absolute Scalar Magnetometers first in-orbit results. *Acta Astronaut.* **2016**, *121*, 76–87. [CrossRef]
11. Knudsen, D.J.; Burchill, J.K.; Buchert, S.C.; Eriksson, A.I.; Gill, R.; Wahlund, J.E.; Åhlen, L.; Smith, M.; Moffat, B. Thermal ion imagers and Langmuir probes in the Swarm electric field instruments. *J. Geophys. Res. Space Phys.* **2017**, *122*, 2655–2673. .: 10.1002/2016JA022571. [CrossRef]
12. Buchert, S.; Zangerl, F.; Sust, M.; André, M.; Eriksson, A.; Wahlund, J.E.; Opgenoorth, H. SWARM observations of equatorial electron densities and topside GPS track losses. *Geophys. Res. Lett.* **2015**, *42*, 2088–2092. [CrossRef]
13. Xiong, C.; Stolle, C.; Park, J. Climatology of GPS signal loss observed by Swarm satellites. *Ann. Geophys.* **2018**, *36*, 679–693. [CrossRef]
14. Visser, P.; Doornbos, E.; van den IJssel, J.; Teixeira da Encarnação, J. Thermospheric density and wind retrieval from Swarm observations. *Earth Planets Space* **2013**, *65*, 1319–1331. [CrossRef]
15. Gjerloev, J.W. The SuperMAG data processing technique. *J. Geophys. Res. (Space Phys.)* **2012**, *117*, A09213. [CrossRef]
16. Newell, P.T.; Gjerloev, J.W. Evaluation of SuperMAG auroral electrojet indices as indicators of substorms and auroral power. *J. Geophys. Res. (Space Phys.)* **2011**, *116*, A12211. [CrossRef]
17. Davis, T.N.; Sugiura, M. Auroral electrojet activity index AE and its universal time variations. *J. Geophys. Res. (1896–1977)* **1966**, *71*, 785–801. [CrossRef]
18. Bergin, A.; Chapman, S.C.; Gjerloev, J.W. AE, D_{ST}, and Their SuperMAG Counterparts: The Effect of Improved Spatial Resolution in Geomagnetic Indices. *J. Geophys. Res. (Space Phys.)* **2020**, *125*, e27828. [CrossRef]
19. Richmond, A.D. Ionospheric Electrodynamics Using Magnetic Apex Coordinates. *J. Geomagn. Geoelectr.* **1995**, *47*, 191–212. [CrossRef]
20. Rodger, A.S.; Moffett, R.J.; Quegan, S. The role of ion drift in the formation of ionisation troughs in the mid- and high-latitude ionosphere—A review. *J. Atmos. Terr. Phys.* **1992**, *54*, 1–30. [CrossRef]
21. Zou, S.; Moldwin, M.B.; Ridley, A.J.; Nicolls, M.J.; Coster, A.J.; Thomas, E.G.; Ruohoniemi, J.M. On the generation/decay of the storm-enhanced density plumes: Role of the convection flow and field-aligned ion flow. *J. Geophys. Res. (Space Phys.)* **2014**, *119*, 8543–8559. [CrossRef]
22. Foster, J.C.; Erickson, P.J.; Coster, A.J.; Goldstein, J.; Rich, F.J. Ionospheric signatures of plasmaspheric tails. *Geophys. Res. Lett.* **2002**, *29*, 1623. [CrossRef]
23. Huang, N.E.; Shen, Z.; Long, S.R.; Wu, M.C.; Shih, H.H.; Zheng, Q.; Yen, N.C.; Tung, C.C.; Liu, H.H. The empirical mode decomposition and the Hilbert spectrum for nonlinear and non-stationary time series analysis. *Proc. R. Soc. London. Ser. A Math. Phys. Eng. Sci.* **1998**, *454*, 903–995. [CrossRef]
24. Maheshwari, S.; Kumar, A. Empirical mode decomposition: Theory & applications. *Int. J. Electron. Eng.* **2014**, *7*, 873–878.
25. Rehman, N.; Mandic, D.P. Multivariate empirical mode decomposition. *Proc. R. Soc. A Math. Phys. Eng. Sci.* **2010**, *466*, 1291–1302. [CrossRef]
26. Tanaka, T.; Mandic, D.P. Complex empirical mode decomposition. *IEEE Signal Process. Lett.* **2007**, *14*, 101–104. [CrossRef]
27. Altaf, M.U.B.; Gautama, T.; Tanaka, T.; Mandic, D.P. Rotation invariant complex empirical mode decomposition. In Proceedings of the 2007 IEEE International Conference on Acoustics, Speech and Signal Processing-ICASSP'07, Honolulu, HI, USA, 15–20 April 2007; IEEE: Piscataway, NJ, USA, 2007; Volume 3, pp. III–1009 .
28. Rehman, N.; Mandic, D.P. Empirical mode decomposition for trivariate signals. *IEEE Trans. Signal Process.* **2009**, *58*, 1059–1068. [CrossRef]
29. Niederreiter, H. *Random Number Generation and Quasi-Monte Carlo Methods*; SIAM: Philadelphia, PA, USA, 1992.

30. Alberti, T.; Giannattasio, F.; De Michelis, P.; Consolini, G. Linear versus nonlinear methods for detecting magnetospheric and ionospheric current systems patterns. *Earth Space Sci.* **2020**, *7*, e2019EA000559. [CrossRef]
31. Mandic, D.P.; Goh, V.S.L. *Complex Valued Nonlinear Adaptive Filters: Noncircularity, Widely Linear and Neural Models*; John Wiley & Sons: Chichester, West Sussex, UK, 2009.
32. Flandrin, P.; Rilling, G.; Goncalves, P. Empirical Mode Decomposition as a Filter Bank. *IEEE Signal Process. Lett.* **2004**, *11*, 112–114. [CrossRef]
33. Wu, Z.; Huang, N.E. A study of the characteristics of white noise using the empirical mode decomposition method. *Proc. R. Soc. Lond. Ser. A* **2004**, *460*, 1597–1611. [CrossRef]
34. Chisham, G.; Lester, M.; Milan, S.; Freeman, M.; Bristow, W.; Grocott, A.; McWilliams, K.; Ruohoniemi, J.; Yeoman, T.; Dyson, P.L.; et al. A decade of the Super Dual Auroral Radar Network (SuperDARN): Scientific achievements, new techniques and future directions. *Surv. Geophys.* **2007**, *28*, 33–109. [CrossRef]
35. Ruohoniemi, J.; Baker, K. Large-scale imaging of high-latitude convection with Super Dual Auroral Radar Network HF radar observations. *J. Geophys. Res. Space Phys.* **1998**, *103*, 20797–20811. [CrossRef]
36. Cousins, E.; Shepherd, S. A dynamical model of high-latitude convection derived from SuperDARN plasma drift measurements. *J. Geophys. Res. Space Phys.* **2010**, *115*, A12329. .: 10.1029/2010JA016017. [CrossRef]
37. Nath, S.; Chetia, B.; Kalita, S. Ionospheric TEC prediction using hybrid method based on ensemble empirical mode decomposition (EEMD) and long short-term memory (LSTM) deep learning model over India. *Adv. Space Res.* **2023**, *71*, 2307–2317. [CrossRef]
38. Zhukov, A.V.; Yasyukevich, Y.V.; Bykov, A.E. GIMLi: Global Ionospheric total electron content model based on machine learning. *GPS Solut.* **2021**, *25*, 19. [CrossRef]

Communication

A New Algorithm for Ill-Posed Problem of GNSS-Based Ionospheric Tomography

Debao Wen [1,*][ID], Kangyou Xie [1], Yinghao Tang [1], Dengkui Mei [2], Xi Chen [1] and Hanqing Chen [1]

1 School of Geography and Remote Sensing, Guangzhou University, Guangzhou 510006, China
2 School of Geodesy and Geomatics, Wuhan University, Wuhan 430072, China
* Correspondence: wdbwhigg@gzhu.edu.cn

Abstract: Ill-posedness of GNSS-based ionospheric tomography affects the stability and the accuracy of the inversion results. Truncated singular value decomposition (TSVD) is a common algorithm of ionospheric tomography reconstruction. However, the TSVD method usually has low inversion accuracy and reconstruction efficiency. To resolve the above problem, a truncated mapping singular value decomposition (TMSVD) algorithm is presented to improve the reconstructed accuracy and computational efficiency. To authenticate the effectiveness and the advantages of the TMSVD algorithm, a numerical test scheme is devised. Finally, ionospheric temporal–spatial variations of the selected reconstructed region are studied using the GNSS observations under different geomagnetic conditions. The reconstructed results of TMSVD can accurately reflect semiannual anomalies, diurnal variations, and geomagnetic storm effects. In contrast with the ionosonde data, it is found that the reconstructed profiles of the TMSVD method are more consistent with than those of the IRI 2016. The study suggests that TMSVD is an efficient algorithm for the tomographic reconstruction of ionospheric electron density (IED).

Keywords: ill-posed problem; ionospheric electron density; algorithm; computerized ionospheric tomography

Citation: Wen, D.; Xie, K.; Tang, Y.; Mei, D.; Chen, X.; Chen, H. A New Algorithm for Ill-Posed Problem of GNSS-Based Ionospheric Tomography. *Remote Sens.* **2023**, *15*, 1930. https://doi.org/10.3390/rs15071930

Academic Editor: Fabio Giannattasio

Received: 5 March 2023
Revised: 31 March 2023
Accepted: 2 April 2023
Published: 4 April 2023

1. Introduction

Ionosphere is an ionized component of the atmosphere over the earth. It is well known that the variation mechanism of the ionosphere is complex. The ionospheric delay phenomenon will occur when the electromagnetic signal penetrates the ionosphere, and the ionospheric delay error is an important error source in the fields of communication, satellite navigation and positioning, and radio science and space geodesy [1–3]. Therefore, it is necessary to grasp the temporal–spatial variation rules of the ionosphere. Fortunately, the constructions of global navigation satellite systems (GNSS) provide a promising means for the ionosphere sounding. Due to the real-time, continuous operation and global coverage characteristics of GNSS, GNSS has unique advantages over other ionospheric sounding techniques [4–6].

Vertical total electron content (VTEC) and ionospheric electron density (IED) are the frequently used ionospheric sounding parameters [7,8]. In general, VTEC can be obtained by using the thin-layer ionospheric model, which ignores the vertical structure of the ionosphere. Therefore, VTEC is usually used to describe the horizontal variations of the ionosphere [9–11]. To obtain high accuracy ionospheric delay correction, it is necessary to reconstruct a three-dimensional IED distribution.

Combining computerized ionospheric tomography (CIT) technique with simulated GNSS observations, Kunitsyn et al. have confirmed the possibilities of three-dimensional IED reconstruction [12]. However, ill-posedness of GNSS-based CIT is very prominent due to the nonuniformity and the sparsity of ground observation stations distribution [8,13–17]. To solve the above problem, some tomographic methods have been developed. In general, the methods are usually divided into iterative and noniterative algorithms. Algebraic reconstruction technique (ART), multiplicative ART, and simultaneous iterative reconstruction technique (SIRT) are the typical representative of the iterative algorithms. Andreeva et al. introduced ART to study the ionospheric disturbance in Alaska during a geomagnetic storm at the end of October 2003 [18]. To overcome the deficiency of the conventional iterative algorithms, some constrained and improved iterative algorithms are proposed by some scholars [19–22]. In noniterative algorithms, singular value decomposition (SVD) and truncated SVD (TSVD) are usually adopted to reconstruct the IED distributions [23–27]. However, SVD and TSVD methods fail to obtain high accuracy solution since slant TEC (STEC) observations are contaminated by the discretized error and GNSS observation noise. Various approaches have been presented to overcome the problem that can be experienced with SVD and TSVD. In the approaches, modified SVD [28] and truncated generalized SVD [29] are the two typical algorithms. Although these approaches have their advantages over SVD and TSVD, they suffer from the difficulty of the regularization operator selection or the heavy computational effort. To overcome the disadvantage of the above-described algorithms, the truncated mapping SVD (TMSVD) method is proposed to obtain high-accuracy inversion results. The proposed TMSVD amalgamates information about the properties of the anticipated solution into the solutions process through an initial mapping of the IED tomographic reconstruction. The reconstructed accuracy and efficiency can be improved using the TMSVD. The numerical simulation results verify that the TMSVD is feasible to improve the reconstructed accuracy and computational efficiency. Finally, the TMSVD methods are successfully applied to reconstruct the IED images based on the actual GNSS observations. The error statistics and the comparisons of the vertical profiles further verify the advantages of the TMSVD.

2. Materials and Methods

2.1. Tomographic Theory

As is well known, GNSS-based CIT uses the input STEC to reconstruct the IED distribution [30]. The relation between STEC and IED can be represented using the following equation:

$$y_i = \int_p Ne(l)dl \tag{1}$$

where $Ne(l)$ is the reconstructed IED distribution. y_i represents the STEC along ith ray path p_i. The actual IED Ne can be approximated using a basis function $f_{ij}(l)$. Then Ne is written as:

$$Ne(l) = \sum_{j=1}^{n} x_j f_{ij}(l) \tag{2}$$

where x_j is the IED within the jth voxel. Substituting Equation (2) into Equation (1), Equation (1) is reformulated as:

$$y_i = \int_p \sum_{j=1}^{n} x_j f_{ij}(l) \, dl = \sum_{j=1}^{n} x_j \int_p f_{ij}(l) \, dl \tag{3}$$

The path integral of $f_{ij}(l)$ represents the length of the *i*th ray path traversing the *j*th voxel, which will be defined as A_{ij}.

$$A_{ij} = \int_p f_{ij}(l) \, dl \tag{4}$$

An indicator function is selected as the basis function, which is represented as:

$$f_{ij}(l) = \begin{cases} 1 & j\text{th voxel intersected by } p_i \\ 0 & \text{otherwise} \end{cases} \tag{5}$$

Then the shorter path integrals become

$$\sum_{j=1}^{n} A_{ij}x_j = y_i \qquad i = 1, 2, \cdots, m \qquad (6)$$

Considering the discretized error and GNSS observation noise, the matrix expression of Equation (6) is as follows:

$$A_{m \times n} x_{n \times 1} + e_{m \times 1} = y_{m \times 1} \qquad (7)$$

where n is the number of the discretized voxels in the reconstructed area, m is the number of the input STEC y is a column vector of the m known STEC values, A is the coefficient matrix, and x is the vector consisting of all the unknown IED in all the voxels [1].

2.2. Tomographic Method

To overcome the encountered difficulties of TSVD, the orthogonal mappings of the coefficient matrix A and the input STEC vector y are first performed, and then the TSVD is used to determine the approximation of the mapping problem. The above procedure is named TMSVD. The mapping divides the subspace of the inversion results into two sections. One section is obtained from the user; the other is computed using TSVD. The use-supplied section improved the reconstructed accuracy by incorporating prior information of the ionosphere, which is obtained from IRI 2016 model.

A subspace ω of $S_{n \times l}$ is first chosen, and the columns of the matrix W constitute an orthogonal basis of ω. The following formula can be obtained by introducing QR decomposition.

$$AW = QR \qquad (8)$$

where $W \in S_{n \times l}$, $Q \in S_{m \times l}$, $R \in S_{l \times l}$. In this case, Q has orthogonal columns and R is upper triangular. The selection of the subspace ω is to ensure that AW is not rank deficiency. Therefore, the matrix R is zero strangeness. Innovating the following orthogonal mapping factors:

$$\Psi_W = WW^T, \quad \Psi'_W = I - \Psi_W, \quad \Psi_Q = QQ^T, \quad \Psi'_Q = I - \Psi_Q$$

Then the solution x of Equation (3) can be divided into two sections, which can be represented as:

$$\begin{cases} x = x' + x'' \\ x' = \Psi_W x \\ x'' = \Psi'_W x \end{cases} \qquad (9)$$

The decomposition of Equation (3) can be written as:

$$\Psi_Q Ax' + \Psi_Q Ax'' = \Psi_Q y \qquad (10)$$

In general, $\Psi'_Q A \Psi_W = 0$, so the following expression can be obtained:

$$\Psi'_Q Ax'' = \Psi'_Q y \qquad (11)$$

TSVD is innovated to fulfill the mapping of Equation (11). Since $\Psi'_Q A \Psi'_W = \Psi'_Q A$, the TSVD of $\Psi'_Q A$ can be performed. The approximate x''_k of Equation (11) is obtained. The k value is determined using deviation principle. Then the approximate solution x'_k of Equation (10) is computed. The final expression can be represented as:

$$Rz'_k = Q^T(y - Ax''_k) \qquad (12)$$

The solution z'_k can be obtained, and then the solution x'_k is inverted using the following equation:

$$x'_k = W z'_k \tag{13}$$

The final solution of IED can be expressed as:

$$x_k = x'_k + x''_k \tag{14}$$

For TMSVD, the regularization is only carried out in the subspace ω. Therefore, the subspace is selected in order that ill-posedness of the matrix $R = Q^T A W$ can be circumvented. For the ill-posed problem of discrete tomographic inversion, this condition is easily satisfied when ω represents smooth functions.

Let x''_k represents the approximate solution of Equation (11). The linear system of Equation (12) is exactly solved for z'_k. The solutions of x'_k and x_k are solved using Equations (13) and (14), respectively. Then:

$$\|y - A x_k\| = \|\Psi' y - \Psi' A x''_k\| = \|y - A x''_k\| \tag{15}$$

The solution of high accuracy can be determined since Equation (12) is not ill-posed.

3. Results

3.1. Numerical Simulation

To confirm the performance of TMSVD, a numerical scheme is first devised. Since TMSVD is the improvement of the common TSVD, the TSVD is introduced to compare with the new algorithm. The simulated process is as follows.

A numerical scheme is devised to illustrate the advantages of the TMSVD in comparison with the TSVD. In the numerical simulation, the true values of IED distribution are generated from IRI 2016 model. The selected time period is 05:30–06:00 UT, 30 October 2020. The latitudinal range is 30°–36°N, and the longitudinal range is 116°–122°E. In vertical direction, the altitudinal range is 100–700 km in steps of 10 km. The discretized intervals are 0.5° in latitude and longitude. Thus, the reconstructed region is divided into 4320 voxels. For the test of TMSVD, the space coordinates of the GNSS observation stations and the observed GNSS satellites are used to construct the matrix A.

To evaluate the advantage of the TMSVD to the common TSVD, the mean absolute error (MAE) and the root mean square error (RMSE) of the two algorithms can be calculated using Equations (16) and (17), respectively [8].

$$\text{MAE} = \sum_{j=1}^{n} \left| x_j^{true} - x_j^{tomo} \right| \Big/ n \tag{16}$$

$$\text{RMSE} = \sqrt{\sum_{j=1}^{n} \left(x_j^{true} - x_j^{tomo} \right)^2 \Big/ n} \tag{17}$$

where x_j^{true} is the simulated IED true value of the jth voxel, and x_j^{tomo} represents the tomographic solution of the two algorithms.

Using the devised scheme, the two algorithms are used to reconstruct the IED distribution of the selected geographic region. Figure 1 illustrates the comparisons between the tomographic solutions of the above methods and the simulated IED true values. The comparisons confirm that the reconstructed IED distributions of TMSVD has better agreement with IED true values than those of TSVD.

The reconstructed error between the inversion results of the two algorithms and the simulated true values is computed. Figure 2 shows the error statistical diagram of the two algorithms. Figure 2a illustrates that the maximum error absolute value is 3.87×10^{10} el/m^3, and the error absolute values of about 85% voxels are less than 2×10^{10} el/m^3. Figure 2b shows that the maximum error absolute value of the error is 1.16×10^{11} el/m^3, and the error absolute values of only 27% voxels are less than 2×10^{10} el/m^3. The error statistics validate the accuracy of TMSVD as being higher than that of the TSVD. According to Equations (16) and (17), the MAE and RMSE of two algorithms is obtained. The MAE of TMSVD is 1.01×10^{10} el/m^3, and the RMSE is 1.54×10^8 el/m^3. However, the MAE of TSVD is 4.45×10^{10} el/m^3, and the RMSE of TSVD is 6.8×10^8 el/m^3. The above facts validate that the TMSVD is superior to the common TSVD in performing the tomographic reconstruction of IED distributions.

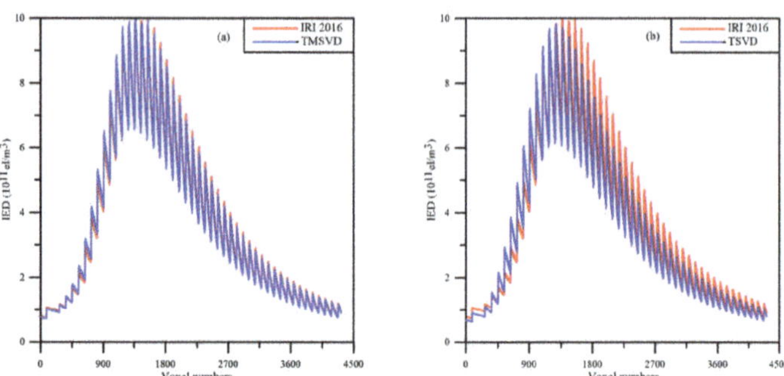

Figure 1. Comparisons of the tomographic results of two algorithms with the simulated IED true values. (**a**) TMSVD; (**b**) TSVD.

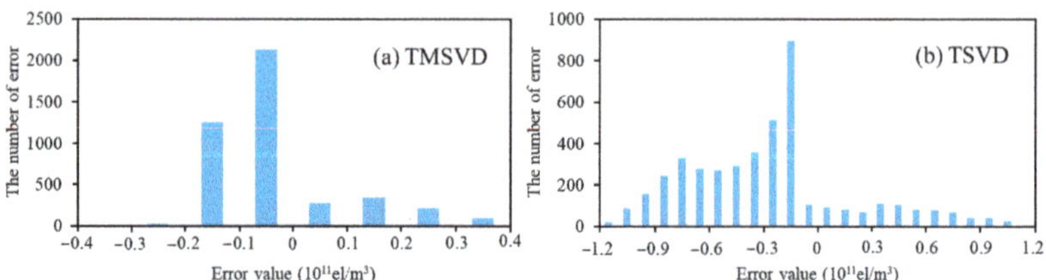

Figure 2. Error statistics of the TMSVD and the TSVD. (**a**) TMSVD; (**b**) TSVD.

3.2. Test of TMSVD Based on GNSS Data

To further test the performance of the TMSVD method, the GNSS observations is applied to reconstruct three dimensional IED distribution of the Hunan province. The sample interval of GNSS data is 30 s. The geographic positions of the selected GNSS ground stations are shown in Figure 3. To avoid the boundary distortion effect of the reconstructed imaging, the latitudinal and the longitudinal ranges are enlarged. For the selected reconstructed area, the latitude ranges from 24°N to 31°N with the step of 0.5°, the longitude ranges from 108°E to 115°E with the step of 1°, and the altitude ranges from 100 km to 1000 km with the spatial resolution of 50 km. Considering the small variation magnitude of IED between 650 km and 1000 km, the altitude ranges from 100 to 650 km when the IED distributions are reconstructed. In this work, the temporal resolution is 30 min.

Using the proposed TMSVD method, we investigate the IED variation rules under the conditions of geomagnetic quiet and disturbance. Figure 4 illustrates three dimensional IED distributions during geomagnetic quiet days on 4 January, 17 March, 17 June, and 18 September 2022. Figure 4 shows that the peak height of the ionosphere is 250 km on 4 January 2022. In the other three days, the peak heights appear at the altitude of 350 km. The reconstructed images capture the IED varied trends in the altitudinal direction.

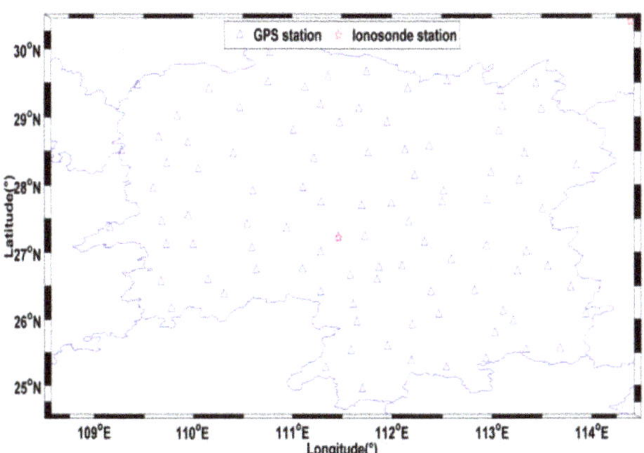

Figure 3. Locations of the selected GNSS and ionosonde stations.

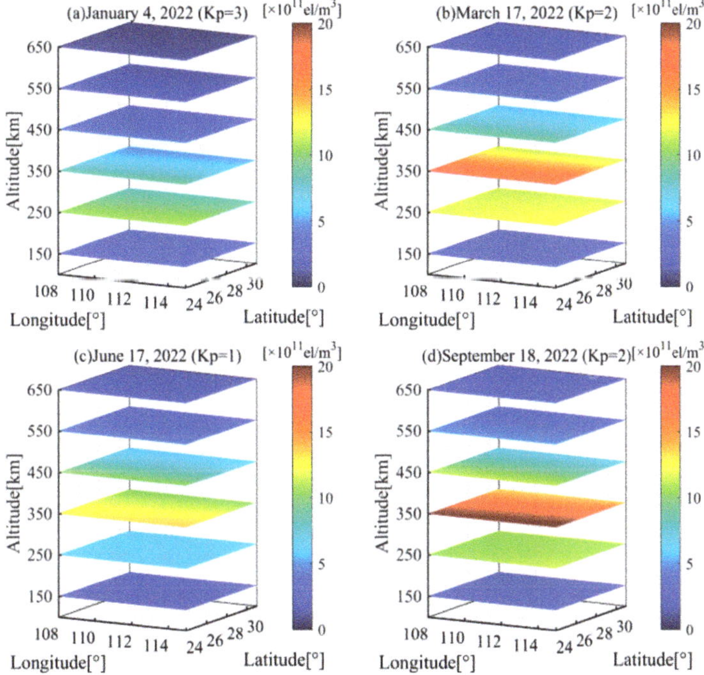

Figure 4. The horizontal section of the reconstructed IED distributions using TMSVD at the altitude of 150–650 km. (**a**) 5:00 UT on 4 January 2022; (**b**) 5:00 UT on 17 March 2022; (**c**) 5:00 UT on 17 June 2022; (**d**) 5:00 UT on 18 September 2022.

Figure 5 illustrates the vertical section at the latitude of 24°–31°N during the same time periods as Figure 4. It shows that the IED values in north Hunan are greater than those in south Hunan. In the meanwhile, the seasonal variations are unlocked. Comparing the reconstructed images in spring and fall with those in winter and summer in Figures 4 and 5, the IED values on 17 March and 18 September are greater than those on 4 January and 17 June 2022. The reconstructed results reflect the semiannual anomaly of three-dimensional ionospheric variations.

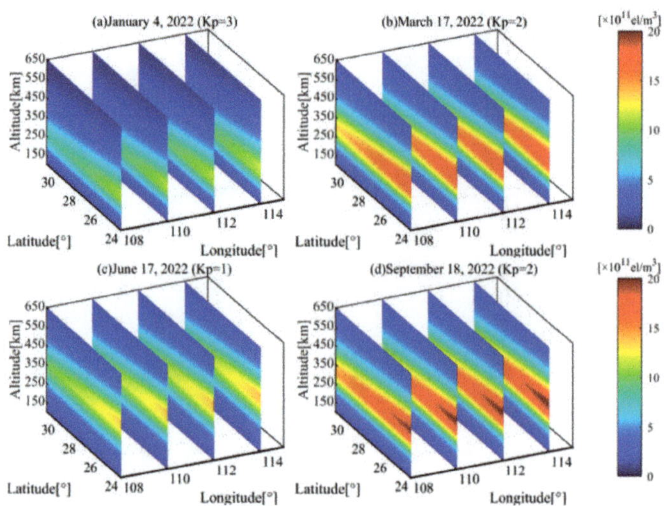

Figure 5. The three-dimensional IED distributions reconstructed by TMSVD at 05:00 UT. (**a**) 4 January 2022; (**b**) 17 March 2022; (**c**) 17 June 2022; (**d**) 18 September 2022.

Figure 6 reflects the diurnal variations of the ionosphere at the cross section of 350 km on 25 August 2018. According to the longitudinal distribution of the IED over the Hunan province, an apparent difference is shown. The IED values in the east is greater than those in the west during 01:00 UT–05:00 UT, and then the IED values in the west are greater than those in the east. However, during 21:00 UT–23:00 UT, the longitudinal distributions of the IED return to the state of the time periods of 01:00 UT–05:00 UT. In the latitudinal direction, the IED values in the north are greater than those in the south between 13:00 UT and 21:00 UT. Nevertheless, the IED values in the low-latitude region are greater than those in the high-latitude region in other time periods. As time goes on, the IED values gradually increase, and the IED value reaches its maximum at 07:00 UT. Subsequently, the electron density distributions start to decrease; the minimum IED value occurs at 21:00 UT (5:00 local time). The diurnal variations rules are identical to the Earth's rotation phenomenon. This also coincides with the alternation of day and night.

According to the statistics analysis, 4.15 Gb computer memory and 1400 CPU seconds are usually required when the TSVD is applied to reconstruct the diurnal variation in the IED distributions, whereas the TMSVD required 2.36 Gb computer memory and 650 CPU seconds. The statistics validate that the reconstructed efficiency of the TMSVD is higher than that of the TSVD.

An ionospheric storm occurred on 26 August 2018. The Kp index reached 7+ at 8:00 UT. To verify the ability of the TMSVD method to capture ionospheric storm, the above storm is selected as the test case. Figure 7d,e shows the IED distributions obtained from the TMSVD and the IRI 2016 model at 8:00 UT on 26 August 2018, respectively. Figure 7f reveals the difference between the results of the TMSVD and those of IRI 2016 model. Figure 7f manifests that the IED inversion results of the TMSVD are greater than those of the IRI 2016 model. Considering the quiet ionospheric activity on 25 August 2018, the three-dimensional IED distributions of the day are introduced to compare with those of the selected storm day. Figure 7a,b represents the reconstructed images of the TMSVD and IRI 2016 model, respectively. Figure 7c represents the difference between Figure 7a,b. Figure 7c shows that the reconstructed IED distributions become lower than the simulated IED values by the IRI 2016 model. Comparing Figure 7d with Figure 7a, the IED values of the storm day evidently increase. It exhibits the positive storm phase effect. Figure 7f shows that the positive storm phase covers the selected latitude range. However, the results obtained from the IRI 2016 model displays no visible difference on 25–26 August 2018.

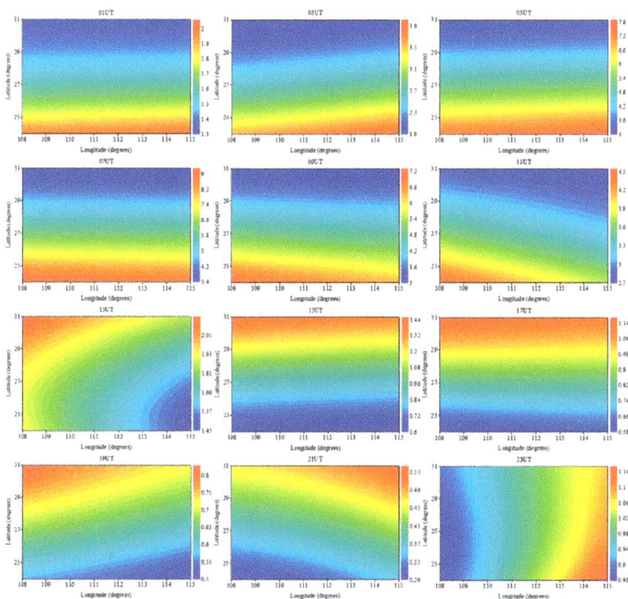

Figure 6. The IED diurnal variation at the altitude of 350 km on 25 August 2018. The unit of IED is 10^{11} el/m^3.

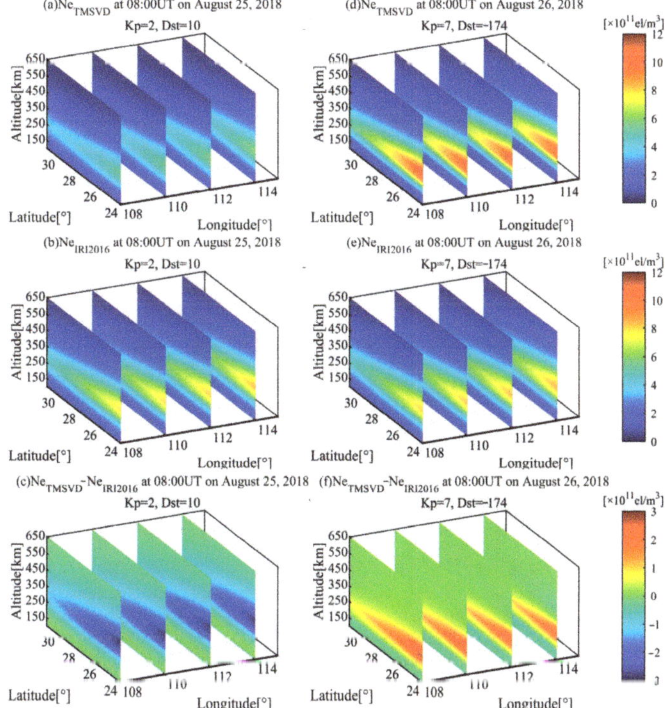

Figure 7. The three-dimensional IED distributions reconstructed using TMSVD and the IRI 2016 model at 8:00 on 25–26 August 2018.

As illustrates in Figure 8, the comparisons of the vertical IED profiles are made at the time mentioned in Figure 7. In the comparisons, the recorded data of ionosonde located at Shaoyang is introduced. In general, ionosonde diagnoses only the lower part of the ionosphere. In this work, the type of profile above the F2 layer peak is derived from the Chapman model. The comparisons show that the vertical profiles reconstructed by TMSVD are identical to those obtained from ionosonde in Shaoyang. The fact validates that the TMSVD is effective and superior to the IED reconstruction under different geomagnetic conditions.

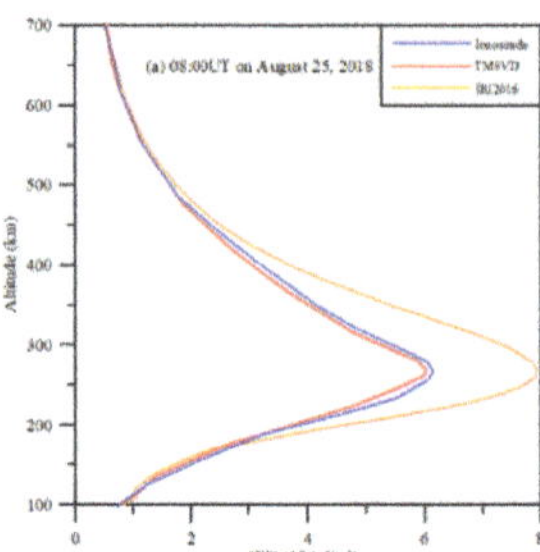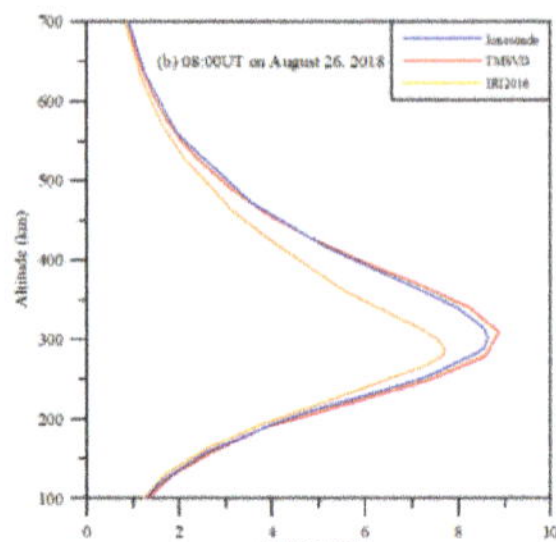

Figure 8. The comparisons of the IED profiles reconstructed by the TMSVD with those obtained from the IRI 2016 and ionosonde in Shaoyang.

4. Conclusions

This work develops a new algorithm to reconstruct the ionospheric structure by using the GNSS observation in the Hunan province. Two experimental schemes are devised to validate the advantages of the TMSVD method. The experiments confirm that the inversion accuracy and efficiency is obviously improved by the TMSVD method. Seasonal characteristics were first studied on 4 January, 17 March, 17 June 2022, and 18 September 2022. The reconstructed results of the TMSVD can accurately capture the semiannual anomalies and the diurnal variations. Compared to the corresponding values obtained from the IRI 2016 model, the reconstructed IED values of the TMSVD have an obvious improvement. Finally, a strong ionospheric storm is selected as a test case. The test results show that the TMSVD method can effectively capture the ionospheric structure during an ionospheric storm. However, the IRI 2016 model cannot reflect the IED variations under the condition of ionospheric disturbance.

Although the TMSVD can effectively reconstruct the three-dimensional IED distributions under different geomagnetic activities, the algorithm has difficulty in reconstructing the polar cap absorption event. In future, the TMSVD will be extended to reconstruct the solar flare and equatorial anomaly. In addition, the study of ionospheric activities caused by geohazards can be carried out.

Author Contributions: Methodology, D.W.; software, Y.T. and K.X.; validation, X.C. and D.M.; formal analysis, D.W.; writing—original draft preparation, D.W.; writing—review and editing, H.C.; funding acquisition, D.W. and H.C. All authors have read and agreed to the published version of the manuscript.

Funding: This research was funded by the National Nature Science Foundation of China, grant number 42070430, the Natural Science Foundation of Guangdong Province of China, grant number 2022A1515011039.

Data Availability Statement: The readers can obtain the data from the corresponding author.

Acknowledgments: All authors would like to thank the editors and anonymous reviewers for their valuable comments and suggestions which improved the quality of the manuscript.

Conflicts of Interest: The authors declare no conflict of interest.

References

1. Wen, D.; Mei, D.; Chen, H. Three-step tomographic algorithm for ionospheric electron density reconstruction. *IEEE Trans. Geosci. Remote Sens.* **2022**, *60*, 5801408. [CrossRef]
2. Samardjiev, T.; Bradley, P.A. Ionospheric mapping by computer contouring techniques. *Electron. Lett.* **1993**, *29*, 1794–1995. [CrossRef]
3. Goodman, J.M. Operational communication systems and relationships to the ionosphere and space weather. *Adv. Space Res.* **2005**, *36*, 2241–2252. [CrossRef]
4. Song, R.; Hattori, K.; Zhang, X.; Yoshino, C. The three-dimensional ionospheric electron density imaging in Japan using the approximate Kalman filter algorithm. *J. Atmos. Sol.-Terr. Phy.* **2021**, *219*, 105628. [CrossRef]
5. Yao, Y.; Tan, J.; Kong, J.; Zhang, L.; Zhang, S. Application of hybrid regularization method for tomographic reconstruction of millatitude ionospheric electron density. *Adv. Space Res.* **2013**, *12*, 2215–2225. [CrossRef]
6. Wen, D.; Mei, D. Ionospheric TEC disturbances over China during the strong geomagnetic storm in September 2017. *Adv. Space Res.* **2020**, *65*, 2529–2539. [CrossRef]
7. Zheng, D.; Zheng, H.; Wang, Y.; Nie, W.; Li, C.; Ao, M.; Hu, W.; Zhou, W. Variable pixel size ionospheric tomography. *Adv. Space Res.* **2017**, *59*, 2969–2986. [CrossRef]
8. Wen, D.; Mei, D.; Du, Y. Adaptive smoothness constraint ionospheric tomography algorithm. *Sensors* **2020**, *20*, 2404. [CrossRef]
9. Chen, C.H.; Saito, A.; Lin, C.H.; Yamamoto, M.; Suzuki, S.; Seemala, G.K. Medium-scale traveling ionospheric disturbances by three-dimensional ionospheric GPS tomography. *Earth Planets Space* **2016**, *68*, 32. [CrossRef]
10. Kong, J.; Li, F.; Yao, Y.; Wang, Z.; Peng, W.; Zhang, Q. Reconstruction of 2D/3D ionospheric disturbances in high-latitude and Arctic regions during a geomagnetic storm using GNSS carrier TEC: A case study of the 2015 great storm. *J. Geodesy* **2019**, *93*, 1529–1541. [CrossRef]
11. Wen, D.; Mei, D.; Du, Y. Imaging the Three-Dimensional Ionospheric Structure with a Blob Basis Functional Ionospheric Tomography Model. *Sensors* **2020**, *20*, 2182. [CrossRef] [PubMed]
12. Kunitsyn, V.E.; Andreeva, E.S.; Razinkov, O.G. Possibilities of the near-space environment radio tomography. *Radio Sci.* **1997**, *32*, 1953–1963. [CrossRef]
13. Li, H.; Yuan, Y.; Yan, W.; Li, Z. A constrained ionospheric tomography algorithm with smoothing method. *Geomat. Inf. Sci. Wuhan Univ.* **2013**, *38*, 412–415. [CrossRef]
14. Hobiger, T.; Kondo, T.; Koyama, Y. Constrained simultaneous algebraic reconstruction technique (C-SART)—A new and simple algorithm applied to ionospheric tomography. *Earth Planets Space* **2008**, *60*, 727–735. [CrossRef]
15. Chen, B.; Wu, L.; Dai, W.; Luo, X.; Xu, Y. A new parameterized approach for ionospheric tomography. *GPS Solut.* **2019**, *23*, 96. [CrossRef]
16. Prol, F.S.; Camargo, P.O. Ionospheric tomography using GNSS: Multiplicative algebraic reconstruction technique applied to the area of Brazil. *GPS Solut.* **2016**, *20*, 807–814. [CrossRef]
17. Hirooka, S.; Hattori, K.; Takeda, T. Numerical validations of neural-network based ionospheric tomography for disturbed ionospheric conditions and sparse data. *Radio Sci.* **2011**, *46*, 1–13. [CrossRef]
18. Andreeva, E.S.; Kunitsyn, V.E.; Leonyeva, E.A.; Fedyunin, Y.N. The ionosphere over Alaska during a storm period in October 2003: Radio tomography and data obtained with GAIM/IFM ionospheric models. *Mosc. Univ. Phys. Bull.* **2009**, *64*, 84–88. [CrossRef]
19. Liu, S.Z.; Wang, J.X.; Gao, J.Q. Inversion of ionospheric electron density based on a constrained simultaneous iteration reconstruction technique. *IEEE Trans. Geosci. Remote Sens.* **2010**, *48*, 2455–2459. [CrossRef]
20. Wen, D.B.; Liu, S.Z.; Tang, P.Y. Tomographic reconstruction of ionospheric electron density based on constrained algebraic reconstruction technique. *GPS Solut.* **2010**, *14*, 375–380. [CrossRef]
21. Yao, Y.B.; Tang, J.; Chen, P.; Zhang, S.; Chen, J.J. An improved iterative algorithm for 3-D ionospheric tomography reconstruction. *IEEE Trans. Geosci. Remote Sens.* **2014**, *52*, 4696–4706. [CrossRef]
22. Zheng, D.Y.; Yao, Y.B.; Nie, W.F.; Yang, W.T.; Hu, W.S.; Ao, M.S.; Zheng, H.W. An improved iterative algorithm for ionospheric tomography by using the automatic search technology of relaxation factor. *Radio Sci.* **2018**, *53*, 1051–1066. [CrossRef]
23. Ou, M.; Zhen, W.; Yu, X.; Xu, J.; Deng, Z. A computerized ionospheric tomography algorithm based on TSVD regularization. *Chin. J. Radio Sci.* **2014**, *29*, 345–352.
24. Dymond, K.F.; Budzien, S.A.; Hei, M.A. Ionospheric thermospheric UV tomography: 1. Image space reconstruction algorithms. *Radio Sci.* **2017**, *52*, 338–356. [CrossRef]
25. Razin, M. Ionospheric electron density reconstruction using TSVD methods over Iran. *Curr. Sci.* **2010**, *79*, 1547–1556.
26. Raymund, T.D. Comparison of several ionospheric tomography algorithms. *Ann. Geophys.* **1994**, *13*, 1254–1262. [CrossRef]
27. Bhuyan, K.; Bhuyan, P.K. International reference ionosphere as a potential regularization profile for computerized ionospheric tomography. *Adv. Space Res.* **2007**, *39*, 851–858. [CrossRef]
28. Hansen, P.C.; Sekii, T.; Shibahashi, H. The modified truncated SVD method for regularization in general form. *SIAM J. Sci. Stat. Comput.* **1992**, *13*, 1142–1150. [CrossRef]

29. Bhuyan, K.; Singh, S.B.; Bhuyan, P.K. Tomogrpahic reconstruction of the ionosphere using generalized singular value decomposition. *J. Geog. Infor. Techn.* **2002**, *83*, 1117–1120.
30. Austen, J.R.; Franke, S.J.; Liu, C.H. Ionospheric imagingusing computerized tomography. *Radio Sci.* **1988**, *23*, 299–307. [CrossRef]

remote sensing

Article

An Explainable Dynamic Prediction Method for Ionospheric *fo*F2 Based on Machine Learning

Jian Wang [1,2,3], Qiao Yu [1], Yafei Shi [1,2,*], Yiran Liu [1] and Cheng Yang [1,2]

1 School of Microelectronics, Tianjin University, Tianjin 300072, China
2 Qingdao Institute for Ocean Technology, Tianjin University, Qingdao 266200, China
3 Shandong Engineering Technology Research Center of Ocean Information Awareness and Transmission, Qingdao 266200, China
* Correspondence: shiyafei@tju.edu.cn

Abstract: To further improve the prediction accuracy of the critical frequency of the ionospheric F2 layer (*fo*F2), we use the machine learning method (ML) to establish an explanatory dynamic model to predict *fo*F2. Firstly, according to the ML modeling process, the three elements of establishing a prediction model of *fo*F2 and four problems to be solved are determined, and the idea and concrete steps of model building are determined. Then the data collection is explained in detail, and according to the modeling process, *fo*F2 dynamic change mapping and its parameters are determined in turn. Finally, the established model is compared with the International Reference Ionospheric model (IRI-2016) and the Asian Regional *fo*F2 Model (ARFM) to verify the validity and reliability. The results show that compared with the IRI-URSI, IRI-CCIR, and ARFM models, the statistical average error of the established model decreased by 0.316 MHz, 0.132 MHz, and 0.007 MHz, respectively. Further, the statistical average relative root-mean-square error decreased by 9.62%, 4.05%, and 0.15%, respectively.

Keywords: ionospheric *fo*F2; machine learning; dynamic prediction

Citation: Wang, J.; Yu, Q.; Shi, Y.; Liu, Y.; Yang, C. An Explainable Dynamic Prediction Method for Ionospheric *fo*F2 Based on Machine Learning. *Remote Sens.* **2023**, *15*, 1256. https://doi.org/10.3390/rs15051256

Academic Editor: Fabio Giannattasio

Received: 21 January 2023
Revised: 21 February 2023
Accepted: 22 February 2023
Published: 24 February 2023

1. Introduction

The ionosphere is the ionized part of the atmosphere around the earth, and is essential to the near-Earth space environment. Because the ionosphere is a time-varying dispersive channel, the radio path loss, delay dispersion, noise, and interference transmitted through the ionosphere constantly change with frequency, location, season, and day and night [1]. The ionosphere is divided into the D layer, E layer, F1 layer, and F2 layer, among which the changes in the ionospheric F2 layer and the radio propagation mechanism and effect in this layer are more complex, which is one of the crucial components of a wireless channel [2]. The critical frequency of the ionospheric F2 layer (*fo*F2), which is a key parameter, plays a vital role in radio-wave propagation and is widely used in military and civilian applications such as high-frequency (HF) communication [3], navigation timing [4], direction finding and positioning [5], spectrum management [6], radar detection [7], and disaster warning [8,9]. The *fo*F2 can be directly or indirectly observed by vertical and oblique ionospheric detectors [10]. Without ionospheric sounding stations, these parameters can be obtained from models.

At present, ionospheric reference models are often used to analyze ionospheric characteristics, among which the International Reference Ionospheric (IRI) model is a well-known empirical model and recommended international standard. The IRI model can provide a prediction of the ionospheric characteristic parameters, such as the height, critical frequency, propagation factor, and other parameters of the quiet period [11]. The IRI model has evolved into many important versions [12], among which IRI-2016 is the latest updated version of the IRI model [13]. In addition, experts and scholars are constantly improving the ionospheric model according to the development of ionospheric observations and new innovations. Regional ionospheric models generally give better results and agreement with observations than the global ionospheric model. Therefore, in recent years, the scientific community has paid more attention to the construction of regional models rather than global ionospheric models, because regional models have more accurate simulations of the ionosphere within the modeled region [14]. Therefore, it has become a trend in ionospheric modeling to continuously improve regional ionospheric models' prediction accuracy. Typical regional ionospheric models include the Asia–Oceania method [15] and its revised version [16], the Indian regional model [17], the South American model [18], the Antarctic model [19], the Arctic model [20], the simplified ionospheric regional model (SIRM) and its updated version for the European region [21], the Chinese regional model [22], the Asian model [23], the African model [24], the East Asian model [25,26], the Korean Peninsula model [27], the European–African regional model [28], the mid-latitude model [29], the Pakistan regional model [30], and the African equatorial regional model [31]. In addition, the study of ionospheric prediction models for single stations in specific areas has attracted more and more attention. A large number of single-station ionospheric prediction models have been developed, such as those in Hyderabad and Bangalore [32] in India, Jeju island in Korea [33], Turkey [34], Darwin [35] in Australia, China [36], Karachi, Pakistan [37], and Graham, South Africa [38].

Meanwhile, with the rapid development of computer technology, artificial intelligence algorithms such as statistical machine learning (SML), artificial neural networks (ANN), and deep learning (DL) have been introduced into the modeling of ionospheric parameters. They have improved the foF2 prediction accuracy [39]. For example, Bai et al. [40] proposed a method based on an extreme learning machine when predicting foF2 at the Darwin station in Australia, and RMSE increased by at least 27.2% and 33.3% compared with the two types of IRI models. Fan et al. [41] proposed using the Elman neural network model to predict short-term foF2 data in Wuhan. Kim et al. [29] used long short-term memory (LSTM) to predict foF2 and hmF2 during geomagnetic explosions in mid-latitude regions. Rao et al. [42] used the bidirectional long short-term memory (Bi-LSTM) model to predict foF2 during Hyderabad geomagnetic storms. They compared it with the IRI-2016 model; the Bi-LSTM performs well in seasonal changes and geomagnetic events. Bi et al. [43] used the Informer architecture model to predict foF2 at the Beijing railway station, and compared with the IRI-2016, LSTM, and Bi-LSTM models; the prediction performance of foF2 during the calm period and the geomagnetic storm was improved to varying degrees. Ameen et al. [30] used the ANN method to predict the foF2 of Pakistan's regional magnetic calm period. Compared with the two types of IRI models, the RMSE decreased by 0.945 MHz and 0.97 MHz, respectively. Sivavaraprasad et al. [32] proposed a hybrid model based on a nonlinear autoregressive neural network to predict ionospheric electron content. They compared it with the IRI-2016, autoregressive moving average, and neural network models, and the results show that the new model has the best prediction performance. The prediction of ionospheric parameters by using bidirectional Bi-LSTM [44] and median and LSTM methods [45] is successful in predicting ionospheric parameters before earthquakes. A method based on fuzzy neural networks is also used to detect ULF waves, and encouraging results have also been achieved [46]. In addition, the empirical orthogonal function (EOF) method has been introduced to analyze ionospheric parameters such as foF2 [47], the peak height of the F2 layer [48], the propagation factor of the F2 layer [26], and total electron content [49], and some of the orthogonal function results have been included in the IRI model [49]. These studies aim to improve the prediction accuracy of ionospheric characteristic parameters by constructing new models or improving existing models with new data.

To further improve the prediction accuracy of foF2, this paper takes Kokubunji station as an example to propose an ML-based explainable dynamic prediction method according to the relationships of foF2 and the solar activity index. The structure of this paper is as follows: The Section 2 proposes the idea of establishing the model, which provides the basis for the implementation of the subsequent modeling. The Section 3 describes the collected data and the primary processing process, which provides data for machine learning. The Section 4 is the paper's focal point, which systematically explains the process of model mapping selection and parameter determination. Finally, the model in this paper is compared with the IRI-2016 model and the Asia Regional *fo*F2 model (ARFM) [23].

2. Modeling Idea

Machine learning (ML) is an interdisciplinary subject that constructs a probabilistic statistical model based on data and uses the model to predict and analyze data [50]. Compared with deep learning and black box algorithms, the data processing ideas of machine learning are simple and easy to understand, and the processing process is straightforward. Moreover, the final model parameters determined by the SML method have transparent and explainable meanings [2]. Therefore, SML is widely used in the modeling of ionospheric parameters.

As shown in Figure 1, using machine learning to reconstruct the ionospheric parameter *fo*F2 model is a process that takes data as the core, and focuses on the model, algorithm, and strategy. The data, model, algorithm, and strategy correspond to the four questions that SML needs to solve, namely: how to determine what data is needed, how to choose the model, how to determine the model, and how to evaluate the model. These questions are outlined in the following:

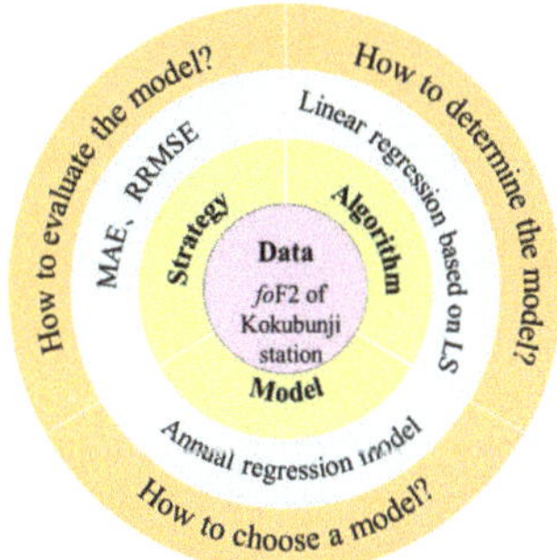

Figure 1. Three elements needed for modeling and four problems needed to be solved based on SML.

(1) What data is needed? The required data are the *fo*F2 of Kokubunji, solar activity index, month, and universal time (UT). *fo*F2 is the model output variable. Solar activity index, month, and UT are model input variables;

(2) How to choose the model? Model choice is the first step of SML, which can be interpreted as a mapping or function. SML can find a specific optimum model in the set of all hypothesized model spaces. Before learning, the assumption modeling sets should be predetermined; this determines the scope of SML. According to the temporal characteristics of model output variable *fo*F2, *fo*F2 is correlated with solar activity index, year, month, and universal time; the hypothesis space of the multiparametric model is determined as in previous research [51];

(3) How to determine the model? We determine a calculation method to measure the relationship between target dependent and independent variables. Regression analysis is a typical method of supervised learning. The optimal model is determined by the least squares (LS) method;

(4) How to evaluate the model? To find the optimal model solution, it is necessary to define a general evaluation criterion in the set of all hypothesized model spaces to find the optimal model solution. In this paper, the mean absolute error (MAE) and the relative root-mean-square error (RRMSE) are used as a general evaluation standard to evaluate the model. Given the prominent time-varying characteristics of the ionosphere, absolute root-mean-square error may lead to deviations in error statistics. Therefore, the RRMSE is adopted to ensure that training learning can stably reflect the changing characteristics of errors.

In short, the model built in this paper is a long-term dynamic prediction model of the monthly median value of *fo*F2 based on the Kokubunji station. Moreover, the model is determined by the RRMSE. In the end, the observations, the IRI model, and the ARFM model are used to verify the reliability, validity, robustness of the established model. The following describes the process of data acquisition, model building, and model verification based on the learning process of SML.

Figure 2 shows the modeling process using SML as follows:

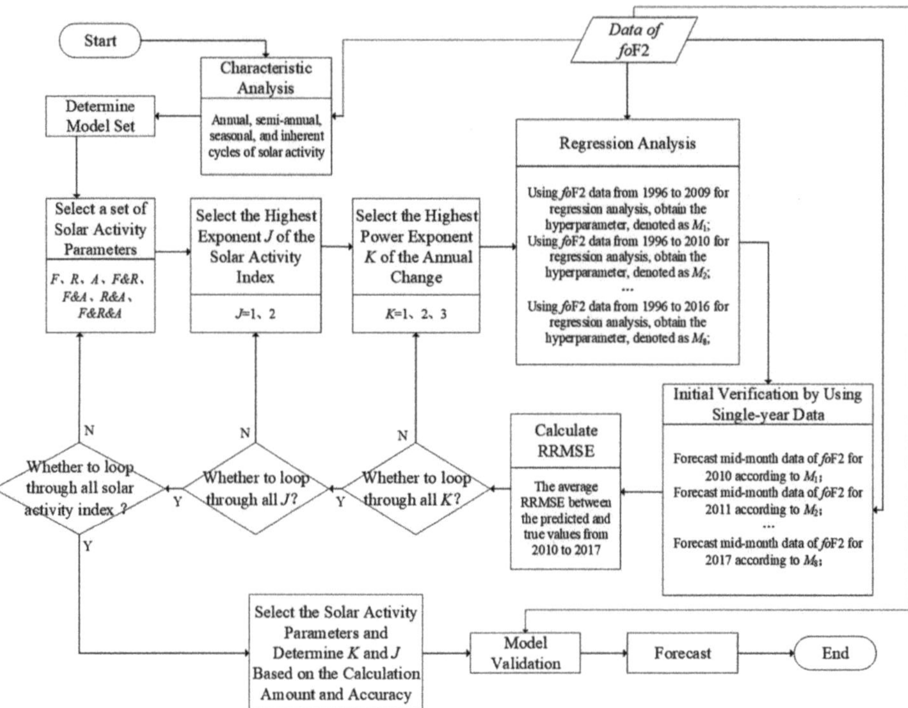

Figure 2. Flow chart of long-term dynamic prediction model of *fo*F2 based on SML.

(1) Analyze the time variation characteristics of *fo*F2 and determine the model set according to the annual, semi-annual, seasonal, and inherent cycle variation rules of solar activity;

(2) Select a set of solar activity parameters, the highest exponent *J* of the solar activity index, and the highest exponent *K* of annual change;

(3) Conduct regression analysis based on the data of *fo*F2 and conduct preliminary verification using single-year data. The specific process is as follows: First, use *fo*F2 data from 1996 to 2009 for regression analysis, obtain the hyperparameter, denoted as M_1, and forecast mid-month data of *fo*F2 for 2010 according to M_1. Secondly, using *fo*F2 data from 1996 to 2010 for regression analysis, obtain the hyperparameter, denoted as M_2, and forecast mid-month data of *fo*F2 for 2011 according to M_2, and so on, using *fo*F2 data from 1996 to 2016 for regression analysis, obtaining the hyperparameter, denoted as M_8, and forecasting the mid-month data of *fo*F2 for 2017 according to M_8;

(4) Calculate and record the mean RRMSE between the predicted and actual median values of *fo*F2 from 2010 to 2017;

(5) Judge, in turn, whether *K*, *J*, and all solar activity index combinations have been traversed. If the traversal is complete, proceed to the next step; if not, proceed to the next round of traversal until the traversal is completed;

(6)   Select the solar activity parameters and determine K and J based on the calculation amount and accuracy;

(7)   Validate the model. The effectiveness and reliability of the model are proven by comparing it with the IRI-URSI, IRI-CCIR, and ARIM models.

3. Data Collection

3.1. Processing Data of foF2

The *fo*F2 data acquisition system is a general-purpose vertical-incidence sounding ionosonde [52], which has the advantages of easy installation, unattended operation, support for remote operation, and low maintenance costs [53]. The main functional performance parameters of the system include: (1) peak transmitting power—usually less than 5000 W; (2) rhe frequency range is wide. The coverage range is 2 MHz to 30 MHz; (3) sweep resolution: 25 kHz, 50 kHz, 75 kHz, or custom. As an example, *fo*F2 data of Kokubunji station (35.7°N, 139.5°E) are used here for training and learning of the proposed model and parameters [54]. At the same time, we collected the corresponding solar activity parameters [55]. The selected observation station lies in the mid-latitude region and the data collected are shown in Figure 3. According to Figure 3, *fo*F2 data collected from this station from 1996 to 2020 are complete, and the modeling selects a sampling interval of 1 h.

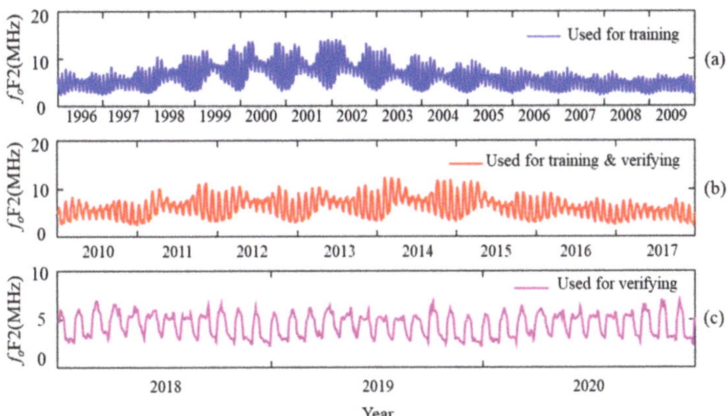

Figure 3. The data of *fo*F2 of Kokubunji station: (**a**) used for training; (**b**) used for training and verifying; and (**c**) used for verifying.

3.2. Deal with the Solar Activity Index

Usually, the solar activity index is used to describe the intensity of solar activity, described as follows: (1) the flux of solar radio waves with a wavelength of 10.7 cm (F10.7, which is labeled F), which is influenced by the corona and the upper atmosphere of the chromosphere [56]; (2) the number of sunspots (R), which is influenced by the photosphere and the lower chromosphere; (3) emission lines in the sun's Lyman-alpha band (Lyman-α, labeled A) are the most vital lines in the ultraviolet band. Among them:

(1)   The solar 10.7 cm radio radiation flux is emitted from the outer chromosphere of the sun's atmosphere and part of the bottom of the corona (inner corona). The unit of $F10.7$ is flux, and 1 sfu = 10^{-22} Wm^{-2}Hz^{-1} [55]. $F10.7$ is a solar activity index closely related to solar activity [57], and is widely used in solar and upper atmosphere empirical models, such as IRI [58,59] and NRLMSISI-00 [60]. Observations of $F10.7$ date back to 1947 and have a history of about 70 years, or about six solar cycles. This paper uses the 12-month sliding average of F, denoted as F_{12}. F10.7 data can be downloaded from the database of the National Oceanic and Atmospheric Administration (NOAA) [61];

(2)   Sunspot numbers are swirls of air caused by solid magnetic activity in the sun's photosphere [62]. Since the first solar cycle in 1755, sunspot data have been observed

for about 24 solar cycles. Because of its prolonged observation, the sunspot number R is the most widely used solar index. This paper uses the 12-month sliding average of R, denoted as R_{12}. The daily values provided by the SILSO database were considered [63];

(3) Hydrogen emission at 121.6 nm represents the most robust single line in the ultraviolet band. It has been investigated in recent years by rockets, Atmospheric Explorer series satellites, the Solar Mesospheric Explorer, the Upper Atmosphere Research Satellite, Thermospheric Ionospheric Mesospheric Energy Dynamics, and the Solar Radiation and Climate Experiment mission [57]. In this paper, we use the 12-month sliding average of A, which is denoted as A_{12}. Detailed information about this composite dataset and free downloadable data are available at the website [64].

In the above, the formula for calculating the 12-month sliding average is as follows:

$$M_{12} = \frac{1}{12}\left[\sum_{i=n-5}^{n+5} \overline{M}_i + \frac{1}{2}(\overline{M}_{n-6} + \overline{M}_{n+6})\right], \tag{1}$$

where M is the solar activity index, $\overline{M}$ is the monthly average of the solar activity index, and n is the month of the required solar index.

The changes in solar activity indexes F_{12}, R_{12}, and A_{12} over time are shown in Figure 4. As can be seen, there are two complete cycles of solar activity between 1996 and 2020.

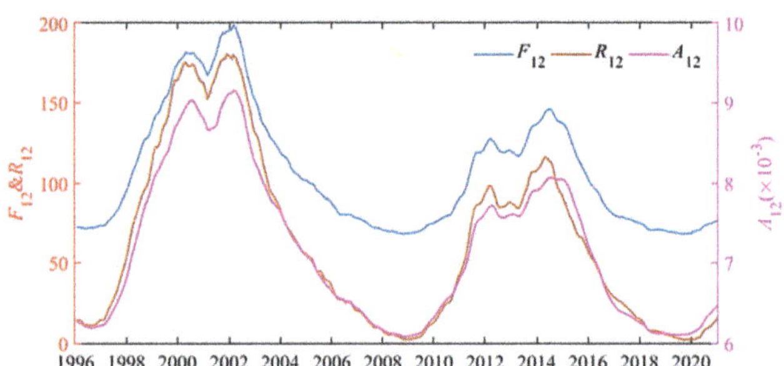

Figure 4. Figure of solar activity indexes over time.

4. Model Construction

4.1. Data Analysis

According to the data collected above, Figure 5 shows the correlation coefficients (CC) between the monthly median value of *f*oF2 of Kokubunji station from 1996 to 2020 and the three types of solar activity index (F_{12}, R_{12}, and A_{12}) by month and hour. As shown in Figure 5, the values of the three groups' CC are mostly between 0.8 and 1, indicating that the median value of *f*oF2 significantly correlates with the three types of solar activity index. In addition, the variation trends of the three groups' CC are similar, but there are differences in some minor details. Among the three groups of CC data, in November, when UT = 19, values are all the smallest, but specifically, they are 0.79, 0.81, and 0.83 respectively.

In summary, it is considered that *f*oF2 is directly related to F_{12}, R_{12}, or A_{12}, but there are specific differences in different periods.

4.2. Model Set

According to the atmosphere–ionosphere–magnetosphere coupling mechanism [65–67] and the previous section's analysis, this paper takes the correlation between *f*oF2 and the above three types of solar activity indices [68] and the annual, semi-annual, seasonal, and more subtle changes of ionospheric characteristic parameters [1] into account. Therefore, the training model is selected for the given geographical coordinates and local time. The general formula for defining the harmonic mapping between *fo*F2 and solar cycle variation parameters, year, season, and month is shown in Equation (2):

$$fo\text{F2}(p,m) = \sum_{k=0}^{K}\sum_{j=0}^{J}\left[\gamma_{k,j}p^j \cdot \cos(2\pi km/12) + \eta_{k,j}p^j \cdot \sin(2\pi km/12)\right], \tag{2}$$

where p represents F_{12}, R_{12}, or A_{12}.

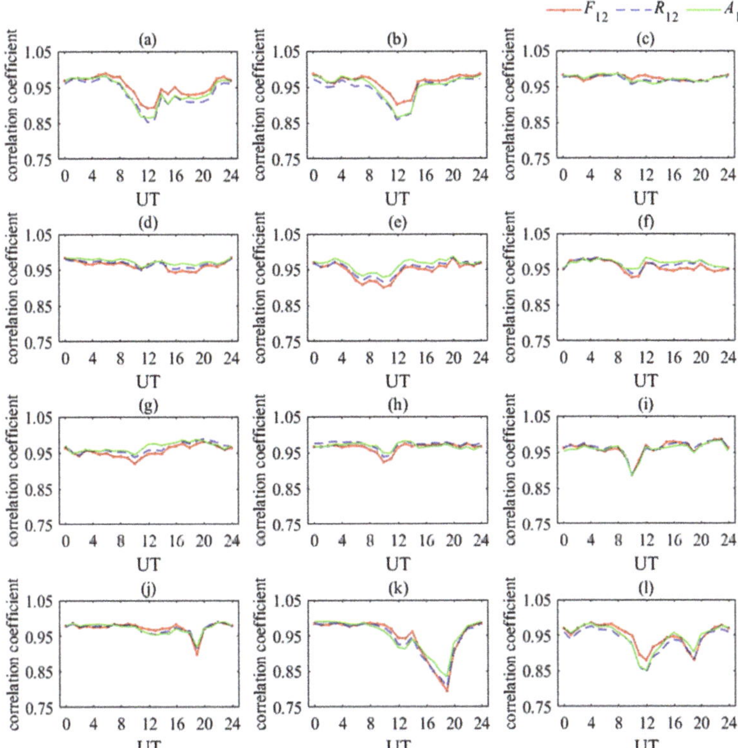

Figure 5. The correlation coefficients between the monthly median value of *fo*F2 and solar activity indices by month: (**a**) Jan; (**b**) Feb; (**c**) Mar; (**d**) Apr; (**e**) May; (**f**) Jun; (**g**) Jul; (**h**) Aug; (**i**) Sept; (**j**) Oct; (**k**) Nov; (**l**) Dec.

Further, the above formula can be converted to Equation (3):

$$fo\text{F2}(F_{12}, R_{12}, A_{12}, m) = \bigcup\Big|\bigcap_{i=1}^{3}\sum_{j=0}^{J}\sum_{k=0}^{K}\left[\gamma_{i,k,j}p_i^j \cdot \cos(2\pi km/12) + \eta_{i,k,j}p_i^j \cdot \sin(2\pi km/12)\right], \tag{3}$$

In Equations (2) and (3), m is an integer representing the month; in the trigonometric functions, the harmonic number k represents the periodic variation characteristics of year, half-year, season, and month. $k = 1, 2, 3,$ and 4, and represents twelve months, six months, three months, and one month, respectively, where the denominator 12 represents the maximum value of the time cycle and corresponds to 12 months of a year. $k = 0$ represents a constant term. Considering that $K = 4$ is relative to $K = 3$, the increase in calculation amount does not significantly reduce the error [26], so we choose $k = 1, 2,$ and 3 here. j represents the variation in the solar activity index, represented by functions of order 1, 2, 3, or higher. The highest harmonic frequency (K and J) can be determined by regression analysis. The coefficient is calculated from the observational data and the corresponding solar activity index.

4.3. Determined Parameters

Figure 6 shows that the model is established by dynamic enhanced prediction. First, the data set is divided into a complete training data set and a full verification data set. The complete training dataset includes *fo*F2 data from 1996 to 2009, the first validation dataset includes *fo*F2 data from 2010 to 2017, and the test dataset includes *fo*F2 data from 2018 to 2020.

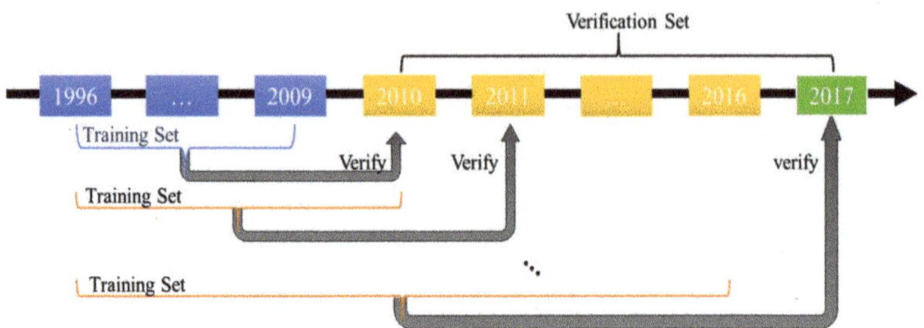

Figure 6. Schematic diagram of the dynamic enhanced prediction model.

First, the data from 1996 to 2009 are used as training data, and the data from 2010 are used as initial validation data to obtain the prediction results of *fo*F2 in 2010. Second, we used data of *fo*F2 from 1996 to 2010 as training data and data from 2011 as initial validation data to obtain prediction results of *fo*F2 in 2011, and so on, until the data from 1996 to 2016 are used as training data and the data from 2017 as validation data; thus, the prediction results of *fo*F2 in 2017 are obtained. Finally, we calculate the RRMSE between the predicted *fo*F2 and the real *fo*F2 from 2010 to 2017. Then the final prediction model is selected from the model set based on the amount of calculation and RRMSE.

Among them, the calculation amount can be determined by Equation (4) according to Equation (3):

$$CM_s = c \cdot [(3J + K)f_+ + 4(J + 1)f_\times + (J + 1)f_2 + (K + 1)f_\Delta], \tag{4}$$

where c is the number of solar activity indices, f_+ represents the number of addition operations, $f_\times$ represents the number of multiplication operations, f_2 represents the number of square operations, and f_Δ represents the number of sine or cosine operations. If the four types of operation times are assumed to be equal, then Equation (4) can be simplified to the form shown in Equation (5):

$$CM_s = c \cdot (8J + 2K + 6) \cdot f, \tag{5}$$

where f represents $f_+, f_\times, f_2,$ and f_Δ.

It can be seen from Equation (5) that the calculated amount is proportional to the number of solar activity parameters, and the change with J is more drastic than that with K.

At the same time, the RRMSE is selected as the evaluation strategy and expressed as:

$$\text{RRMSE} = \sqrt{\frac{1}{N}\sum_{i=1}^{N}\left(\frac{foF2'_i - foF2_i}{foF2_i}\right)^2}, \tag{6}$$

where $foF2'_i$ is the prediction of the model, $foF2_i$ is the measured value, and N is the total data volume.

Figure 7 shows the radar diagram of the RRMSE and computation amounts for different modeling orders J and K obtained by SML. In the figure, 10, 20, 30, etc., can not only represent the amount of computation, but also the value of $4 \times$ RRMSE (%). The analysis found that:

(1) When the value of J is the same, $K = 2$ has an increase compared with $K = 1$, but the RRMSE has a significant decrease; $K = 3$ has an increase compared with $K = 2$, but the RRMSE does not have a significant decrease;

(2) When the values of J and K are the same, the calculation amount of a single solar activity parameter is less than that of multiple solar activity parameters. However, the RRMSE is similar to or even smaller than that obtained with multiple solar activity parameters;

(3) Under the same conditions, the regression prediction effect using a single solar activity parameter R_{12} is worse than that using F_{12} or A_{12}.

According to the above analysis results, when $J = 1, 2$ and $K = 1, 2$, respectively, the calculation amounts required when using solar activity indices F and A for modeling prediction and the RRMSE are calculated. A comparison showed that the increased percentage of the calculation amount and the decreased percentage of the RRMSE when $J = 1$ and $K = 1$ were used for modeling.

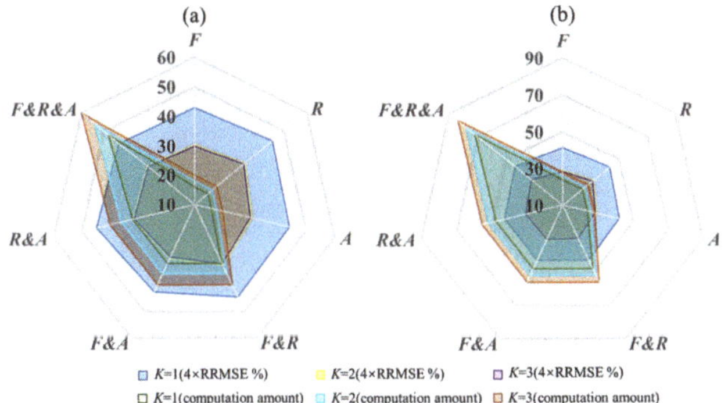

Figure 7. Radar map of calculation amount and $4 \times$ RRMSE(%) under different solar activity parameters J and K: (**a**) $J = 1$; (**b**) $J = 2$.

As shown in Figure 8, J and K are both determined to be 2 by combining RRMSE values and considering engineering complexity.

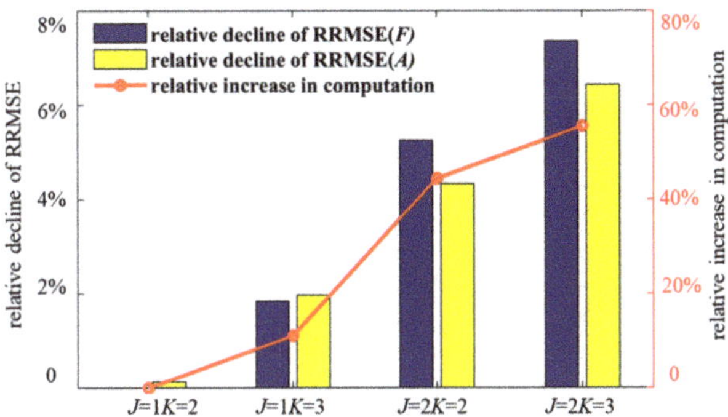

Figure 8. Schematic diagram of calculated increase percentage and RRMSE decrease percentage under different solar activity parameters and J and K conditions.

According to the determined $J = 2$ and $K = 2$, the time regression prediction equation can be explicitly expressed as:

$$\begin{aligned}
\hat{f}_o\text{F2}(F, m) &= \sum_{k=0}^{2} \sum_{j=0}^{2} \left[\gamma_{k,j} F^j \cdot \cos\left(\tfrac{2\pi km}{12}\right) + \eta_{k,j} F^j \cdot \sin\left(\tfrac{2\pi km}{12}\right) \right] \\
&= \left(\gamma_{0,0} + \gamma_{0,1}F + \gamma_{0,2}F^2\right) + \left(\gamma_{1,0} + \gamma_{1,1}F + \gamma_{1,2}F^2\right)\cdot\cos\left(\tfrac{2\pi m}{12}\right) + \left(\gamma_{2,0} + \gamma_{2,1}F + \gamma_{2,2}F^2\right)\cdot\cos\left(\tfrac{2\pi m}{6}\right) + \\
&\quad \left(\eta_{0,0} + \eta_{0,1}F + \eta_{0,2}F^2\right) + \left(\eta_{1,0} + \eta_{1,1}F + \eta_{1,2}F^2\right)\cdot\sin\left(\tfrac{2\pi m}{12}\right) + \left(\eta_{2,0} + \eta_{2,1}F + \eta_{2,2}F^2\right)\cdot\sin\left(\tfrac{2\pi m}{6}\right)
\end{aligned} \tag{7}$$

According to Equation (7), the observation data of Kokubunji station are used for LS regression analysis, and the corresponding $\eta_{k,j}$, $\gamma_{k,j}$, $\eta'_{k,j}$, and $\gamma'_{k,j}$ of the station can be obtained. Figure 9 shows the distribution of coefficients $\gamma_{k,j}$ and $\eta_{k,j}$ using Kokubunji station in 2017 as an example. Coordinate marks 1~18 in the figure correspond to $\gamma_{0,0}$, $\gamma_{0,1}$, $\gamma_{0,2}$, $\eta_{0,0}$, $\eta_{0,1}$, $\eta_{0,2}$, $\gamma_{1,0}$, $\gamma_{1,1}$, $\gamma_{1,2}$, $\eta_{1,0}$, $\eta_{1,1}$, $\eta_{1,2}$, $\gamma_{2,0}$, $\gamma_{2,1}$, $\gamma_{2,2}$, $\eta_{2,0}$, $\eta_{2,1}$, and $\eta_{2,2}$ respectively.

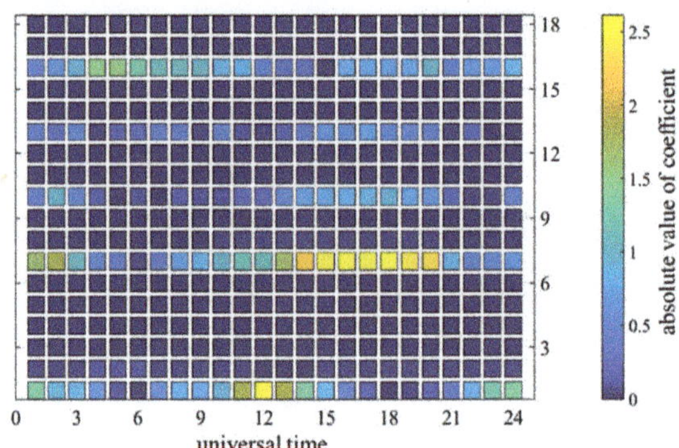

Figure 9. Function reconstruction parameters.

5. Model Verification

To verify the predictive accuracy of the proposed model (identified as PROP), the predictive performance of the new model is compared with that of IRI-2016 [69] and ARFM [23]. The steps to obtain the prediction of *f*oF2 from the PROP model are as follows:

(1) According to Equation (7), the median data of *f*oF2 from 1996 to 2017, 1996 to 2018, and 1996 to 2019 were used for regression analysis training, and we obtained the three corresponding groups of hyperparameters, denoted as M_9, M_{10}, and M_{11};

(2) According to M_9, M_{10}, and M_{11}, and according to F_{12} of 2018, 2019, and 2020, the monthly median forecast of 2018, 2019, and 2020 is obtained by substituting Equation (7).

Figure 10 compares the predictions of the IRI model, ARFM model, PROP model, and the real monthly median value of *f*oF2 from 2018 to 2020. The difference between the predictions of the IRI model and the real monthly median value of *f*oF2 is significant. The predicted value of the ARFM and PROP models is close to the actual value of the monthly median value of *f*oF2.

To further evaluate the models' accuracy, the statistical mean value of MAE and RRMSE (Equation (6)) are calculated to statistically analyze the error between the median value of *f*oF2 of the prediction model and the median value of the measured value. MAE is calculated by the following formula:

$$\text{MAE} = \frac{1}{N} \sum_{i}^{N} \left| f o\text{F2}'_i - f o\text{F2}_i \right|, \tag{8}$$

where $f o\text{F2}'_i$ is the prediction of the model, $f o\text{F2}_i$ is the measured value, and N is the total data volume.

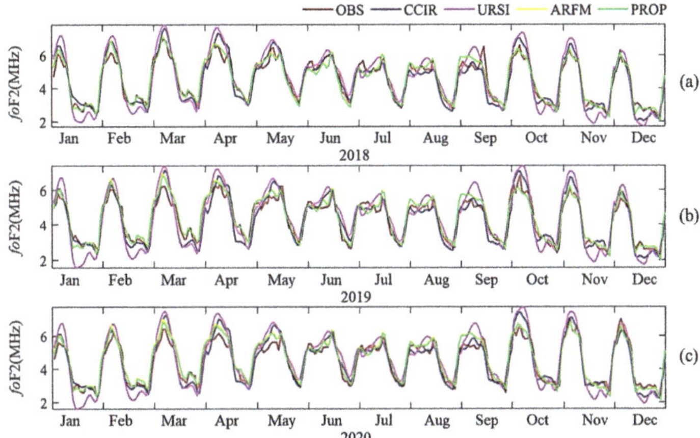

Figure 10. Comparison of IRI, ARFM, and reconstructed model with real value of *fo*F2: (**a**) year of 2018; (**b**) year of 2019; (**c**) year of 2020.

Figure 11 shows the *fo*F2 statistical error diagram of the IRI-URSI, IRI-CCIR, ARFM, and PROP models. From 2018 to 2020, the MAE and RRMSE between the predictions of the IRI model and the observations are more significant than those of the PROP model. However, in 2018, the MAE and RRMSE between the predicted value and the actual value of the ARFM model are slightly smaller than those of the PROP model. Nevertheless, considering the average of three years, the PROP model still has some advantages.

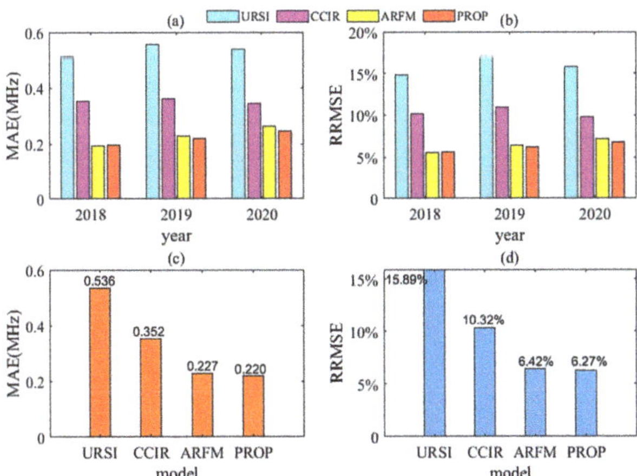

Figure 11. Statistical results of MAE and RRMSE between the predicted value and the real value of IRI, ARFM, and PROP models: (**a**) MAE of each year; (**b**) RRMSE of each year; (**c**) MAE of all years; (**d**) RRMSE of all years.

Specifically, the MAEs corresponding to the IRI-URSI model, IRI-CCIR model, ARFM model, and PROP model are 0.536 MHz, 0.352 MHz, 0.227 MHz, and 0.220 MHz, respectively. Compared with the IRI-URSI, IRI-CCIR, and ARFM models, the statistical average MAE corresponding to *fo*F2 decreased by 0.316 MHz, 0.132 MHz, and 0.007 MHz, respectively. The RRMSEs corresponding to the IRI-URSI model, IRI-CCIR model, ARFM model, and PROP model are 15.89%, 10.32%, 6.42%, and 6.27%, respectively. Compared with the IRI-URSI, IRI-CCIR, and ARFM models, the statistical average RRMSE corresponding to *fo*F2 decreased by 9.62%, 4.05%, and 0.15%, respectively. The reconstruction model PROP in

Kokubunji station has higher precision than the IRI and ARFM models. In other words, the PROP model in this station has realized improvements against the IRI and ARFM models.

6. Conclusions

This paper uses an ML method to propose an interpretable dynamic predictive annual regression model of the ionospheric *fo*F2 median value. The model's main features are as follows: (1) using a dynamic data-updating method for dynamic training prediction; (2) using only one solar activity parameter can achieve high precision prediction and reduce engineering complexity. Compared with the IRI-2016 model and the ARFM model, this model still has some advantages in the prediction performance of *fo*F2 parameters of Kokubunji station. In addition, a goal of future work is to compare the proposed model with other methods such as LSTM and Bi-LSTM. Future research expects to extend the model's applicability to a broader range of stations, regions, and even globally. In addition, this modeling method can be used to predict other ionospheric parameters, such as total electron content and peak height. This method can also be used for natural hazard and space weather monitoring.

Author Contributions: Conceptualization, J.W. and Q.Y.; methodology, J.W. and Q.Y.; software, J.W., Q.Y., Y.S., Y.L. and C.Y.; validation, J.W. and Q.Y.; formal analysis, J.W., Q.Y. and Y.S.; investigation, J.W., Q.Y., Y.S., Y.L. and C.Y.; resources, J.W. and C.Y.; data curation, J.W. and Q.Y.; writing—original draft preparation, J.W. and Q.Y.; writing—review and editing, J.W., Y.S. and C.Y.; visualization, J.W., Q.Y. and C.Y.; supervision, J.W., Q.Y. and Y.S.; project administration, J.W. and C.Y.; funding acquisition, J.W. and C.Y. All authors have read and agreed to the published version of the manuscript.

Funding: This research was funded by the State Key Laboratory of Complex Electromagnetic Environment Effects on Electronics and Information System (No. CEMEE2022G0201).

Data Availability Statement: Not applicable.

Conflicts of Interest: The authors declare no conflict of interest.

References

1. Qian, L.Y.; Burns, A.G.; Solomon, S.C.; Wang, W.B. Annual/semiannual variation of the ionosphere. *Geophys. Res. Lett.* **2013**, *40*, 1928–1933. [CrossRef]
2. Wang, J.; Yang, C.; An, W. Regional Refined Long-term Predictions Method of Usable Frequency for HF Communication Based on Machine Learning over Asia. *IEEE Trans. Antennas Propag.* **2022**, *70*, 4040–4055. [CrossRef]
3. Fagre, M.; Zossi, B.S.; Chum, J.; Yigit, E.; Elias, A.G. Ionospheric high frequency wave propagation using different IRI hmF2 and foF2 models. *J. Atmos. Sol.-Terr. Phys.* **2019**, *196*, 105141. [CrossRef]
4. Swamy, K.C.T.; Sarma, A.D.; Srinivas, V.S.; Kumar, P.N.; Rao, P.V.D.S. Accuracy evaluation of estimated ionospheric delay of GPS signals based on Klobuchar and IRI-2007 models in low latitude region. *IEEE Geosci. Remote Sens. Lett.* **2013**, *10*, 1557–1561. [CrossRef]
5. Erdogan, E.; Seitz, F. High-Resolution Ionosphere Corrections for Single-Frequency Positioning. *Remote Sens.* **2021**, *13*, 12.
6. Wang, J.; Yang, C.; Yan, C. Study on digital twin channel for the B5G and 6G communication. *Radio Sci.* **2021**, *36*, 340–348.
7. Thayaparan, T.; Marchioni, J.; Kelsall, A.; Riddolls, R. Improved Frequency Monitoring System for Sky-Wave Over-the-Horizon Radar in Canada. *IEEE Geosci. Remote Sens. Lett.* **2020**, *17*, 606–610. [CrossRef]
8. Ikuta, R.; Oba, R.; Kiguchi, D.; Hisada, T. Reanalysis of the ionospheric total electron content anomalies around the 2011 Tohoku-Oki and 2016 Kumamoto earthquakes: Lack of a clear precursor of large earthquakes. *J. Geophys. Res. Space Phys.* **2021**, *126*, e2021JA029376. [CrossRef]
9. Santis, A.D.; Perrone, L.; Calcara, M.; Campuzano, S.A.; Cianchini, G.; D'Arcangelo, S.; Mauro, D.D.; Marchetti, D.; Nardi, A.; Orlando, M.; et al. A comprehensive multiparametric and multilayer approach to study the preparation phase of large earthquakes from ground to space: The case study of the June 15 2019, M7.2 Kermadec Islands (New Zealand) earthquake. *Remote Sens. Environ.* **2022**, *283*, 113325. [CrossRef]
10. Wang, J.; Shi, Y.; Yang, C. Investigation of Two Prediction Models of Maximum Usable Frequency for HF Communication Based on Oblique- and Vertical-Incidence Sounding Data. *Atmosphere* **2022**, *13*, 1122. [CrossRef]
11. Bilitza, D. IRI the international standard for the ionosphere. *Adv. Radio Sci.* **2018**, *53*, 1–11. [CrossRef]
12. Bilitza, D.; Mckinnell, L.A.; Reinisch, B.; Rowell, T.F. The international reference ionosphere today and in the future. *J. Geod.* **2011**, *85*, 909–920. [CrossRef]
13. Bilitza, D.; Altadill, D.; Truhlik, V.; Shubin, V.; Galkin, I.; Reinisch, B.; Huang, X. International reference ionosphere 2016: From ionospheric climate to real-time weather predictions. *Space Weather* **2017**, *15*, 418–429. [CrossRef]

14. Wang, J.; Shi, Y.; Yang, C. An Overview and Prospects of Operational Frequency Selecting Techniques for HF Radio Communication. *Adv. Space Res.* **2022**, *69*, 2989–2999. [CrossRef]
15. Sun, X.R. A method of predicting the ionospheric Flayer in the Asia Oceania region. *J. China Inst. Commun.* **1987**, *8*, 153–156.
16. Cao, H.; Sun, X.R. A new method of predicting the ionospheric F2 layer in the Asia Oceania region. *Space Sci.* **2009**, *29*, 502–507. [CrossRef]
17. Bhuyan, P.K.; Chamua, M. An empirical model of electron temperature in the Indian topside ionosphere for solar minimum based on SROSS C2 RPA data. *Adv. Space Res.* **2006**, *37*, 897–902. [CrossRef]
18. Brunini, C.; Meza, A.; Gende, M.; Azpilicueta, F. South American regional ionospheric maps computed by GESA: A pilot service in the framework of SIRGAS. *Adv. Space Res.* **2007**, *42*, 737–744. [CrossRef]
19. An, J.C.; Ning, X.J.; Wang, Z.M.; Zhang, X. Antarctic ionospheric prediction based on spherical cap harmonic analysis and time series analysis. *Wuhan Daxue Xuebao* **2015**, *40*, 677–681.
20. Themens, D.R.; Jayachandran, P.T.; Galkin, I.; Hall, C. The Empirical Canadian High Arctic Ionospheric Model (E-CHAIM): NmF2 and hmF2. *J. Geophys. Res. Space Phys.* **2017**, *122*, 9015–9031. [CrossRef]
21. Perna, L.; Pezzopane, M.; Pietrella, M.; Zolesi, B.; Cander, L.R. An updating of the SIRM model. *Adv. Space Res.* **2017**, *60*, 1249–1259. [CrossRef]
22. Song, R.; Zhang, X.M.; Zhou, C.; Liu, J.; He, J.H. Predicting TEC in China based on the neural networks optimized by genetic algorithm. *Adv. Space Res.* **2018**, *62*, 745–759. [CrossRef]
23. Wang, J.; Ma, J.G.; Huang, X.D.; Bai, H.M.; Chen, Q.; Cheng, H. Modeling of the Ionospheric Critical Frequency of the F2 layer over Asia based on Modified Temporal-Spatial Reconstruction. *Radio Sci.* **2019**, *54*, 680–691. [CrossRef]
24. Okoh, D.; Seemala, G.; Rabiu, B.; Habarulema, J.B.; Jin, S.; Shiokawa, K.; Otsuka, Y.; Aggarwal, M.; Uwamahoro, J.; Mungufeni, P.; et al. A neural network-based ionospheric model over Africa from Constellation Observing System for Meteorology, Ionosphere, and Climate and Ground Global Positioning System observations. *J. Geophys. Res. Space Phys.* **2019**, *124*, 10512–10532. [CrossRef]
25. Wang, J.; Bai, H.; Huang, X.; Cao, Y.; Chen, Q.; Ma, J. Simplified Regional Prediction Model of Long-Term Trend for Critical Frequency of Ionospheric F2 Region over East Asia. *Appl. Sci.* **2019**, *9*, 3219. [CrossRef]
26. Wang, J.; Feng, F.; Bai, H.M.; Cao, Y.B.; Chen, Q.; Ma, J.G. A regional model for the prediction of M(3000)F2 over East Asia. *Adv. Space Res.* **2020**, *65*, 2036–2051. [CrossRef]
27. Jeong, S.-H.; Lee, W.K.; Jang, S.; Kil, H.; Kim, J.-H.; Kwak, Y.-S.; Kim, Y.H.; Hong, J.; Choi, B.K. Reconstruction of the Regional Total Electron Content Maps Over the Korean Peninsula Using Deep Convolutional Generative Adversarial Network and Poisson Blending. *Space Weather* **2022**, *20*, e2022SW003131. [CrossRef]
28. Abuelezz, O.A.; Mahrous, A.M.; Cilliers, P.J.; Fleury, R.; Youssef, M.; Nedal, M.; Yassen, A.M. Neural network prediction of the topside electron content over the Euro African sector derived from Swarm A measurements. *Adv. Space Res.* **2021**, *67*, 1191–1209. [CrossRef]
29. Kim, J.-H.; Kwak, Y.S.; Kim, Y.; Moon, S.-I.; Jeong, S.-H.; Yun, J. Potential of Regional Ionosphere Prediction Using a Long Short-Term Memory Deep-Learning Algorithm Specialized for Geomagnetic Storm Period. *Space Weather* **2021**, *19*, e2021SW002741. [CrossRef]
30. Ameen, M.A.; Tahir, A.; Talha, M.; Khursheed, H.; Siddiqui, I.A.; Iqbal, S.T.; Gul, B. Modelling of foF2 using artificial neural network over Equatorial Ionization Anomaly (EIA) region stations. *Adv. Space Res.* **2022**. [CrossRef]
31. Adebesin, B.O.; Adeniyi, J.O.; Afolabi, P.A.; Ikubanni, S.O.; Adebiyi, S.J. Modelling M(3000)F2 at an African Equatorial Location for Better IRI-Model Prediction. *Radio Science* **2022**, *57*, e2021RS007311. [CrossRef]
32. Sivavaraprasad, G.; Mallika, I.L.; Sivakrishna, K.; Ratnam, D.V. A novel hybrid Machine learning model to forecast ionospheric TEC over Low-latitude GNSS stations. *Adv. Space Res.* **2022**, *69*, 1366–1379. [CrossRef]
33. Moon, S.; Kim, Y.H.; Kim, J.H.; Kwak, Y.S.; Yoon, J.Y. Forecasting the ionospheric F2 parameters over Jeju Station (33.43oN, 126.30oE) by using long short-term memory. *J. Korean Phys. Soc.* **2020**, *77*, 11. [CrossRef]
34. Inyurt, S.; Kashani, M.H.; Sekertekin, A. Ionospheric TEC forecasting using Gaussian process regression (GPR) and multiple linear regression (MLR) in Turkey. *Astrophys. Space Sci.* **2020**, *365*, 99. [CrossRef]
35. Bai, H.M.; Feng, F.; Wang, J.; Wu, T.S. Modeling M(3000)F2 based on Extreme Learning Machine. *Adv. Space Res.* **2020**, *65*, 107–114.
36. Zhao, J.; Li, X.J.; Liu, Y.; Wang, X.; Zhou, C. Ionospheric foF2 disturbance forecast using neural network improved by a genetic algorithm. *Adv. Space Res.* **2019**, *63*, 4003–4014. [CrossRef]
37. Ameen, M.A.; Jabbar, M.A.; Murtaza, G.; Chishtie, F.; Xu, T.; Zhen, W.M.; Atiq, M.; Ali, M.S. Single station modelling and comparison with ionosonde foF2 over Karachi from 1983–2007. *Adv. Space Res.* **2019**, *64*, 2104–2113. [CrossRef]
38. Tshisaphungo, M.; Habarulema, J.B.; McKinnell, L.-A. Modeling ionospheric foF2 response during geomagnetic storms using neural network and linear regression techniques. *Adv. Space Res.* **2018**, *61*, 2891–2903. [CrossRef]
39. Wang, J.; Liu, Y.; Xu, C. The Progress Review and Future Preview of Typical Ionospheric Models. In Proceedings of the International Symposium on Antennas, Propagation and EM Theory, Zhuhai, China, 30 November 2021.
40. Bai, H.M.; Fu, H.P.; Wang, J.; Ma, K.X.; Wu, T.S.; Ma, J.G. A prediction model of ionospheric foF2 based on extreme learning machine. *Radio Sci.* **2018**, *53*, 1292–1301. [CrossRef]
41. Fan, J.Q.; Liu, C.; Lv, Y.J.; Han, J.; Wang, J. A Short-Term Forecast Model of foF2 Based on Elman Neural Network. *Appl. Sci.* **2019**, *9*, 2782

42. Rao, T.V.; Sridhar, M.; Ratnam, D.V.; Harsha, P.B.S.; Srivani, I. A Bidirectional Long Short-Term Memory-Based Ionospheric foF2 and hmF2 Models for a Single Station in the Low Latitude Region. *IEEE Geosci Remote Sens. Lett.* **2022**, *19*, 8005405. [CrossRef]

43. Bi, C.; Ren, P.; Yin, T.; Zhang, Y.; Li, B.; Xiang, Z. An Informer Architecture-Based Ionospheric foF2 Model in the Middle Latitude Region. *IEEE Geosci Remote Sens. Lett.* **2022**, *19*, 1005305. [CrossRef]

44. Tsai, T.C.; Jhuang, H.K.; Ho, Y.Y.; Lee, L.C.; Su, W.C.; Hung, S.L.; Lee, K.H.; Fu, C.C.; Lin, H.C.; Kuo, C.L. Deep learning of detecting ionospheric precursors associated with $M \geq 6.0$ earthquakes in Taiwan. *Earth Space Sci.* **2022**, *9*, e2022EA002289. [CrossRef]

45. Akhoondzadeh, M.; De Santis, A.; Marchetti, D.; Wang, T. Developing a Deep Learning-Based Detector of Magnetic, Ne, Te and TEC Anomalies from Swarm Satellites: The Case of Mw 7.1 2021 Japan Earthquake. *Remote Sens.* **2022**, *14*, 1582. [CrossRef]

46. Georgios, B.; Sigiava, A.-G.; Constantinos, P.; Ioannis, A.; Anastasios, A.; Roger, H. A machine learning approach for automated ULF wave recognition. *J. Space Weather Space Clim.* **2019**, *9*, A13.

47. Aa, E.; Zhang, D.; Xiao, Z.; Hao, Y.-Q.; Ridley, A.J.; Moldwin, M. Modeling ionospheric foF2 by using empirical orthogonal function analysis. *Ann. Geophys.* **2011**, *29*, 1501–1515.

48. Zhang, M.L.; Liu, C.; Wan, W.; Liu, L.; Ning, B. A global model of the ionospheric F2 peak height based on EOF analysis. *Ann. Geophys.* **2009**, *27*, 3203–3212. [CrossRef]

49. Oyeyemi, E.O.; Mckinnell, L.A. A new global F2 peak electron density model for the International Reference Ionosphere (IRI). *Adv. Space Res.* **2008**, *42*, 645–658. [CrossRef]

50. Fokoue, E. Model Selection for Optimal Prediction in Statistical Machine Learning. *N. Am. Math. Soc.* **2020**, *67*, 2. [CrossRef]

51. Santis, D. A Multiparametric Approach to Study the Preparation Phase of the 2019 M7.1 Ridgecrest (California, United States) Earthquake. *Front. Earth Sci.* **2020**, *8*, 540398. [CrossRef]

52. Bibl, K. Sixty years of ionospheric measurements and studies. *Advances in Radio Science* **2004**, *2*, 265–268. [CrossRef]

53. Lan, J.P.; Ning, B.Q.; Zhu, Z.P.; Hu, L.H.; Sun, W.J.; Li, G.Z. Development of agile digital ionosonde and its preliminary observation. *Space Sci.* **2019**, *39*, 167–177.

54. Ionosonde Data in JAPAN. Available online: https://wdc.nict.go.jp/IONO/HP2009/ISDJ/index-E.html (accessed on 3 November 2022).

55. Tapping, K.F. The 10.7cm solar radio flux (F10.7). *Space Weather* **2013**, *11*, 394–406. [CrossRef]

56. Afraimovich, E.L.; Astafyeva, E.I.; Oinats, A.V.; Yasukevich, Y.V.; Zhivetiev, I.V. Global electron content: A new conception to track solar activity. *Ann. Geophys.* **2008**, *26*, 335–344. [CrossRef]

57. Solomon, S.C.; Qian, L.; Burns, A.G. The anomalous ionosphere between solar cycles 23 and 24. *Geophys. Res. Space Phys.* **2013**, *18*, 6524–6535. [CrossRef]

58. Bilitza, D. *International Reference Ionosphere*, 3rd ed.; World Data Center A for Rockets and Satellites: Greenbelt, MA, USA, 1990.

59. Bilitza, D.; Altadill, D.; Zhang, Y.; Mertens, C.; Truhlik, V.; Richards, P.; McKinnell, L.A.; Reinisch, B. The International Reference Ionosphere 2012—A model of international collaboration. *J. Space Weather Space Clim.* **2014**, *4*, A07. [CrossRef]

60. Picone, J.M.; Hedin, A.E.; Drob, D.P.; Aikin, A.C. NRLMSISE-00 empirical model of the atmosphere: Statistical comparisons and scientific issues. *J. Geophys.* **2002**, *107*, 1–16. [CrossRef]

61. National Oceanic and Atmospheric Administration (NOAA). Available online: https://www.ngdc.noaa.gov/stp/space-weather/solar-data/ (accessed on 28 October 2022).

62. Sun, W. *Study on Regional Ionospheric Characteristics Based on Ground-Based GPS and Occultation Technology*; Wuhan University: Wuhan, China, 2015.

63. Sunspot Number. Available online: https://www.sidc.be/silso/datafiles (accessed on 28 October 2022).

64. Data of Hydrogen Emission at 121.6 nm. Available online: https://lasp.colorado.edu/lisird/composite_timeseries.html (accessed on 27 April 2022).

65. Editors, D.; Ouzounov, S.; Pulinets, K.; Hattori, T.P. *Pre-Earthquake Processes: A Multidisciplinary Approach to Earthquake Prediction Studies*; AGU & Wiley: Bristol, UK, 2018; pp. 79–99.

66. Santis, A.D.; Balasis, G.; Pavón-Carrasco, F.G.; Cianchini, G.; Mandea, M. Potential earthquake precursory pattern from space: The 2015 Nepal event as seen by magnetic Swarm satellites. *Earth Planet. Sci. Lett.* **2017**, *461*, 119–126. [CrossRef]

67. Ouzounov, D.; Pulinets, S.; Davidenko, D.; Rozhnoi, A.; Solovieva, M.; Fedun, V.; Dwivedi, B.N.; Rybin, A.; Kafatos, M.; Taylor, P. Transient Effects in Atmosphere and Ionosphere Preceding the 2015 M7.8 and M7.3 Gorkha–Nepal Earthquakes. *Front. Earth Sci.* **2021**, *9*, 757358. [CrossRef]

68. Zhang, X.; Le, G.; Zhang, Y. Phase relationship between the relative sunspot number and solar 10.7 cm flux. *Chin. Sci. Bull.* **2012**, *57*, 2078–2082. [CrossRef]

69. International Reference Ionosphere. Available online: http://IRImodel.org/IRI-2016 (accessed on 18 April 2022).

remote sensing

Article

Statistical Analysis of SF Occurrence in Middle and Low Latitudes Using Bayesian Network Automatic Identification

Jian Feng [1], Yuqiang Zhang [2,*], Shuaihe Gao [3], Zhuangkai Wang [2], Xiang Wang [4], Bo Chen [2], Yi Liu [2], Chen Zhou [2] and Zhengyu Zhao [2]

1   China Research Institute of Radiowave Propagation (CRIRP), Qingdao 266107, China
2   Department of Space Physics, School of Electronic Information, Wuhan University, Wuhan 430072, China
3   National Time Service Center, Chinese Academy of Sciences, Beijing 100045, China
4   School of Artificial Intelligence, Hubei University, Wuhan 430061, China
*   Correspondence: yqzhang_3@stu.xidian.edu.cn

Abstract: Spread-F (SF) is one of the most important types of the ionospheric irregularities as it causes ionospheric scintillation which can severely affect the performance and reliability of communication, navigation, and radar systems. The ionosonde provides the most effective and economical way to study the ionosphere and SF. However, the manual identification of SF from an ionogram is boring and hard work. To automatically identify SF on the ionogram and extend the study of SF into the middle and low latitudes of East Asia, this paper presents a statistical analysis of SF in this region, based on the naïve Bayesian classifier. The results showed that the accuracy of automatic identification reached up to 97% on both the validation datasets and test datasets composed of Mohe, I-Cheon, Jeju, Wuhan, and Sanya ionograms, suggesting that it is a promising way to automatically identify SF on ionograms. Based on the classification results, the statistical analysis shows that SF has a complicated morphology in the middle and low latitudes of East Asia. Specifically, there is a peak of occurrence of SF in the summer in I-Cheon, Jeju, Sanya, and Wuhan; however, the Mohe station has the highest occurrence rate of SF in December. The different seasonal variations of SF might be due to the different geographic local conditions, such as the inland-coastal differences and formation mechanism differences at these latitudes. Moreover, SF occurs more easily in the post-midnight hours when compared with the pre-midnight period in these stations, which is consistent with the previous results. Furthermore, this paper extracts the frequency SF (FSF) index and range SF (RSF) index to characterize the features of SF. The results shows that the most intense FSF/RSF appeared in the height range of 220–300 km/1–7 MHz in these stations, although there are different magnitude extensions on different season in these regions. In particular, strong spread-F (SSF) reached its maximum at the equinox at Sanya, confirming the frequent SSF occurrence at the equinox at the equator and low latitudes. These results would be helpful for understanding the characteristics of SF in East Asia.

Keywords: ionosphere spread-F; naive Bayes classifier; East Asia

Citation: Feng, J.; Zhang, Y.; Gao, S.; Wang, Z.; Wang, X.; Chen, B.; Liu, Y.; Zhou, C.; Zhao, Z. Statistical Analysis of SF Occurrence in Middle and Low Latitudes Using Bayesian Network Automatic Identification. *Remote Sens.* **2023**, *15*, 1108. https://doi.org/10.3390/rs15041108

Academic Editor: Fabio Giannattasio

Received: 11 January 2023
Revised: 15 February 2023
Accepted: 15 February 2023
Published: 17 February 2023

1. Introduction

Spread-F (SF) is one of the most prominent manifestations of ionospheric plasma instability, which is able to influence the performance and reliability of electronic systems, such as satellite, radar, navigation, and communication systems [1,2]. Therefore, it has been a hot topic in space research for several decades. The spatial and temporal characteristics of SF are complicated with time scales ranging from seconds to hours and length scales ranging from centimeters to tens of kilometers. The equatorial spread-F (ESF), which is also nominated as a convective equatorial ionospheric storm (CEIS), has been intensively investigated for more than eight decades since the first observation made by Booker and Wells [3–18]. Rayleigh-Taylor instability (RTI) theory is considered to play a crucial role in the generation of ESF [19–26]. Previous studies demonstrated that factors including electric field, neutral wind, atmospheric gravity wave (AGW), and horizontal gradient all contribute to the ESF onset condition and diverse characteristics [27–32]. Moreover, the geographic location, solar activity, geomagnetic activity, and some other factors might also contribute to the generation and evolution of SF [33–36]. Therefore, the spatial and temporal variation of SF is so complicated that none of the theories that have been put forward could explain them completely, and they need further studies.

In order to detect and characterize SF, different instruments have been used, such as ionosondes, airglow imagers, coherent/incoherent scatter radars, global positioning satellite (GPS) receivers, and rockets [37–39]. Among them, ionosondes provide the most extensive coverage and datasets such as ionograms have been accumulated for nearly one hundred years. Therefore, this provides a convenient way to study the diurnal and seasonal characteristics of SF. However, this requires a considerable expense for the manager to identify SF from a large number of ionograms. To solve this problem, automatic scaling of ionograms has been proposed for more than several decades. Previous studies generally focus on the automatic extraction of the ionospheric parameters from the ionograms, whereas few researchers have studied the automatic identification of SF on ionograms [40–45]. Pillat et al. [46] proposed a method to automatically identify SF on ionograms recorded by the Canadian Advanced Digital Ionosonde (CADI) and the empirical threshold required in the approach to determine whether SF is present or not, resulting in the identification being complicated. In addition, Lan et al. [47] applied the machine learning method of decision trees (DTs) to the automatic identification of SF on ionograms, and the accuracy of identification of SF reached up to 89%. Furthermore, Lan et al. [48] made a comparative study of decision trees, random forests, and convolutional neural networks (CNNs) for spread-F identification, and the results show that CNN is the best method for classifying the performance among the three methods, although it demands a longer training time. Scotto et al. [49] proposed a method to detect ESF automatically and the accuracy rate is 81.55% for SF events. In addition, Rao et al. [50] proposed an auto-detection methodology of sporadic E (Es) and SF events, and the accuracies of the technique were 96.71%, 89.71%, and 93.39% for the detection of Es, range SF (RSF), and strong SF (SSF). Therefore, to identify SF more accurately and extract the related ionospheric parameters, an application of the machine learning approach is necessary.

The mid- and low-latitude SF has been investigated for decades, and the results showed that the SF demonstrated an apparent longitudinal variation [51,52]. Among them, SF in East Asia has attracted much interest, since the low geomagnetic latitude and longitudinal variation nears the $0°$ magnetic declination in East Asia, and the investigation can be divided into two types: single-station investigation and regional investigation. For example, Wang et al. [53] studied the seasonal variation of SF in Hainan (19.4°N, 109°E); Lan et al. [54] investigated the features of SF in Puer (22.7°N, 101.05°E). As for the regional studies, Guo et al. [55] analyzed the SF in Hainan (19.4°N, 109°E), Changchun (43.8°N, 125.3°E), and Urumqi (43.7°N, 87.6°E) and they conducted a comparison study of SF occurrence rates in low and high solar activity years; Wang et al. [56] studied the longitudinal differences of SF at mid-latitude in East Asia; Lan et al. [57] investigated the latitudinal difference in SF in mid and low latitudes based on Chiang Mai (22.7°N, 101.05°E), Puer (22.7°N, 101.05°E), and Leshan (29.6°N, 103.7°E) ionosonde data at Asia longitude sector in the descending phase of the 24th solar cycle. Although the SF in East Asia has been studied by many researchers, the statistical characteristics of SF at other stations, such as the mid-high latitude station of Mohe, might be different due to the different regions and their mechanisms. Therefore, the statistical characteristics of SF in East Asia still need to be further investigated.

In recent years, machine learning has been widely used in the classification, identification, and forecasting in the fields of remote sensing and ionospheric research [58–62]. Among the machine learning algorithms, the naive Bayes classifier is an efficient and effective algorithm based on Bayes' theorem and class conditional independence. In order to identify SF effectively and extend the study of SF in East Asia, this paper presents a statistical analysis of SF occurrence in middle and low latitudes of East Asia based on Bayesian network automatic identification. This study confirms the feasibility and effectiveness of the naive Bayes classifier and shows some new insights about SF in middle and low latitudes. This paper is arranged as follows: Section 2 presents the data and methods used in the work. Section 3 gives a brief introduction of the naive Bayes classifier and the processes of training Bayesian classifiers. The results are shown in Section 4. Section 5 discusses the derived results. Section 6 summarizes the findings.

2. Materials

In this study, the ionograms used for the middle and low latitudes come from Mohe, I-Cheon, Jeju, Wuhan, and Sanya, and they can be accessed through the DIDBase database (http://ulcar.uml.edu/DIDBase, accessed on 16 February 2023). The geographical coordinates and geomagnetic latitudes of the five stations are listed in Table 1. In addition, Figure 1 shows the geographic location of these five ionosondes.

Table 1. The geographical coordinates of the selected stations.

Station	Geographic Latitude/°	Geographic Longitude/°	Geomagnetic Latitude/°
Mohe	52.00	122.52	42.68
I-Cheon	37.14	127.54	28.07
Jeju	33.43	126.30	24.31
Wuhan	30.50	114.40	20.99
Sanya	18.34	109.42	8.82

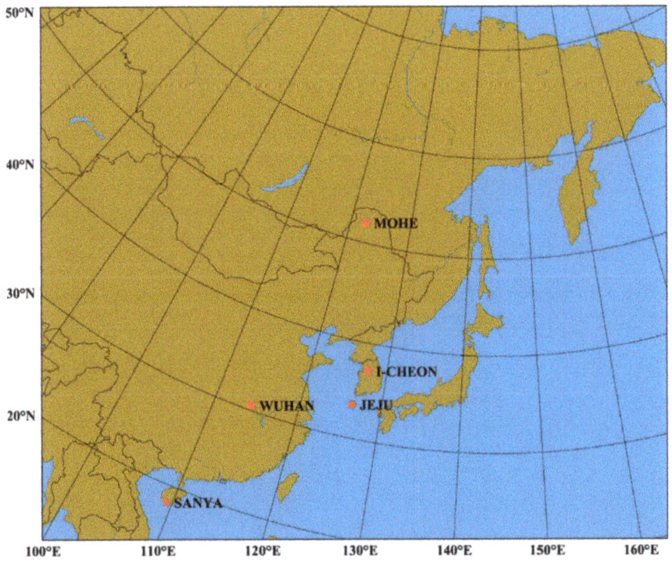

Figure 1. Geographical locations of the selected ionosondes.

Table 2 lists available data of the 5 ionosonde stations, and the diagonals in Wuhan represent that there is no missing data at Wuhan. As shown in Table 2, the data of I-Cheon and Jeju ionosondes are in the period from 2009 to 2013, which are mainly distributed in low solar activity periods, and the data of Mohe, Wuhan, and Sanya ionosondes are in the period from 2013 to 2018 which covers both the high and low solar activity years. Thus, the data availabilities are different for the different ionosondes. In addition, we randomly selected 212/185/191/220/195 samples from the datasets with SF and 212/185/191/221/195 from the dataset without SF in Mohe/I-Cheon/Jeju/Wuhan/Sanya for training, respectively. Moreover, we randomly selected 100/100 samples with/without SF in each of these five stations for testing.

Table 2. Data availability of the selected ionosonde stations during the period of 2009–2018.

Station	Data Period	Data Missing Period
Mohe	2013–2018	January~August in 2013, May~December in year 2018
I-Cheon	2010–2013	November and December in 2013
Jeju	2009–2013	July, November, and December in 2013
Wuhan	2013–2018	/
Sanya	2013–2017	August in 2013, December in 2016, January, March, May, and June in 2017

It is of great importance to note that the Es and the related blanketing multiple reflections could seriously affect the identification of F layers [46]. To eliminate the effect of second-hop Es layers which overlaps the F layer, we applied the method developed by Jiang et al. [63] to detect Es layers and to obtain the virtual height of the Es layer. The Es layer is first automatically scaled, then, the second-hop Es layer is removed using the virtual height parameter of Es layer in this paper. Moreover, we used the method proposed by Lan et al. [48] to filter out the second-hop F layer before training.

3. Methods

3.1. The Naive Bayes Classifier Introduction

The naive Bayes classifier is an efficient and effective machine learning algorithm based on Bayes' theorem and class conditional independence [64]. Bayes' theorem is described as:

$$P(c|x) = \frac{P(x|c)P(c)}{P(x)} \tag{1}$$

where $P(c)$ is the prior probability of c or the probability of the occurrence of class c, $P(x)$ is the independent probability of x or the probability of instance x occurring. $P(c|x)$ denotes the posterior probability, which means the probability of instance x being in class c, and $P(x|c)$ denotes the conditional probability, which means the probability of producing instance x given class c.

For a given set of variables, $X = \{x_1, x_2, \ldots, x_n\}$, we need to know the posterior probability for class C_j among a set of possible outcomes $C = \{c_1, c_2, \ldots, c_n\}$. Because $P(x)$ is constant for all classes in C, we get:

$$p(C_j|x_1, x_2, \ldots, x_n) \propto p(x_1, x_2, \ldots, x_n|C_j)p(C_j) \tag{2}$$

Since all attributes are independent given the value of the class variable, the conditional probability can be decomposed as a product of terms:

$$p(X|C_j) \propto \prod_{k=1}^{n} p(x_k|C_j) \tag{3}$$

Thus, the posterior probability may be rewritten as:

$$p(C_j|X) \propto p(C_j)\prod_{k=1}^{n} p(x_k|C_j) \tag{4}$$

Based on Equation (4), a new case X is classified as C_j if the posterior probability reaches the highest probability.

3.2. Naive Bayesian Classifier Training and Features Extraction

3.2.1. Extract Feature Values

For each ionogram in the training data set, we compute the continuous points in the specified region as feature values. The specified region is defined as 1–16 Mhz in the frequency axis and 200–500 Km in the height axis, and the step frequency is 0.3 Mhz. Therefore, there should be 51 lines in the specified region. However, there are some ionograms which frequency do not exceed 16 Mhz, such as the 2 examples in Figure 2. In this situation, we take the feature values between 1 Mhz and the frequency maximum of the ionogram (14 MHz in Figure 2) as the input and set the feature value between the frequency maximum and 16 Mhz as 0. Taking Figure 2 as an example, we first obtain 39 feature values between 1–14 MHz, and then set the remaining 12 feature values to 0. Figure 2a shows an example of the ionogram without SF and Figure 2b demonstrates another example of the ionogram with SF for comparison. For each line, we search from the bottom of the line (the area whose y-coordinate is set to 200) to the top of the line. During the searching process, we only consider the continuous points, which means that if there are more than 10 white points (no echo trace) after the first no-white point (having echo trace) has been found, the search will stop, and the number of no-white points is the continuous point in this line. It is of great importance to note that the existence of echo trace is determined by comparing the difference between the maximum and minimum values in the array of each pixel (each pixel is composed of a ternary array): if the difference is larger than 50, echo tracing occurs; otherwise, there is no echo tracing in this pixel. After the feature values of an ionogram have been extracted, we give a label to the ionogram. If the SF was shown in this ionogram, the label is 'Y'. Otherwise, the label is 'N'. The feature values and the labels are both included in the processed data. All processed data are recorded in an Excel file.

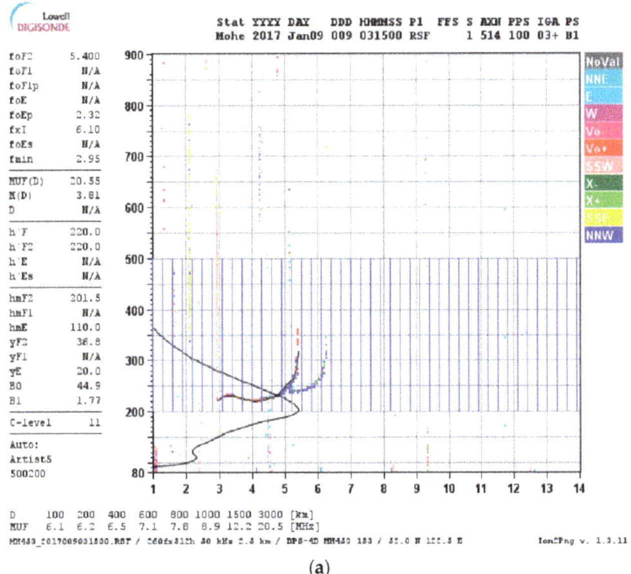

(a)

Figure 2. *Cont.*

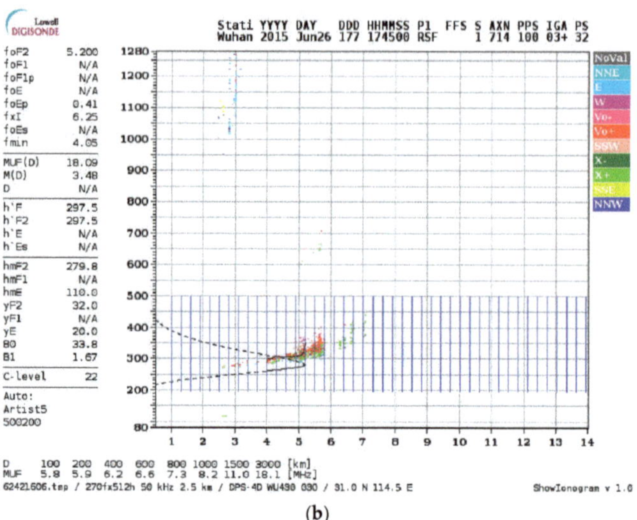

(b)

Figure 2. (**a**) An example of consecutive points (feature value) extraction positions in ionogram without SF in Mohe at 03:15UT on 9 January 2017. (**b**) An example of consecutive points (feature value) extraction position in ionogram with SF in Wuhan at 17:45UT on 26 June 2015.

3.2.2. Training Bayesian Classifier

After recording the processed data in the Excel file, we put the Excel file which contains the processed data into the Weka software. Weka is an efficient data mining software which can perform many data mining tasks and test new methods on data sets. The naive Bayesian classifier is one of the important classifiers in the Weka software. The input of the classifier is 51 feature values of an ionogram. If SF shows in this ionogram, the output of the classifier is 'Y'. If there is no SF, the output is 'N'. Figure 3 displays the interface for training naive Bayesian classifiers using Weka Explorer.

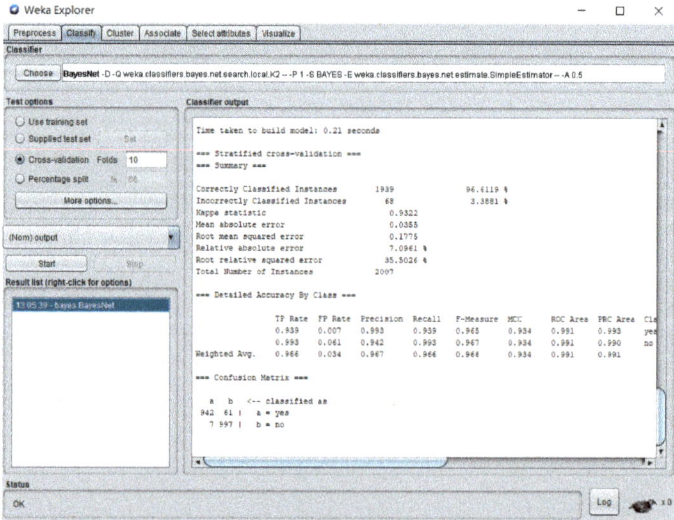

Figure 3. The interface of training naive Bayesian classifier using Weka Explorer.

3.2.3. Extract the Indexes of SF

According to the diffusion characteristics on the vertical incidence ionogram, SF echoes are usually classified into two types: frequency SF (FSF) and range SF (RSF). If the diffusion appears only along the horizontal part of the trace, then it is classified as FSF; if the diffusion is pronounced along the vertical height axis, SF is classified as RSF [65]. In this study, we extract two indexes to characterize the features of SF: one is called the FSF index, the other is called the RSF index. The classifier trained in step two was used to decide whether SF has shown. If there is no SF, the two indexes are set to zero; if there is SF, we used the following steps to obtain these two indexes.

The first step was to extract RSF index. In the process, this paper defined two variables, sum_point and number_point, and initialized them to zero. From the left side to the right side of the ionogram, we extracted the continuous points using the same method as step 2.3.1 described above and added up the numbers to obtain the sum_point. However, each extraction region was separated by 10 pixels in step 2.3.1. In this step, the pixels of each extraction region were adjacent. Every time the extraction region moved one unit forward, the number_point increased by one. However, if one extraction region had no continuous point, number_point remained the same. In the end, the RSF index was equal to number_point/sum_point.x.

Secondly, we extracted the FSF index. Most of the steps were same as the steps in extracting the RSF index. However, the approach to extract continuous points has changed. The extraction of RSF is from the right side to the left side of the ionogram, whereas the extraction of FSF is from the bottom side to the top side. Therefore, the search of continuous points began from the left side of the bottom line to its right side. Every time, the search region was raised by one pixel until the top blue line. Finally, FSF index was equal to the number_point/sum_point.y.

4. Results

There were a total of 2007 samples used for the training of naive Bayes classifier, of which 1003 samples were data with the occurrence of SF and the remaining 1004 samples were the data without SF. In total, 70% of the samples were used for training, and the remaining 30% samples were used for validation. Through the training, the correction rate of the trained Naive Bayes Classifier on the validation datasets was 97.51%. In addition, 1000 samples were selected as the test datasets, of which 500 samples were with the occurrence of SF and the remaining 500 samples were without SF. The correction rate of the trained naive Bayes classifier on the test dataset was 97.60%. Therefore, the naive Bayes classifier introduced an effective and accurate means to automatically identify the occurrence of SF from the ionograms, and it was more precise than previous methods.

Based on the identification results of the naive Bayes classifier, the dependence of SF occurrence rate on the local time and season was investigated to obtain an overall picture of SF distribution at these five stations. Note that the occurrence rate in each bin was defined as the ratio of number of SF to the total number of ionograms. Figure 4 illustrates the occurrence rate (%) of SF at the five stations of different latitudes in East Asia sector. In each subplot of Figure 4, the SF occurrence rate is binned as a function of the local time and month. It can be seen that there is an obvious day-night asymmetry of SF occurrence in these five stations. Specifically, SF basically occurs at local nighttime in these five stations and the occurrence rate of SF at Wuhan is evidently smaller than that at the other four stations. In the nighttime, the occurrence rate of SF also shows an asymmetry; SF occurred more easily in the post-midnight hours when compared with the pre-midnight period, and this feature was most/least evident in Sanya/Mohe, respectively. Moreover, the seasonal dependence showed some similarity and difference in these five stations: there was a peak of occurrence of SF in summer in I-Cheon, Jeju, Sanya, and Wuhan; however, the Mohe station had the highest occurrence rate of SF in December. In addition, the stations with the highest occurrence rate of SF in the summer seemed to have some seasonal differences in the SF incidence: I-Cheon station had the lowest occurrence rate in the winter, while the Jeju station had the lowest occurrence rate in the equinox. On the other hand, the occurrence rate was lowest from October to December at Sanya and there was almost no SF in the equinox and winter at Wuhan.

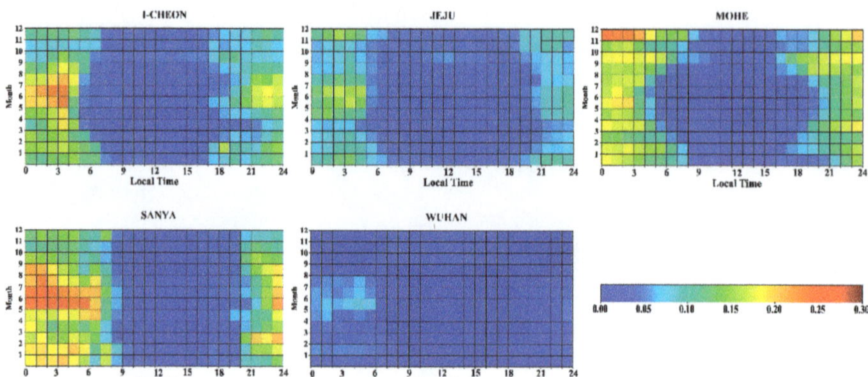

Figure 4. Occurrence rate of SF at Mohe, I-Cheon, Jeju, Wuhan, and Sanya according to the 0–24 LT and January–December division.

Furthermore, we calculated the FSF index and RSF index to characterize the features of SF according to the SF identified with the naive Bayes classifier. Figure 5 shows the FSF index distribution according to the divisions of local time 0:00–24:00LT and heights 200–500 km within a grid of 1 h × 20 km. It can be seen from Figure 5 that the FSF had different morphologies for different stations. Specifically speaking, this occurred mainly in the 220–340 km height in the summer, and most FSF occurred during the post-midnight hours at Mohe. By contrast, it distributed widely within 220–420 km range in the equinox and winter, and it was more symmetrical around midnight in the winter. In I-Cheon, the FSF was mainly distributed in 220–360 km, and it was also more symmetrical around 0:00UT in the winter when compared with the FSF in the equinox and summer. Moreover, it mainly appeared in 220–340 km in summer, while it was widely distributed within the 220–420 km range in the equinox and the 220–440 km range in the winter at Jeju. As for the local time distribution, it showed similar features as Mohe and I-Cheon, and most of the FSF in the summer was distributed in the post-midnight hours. In Wuhan, the FSF mainly occurred in 220–280 km/220–340 km/220–360 km heights in the summer, equinox, and winter, respectively. However, the local time distribution of the FSF was more symmetrical around midnight in the summer when compared with Mohe, I-Cheon, and Jeju. In Sanya, the FSF was mainly distributed in 220–460 km in the summer, which covered a larger range than the four stations above. Moreover, it was distributed mainly in 220–460 km/220–440 km in the equinox and winter, respectively, and the local time distribution was almost symmetrical in the full year. In addition to the above similarities and differences, FSF also had one important common feature in these stations, that is, the most intense FSF tended to appear in the 220–300 km range, although there are different magnitude extensions in different stations. Figure 6 presents the RSF index according to the division of 0:00–24:00 LT and 1–16 MHz within a grid of 1 h × 1 MHz. It is clear that RSF shows different morphologies in the middle and low latitude station. In Mohe, the RSF mainly occurred in the 1–7 MHz interval, and it mainly appeared in the nighttime in the summer, whereas it could also occur in the daytime in the equinox and winter. In I-Cheon, the RSF frequently occurred in 1–12 Mhz in the summer, and it mainly occurred in 1–9 Mhz in the equinox and winter. As for the local time distribution, most of the RSF in the summer is distributed in post-midnight hours and it is more symmetrical around 0:00 LT in the equinox and winter. In Jeju, it mainly occurred in 1–12 Mhz in the summer and 1–8 Mhz in the equinox and winter. Moreover, there are also apparent FSF in the daytime at I-Cheon and Jeju; however, the most intense RSF still occurred during the night-time. In Wuhan, the frequency spread mainly occurred in 1–4 Mhz/1–5 Mhz/1–6 Mhz in the summer/winter/equinox, respectively. By contrast, the RSF index is widely distributed in 1–13 Mhz/1–14 Mhz/1–15 Mhz in the summer/winter/equinox at Sanya, indicating that they are strong Spread-F (SSF) in Sanya. In short, the frequency spread covers the largest frequency range in Sanya and smallest frequency range at Wuhan, and the local time asymmetrical around 0:00 LT is most evident in the summer.

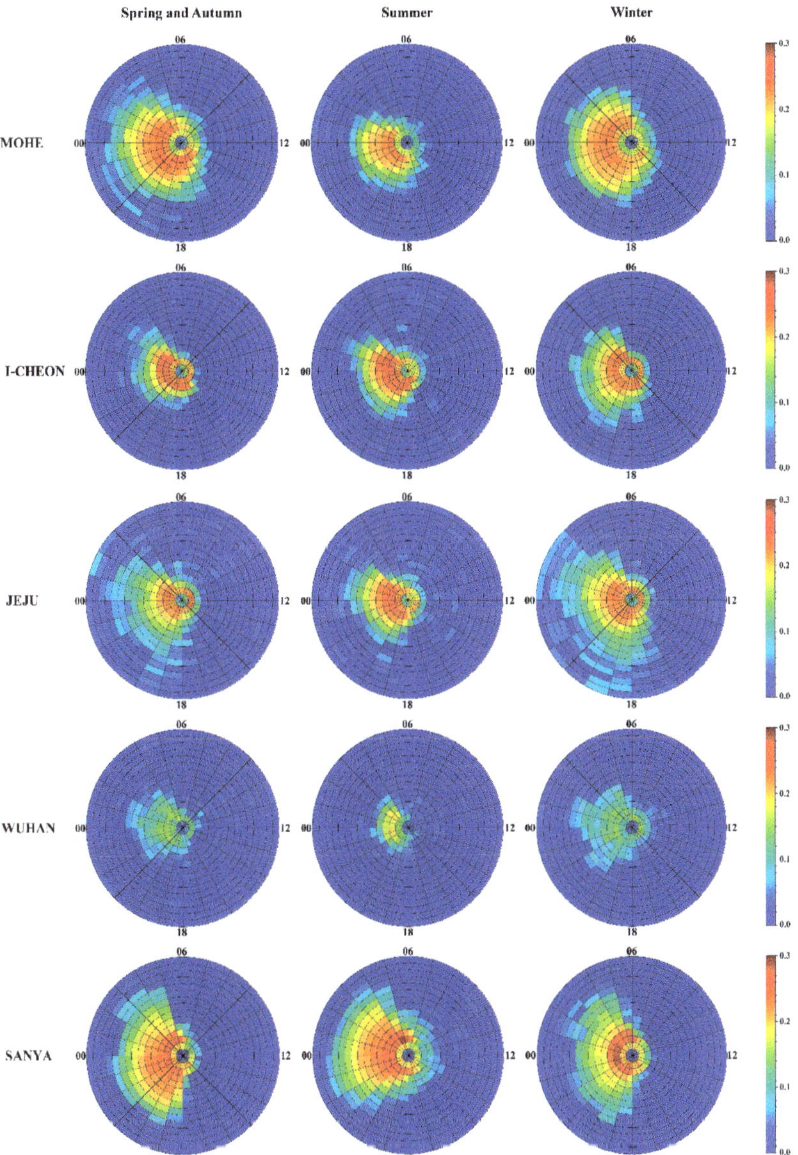

Figure 5. Frequency spread-F (FSF) index distribution in summer, winter, and equinox (spring and autumn) in Mohe, I-Cheon, Jeju, Wuhan, and Sanya, which is binned into the function of local time 0:00–24:00 LT and heights 200–500 km within a grid of 1 h × 20 km.

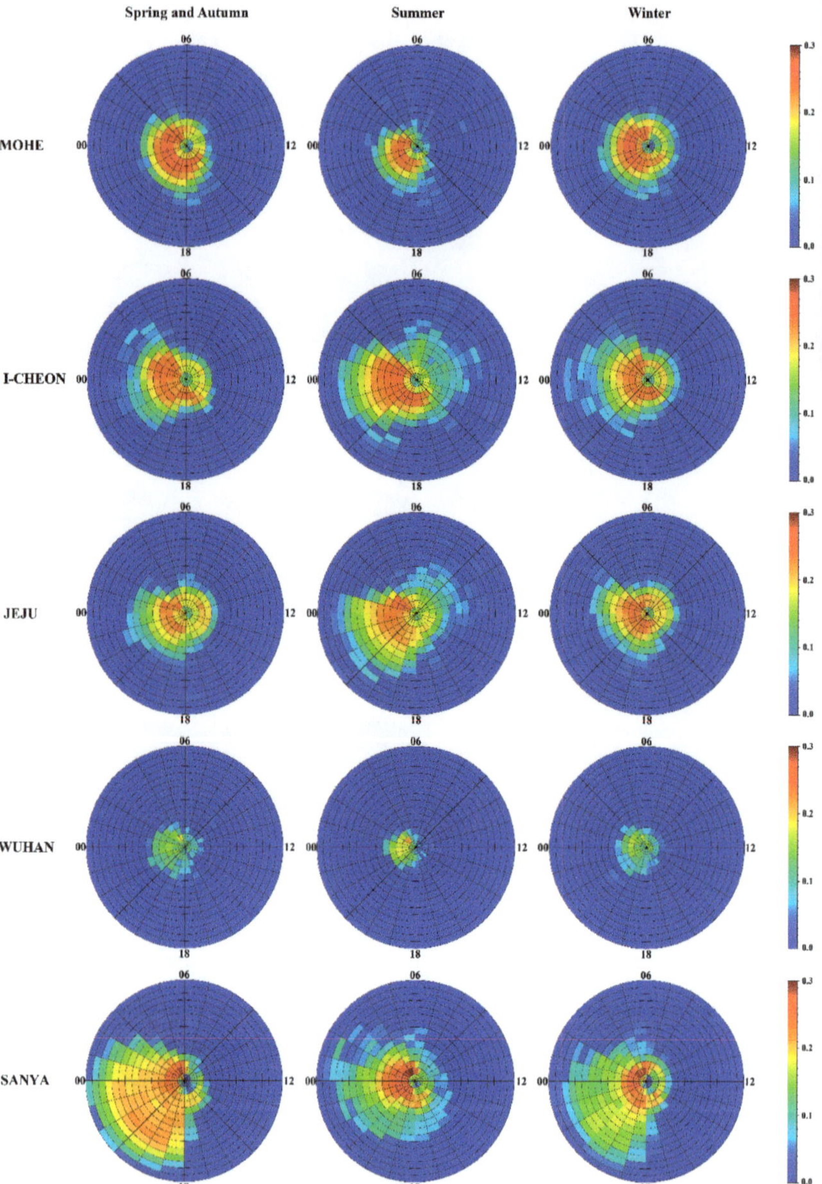

Figure 6. Seasonal distribution of RSF index in Mohe, I-Cheon, Jeju, Wuhan, and Sanya within the same grid as Figure 5.

5. Discussion

The automatic identification of SFs is one of the most challenging tasks in the study of ionospheric irregularities. With the improvement of computer capability and the development of artificial intelligence, more and more machine learning methods such as decision trees (DTs) and long-short temporal memory (LSTM) are applied to the ionosphere research [58–62]. The naive Bayes classifier is an efficient and effective algorithm based on Bayes' theorem and class conditional independence, and we applied it to the automatic identification of SF on the ionogram. The result shows that the accuracy rate of the classifier is 97.51% on the validation dataset, whereas the accuracy rate reached 97.6% on the test dataset. This means that the naive Bayes classifier proposed in this study had no difficulty in classing the ionograms in different geographic locations and performed very well in the identification of SF. However, there are wrong identifications of the ionograms. Pillat et al. [46] have proposed that the possible root causes for spread-F detection failure are bad quality ionograms, the presence in ionogram of E-sporadic (if the E layer second reflection overlaps with the F layer), presence of a satellite F trace, and the presence of a strong noise or lumped points near the F trace. During the pre-processing of training ionograms, we have eliminated the Es second-hop and the second-hop F layer. Therefore, the failure of SF detection could be related to other factors such as bad quality ionograms or lumped points near the F trace. In the case of bad quality ionograms, the F traces were difficult to identify, which makes the method underestimate SF occurrence. In the case of strong noise or lumped points near the F trace, it could make the automatic methods mistake the ionogram as SF type and overestimate SF occurrence. In the future, we would conduct more tests to improve this Bayesian classifier.

Based on the proposed Bayes classifier, this paper conducts the statistical study of SF in five stations: Mohe, I-Cheon, Jeju, Wuhan, and Sanya. The results show that SF has a complicated morphology in different stations. In general, SF occurred most/least frequently in Sanya/Wuhan, and SF mainly occurred in the nighttime. Previous studies have shown that SF has a maximum occurrence rate in the equator and low latitude, and the stations around 31°N have the lowest occurrence rates of SF [56,66]. Sanya is located at equatorial low latitudes, and the occurrence of SF is therefore very high when compared with the mid-latitude station. By contrast, Wuhan is located in the mid latitudes of East Asia with 30.5° latitude and has the lowest occurrence rate of SF. Therefore, the results confirm previous studies' conclusions. In addition, SF occurred most frequently in the summer in I-Cheon, Jeju, Sanya, and Wuhan; however, the Mohe station has the highest occurrence rate of SF in December. Previous studies found that SF has the highest occurrence rate in summer in mid and low latitudes [51,67,68], and SF has two peaks in the summer and winter in low solar activity years [56,69,70]. It is important to note that the latitude of Mohe is 52°N, a latitude higher than most of the mid-latitude stations. Previous studies have confirmed that the latitudinal variation of SF characteristics could be large due to the different geographic local conditions and formation mechanisms [51,56]. Mohe is located inland, whereas Jeju and I-Cheon are in coastal or marine areas. As one of the suggested contributors to the formation of mid-latitude SF, AGW in the ionosphere mostly comes from the lower atmosphere [3]. The meteorological or ground conditions below the ionosphere are different, so there should be some regional characteristics of SF for the inland and coastal ionosphere. Therefore, SF displays a different seasonal variation at Mohe.

SF is generally categorized into two types, frequency SF and range SF, and the echo of FSF/RSF distributes spreading along the frequency/vertical height axes, respectively. Previous studies have shown that these two kinds of SF have different morphologies even in the same place for that the formation mechanisms of them that are different. According to the classification results of the proposed naive Bayes classifier, this paper calculated the FSF index and RSF index. Figure 5 shows that the most intense FSF appeared in the height range of 220–300 km in these stations, although there are different magnitude extensions in different regions. Devasia [71] suggested that 300 km is the threshold of h'F, which has a close relationship with SF, although the data used in their research was from North America. Thus, this feature shown in this study is basically consistent with previous studies. Moreover, the FSF displays different seasonal and local time characteristics in these stations. In Mohe, most FSF occurs during the post-midnight hours in summer, whereas it distributed more symmetrical around midnight in winter. The similar situation applies for the FSF in I-Cheon and Jeju. Wang et al. [56] found that the FSF in the midlatitude of East Asia mainly occurred in the post-midnight period. Lan et al. [57] also drew this conclusion based on the data for three ionosondes in the mid and low latitudes of East Asia, consistent with this study. However, the FSF shows a different morphology in Sanya: it is more symmetrical around midnight in the summer. Sanya is located at equatorial low latitudes, therefore the formation mechanisms of SF is different from other midlatitude station. Although the RSF discussed below is closely related to the equator SF (ESF) and the equator plasma bubble (EPB), the FSF could generated by local ionospheric disturbances induced by atmospheric gravity waves [72]. In addition, it might also be affected by the medium-scale traveling ionospheric disturbances (MSTIDs) induced by Perkins instability during the nighttime [73]. Due to the limitation of data, the generation mechanisms of FSF during pre-midnight period need to be further studied.

Figure 6 shows that the RSF appeared in the frequency range of 1–7 Mhz, although there are different magnitude extensions in different regions. In particular, the RSF covers a large frequency range of 1–13 Mhz in Sanya, and this phenomenon is most apparent for the RSF in the equinox, indicating that the strong RSF occurred most frequently in the equinox in Sanya. Previous studies have defined the strong SF (SSF) as the frequency spread extended to a high frequency (usually greater than 8 MHz) in the ionogram, and they reported that SSF occurred most frequently in the equator and low latitudes [74]. Sanya is located at the equatorial low latitudes and the irregularities in this region have close relationships with the ESF and EPB, which have large-scale plasma density structures, therefore SSF has a maximum occurrence rate at Sanya among these five stations. In addition, SSF occurs most frequently in the equinox at Sanya, consistent with previous results [36,54]. Except the frequent occurrence at night, the RSF also appeared in the daytime in these stations. Li.et al. [75] and Ki et al. [76] suggested that the daytime SF could be a continuation of the nighttime EPB which spread into the low and mid-latitude regions with the ionospheric fountain effect. In addition, the AGW might also be another possible reason for the generation of daytime SF [77,78]. Note that the traveling ionospheric disturbance (TID) induced by typhoon and stronger eastward electric field during the geomagnetic storms might also be the driver of daytime SF [79,80]. As there are many factors such as AGW, Perkins instability, geomagnetic inclination/declination, etc. that could contribute to SF formation, more multi-instrument observations and simulations are needed to provide a statistical picture and study the physical mechanisms in the future. Moreover, the local time asymmetrical around 0:00 LT is most evident in the summer in Mohe, I-Cheon, and Jeju. Previous studies also found that the occurrence rates of mid-latitude SF reached their maximum in the midnight to post-midnight periods during the June solstice, especially at th solar minimum [69,70]. The data used for I-Cheon (2010–2013) and Jeju (2009–2013) are both the data of low solar activity years, and the data used for Mohe also includes the low solar activity years (2013–2018). Therefore, the results confirmed previous conclusions. Note that the FSF index and RSF index in this paper do not represent the actual occurrence of FSF/RSF, it just characterizes the possibility of the occurrence of these two types of SF in the division grid. Therefore, further investigation about the statistical characteristics and mechanisms of these two types of spread-F occurrences are still needed.

6. Conclusions

To solve the problem of the automatic identification of SF on ionograms, this paper presents a method based on naive Bayes classifier. After extracting 51 features on the specified region (1–16 MHz, 200–500 km) on ionograms, we train the Bayesian classifier using Weka Explorer. The results showed that the accuracy of this classifier reached 97% on the Mohe, I-Cheon, Jeju, Wuhan, and Sanya datasets, suggesting it is a promising way to automatically identify SF on ionograms.

Based on the classification results, we made a statistical study of SF in these five stations to expand the research of SF in East Asia. The major conclusions are summarized as follows:

1. Our results showed that SF mainly occurs at the local nighttime and the occurrence rate of SF at Wuhan is evidently smaller than that at the other four stations. In addition, the occurrence rate of SF also shows an asymmetry in the nighttime: SF occurred

more easily in the post-midnight hours when compared with the pre-midnight period, consistent with previous studies.

2. There is a peak of occurrence of SF in the summer in I-Cheon, Jeju, Sanya, and Wuhan; however, the Mohe station has the highest occurrence rate of SF in December. The different seasonal variations of SF might be due to the different geographic local conditions and formation mechanisms at these latitudes: Mohe is located inland at a mid-high latitude, so AGW which comes from the lower atmosphere should be different from that of other stations, contributing to the different seasonal characteristics of SF in these stations.

3. The most intense FSF appeared in the height range of 220–300 km in these stations, although there are different magnitude extensions in different region. Most FSF occurs during the post-midnight hours in the summer and is distributed more symmetrical around midnight in the winter in Mohe, I-Cheon, and Jeju. By contrast, it is more symmetrical around midnight in the summer at Sanya. As Sanya is located in the equatorial low latitudes, local ionospheric disturbances induced by atmospheric gravity waves and medium-scale traveling ionospheric disturbances (MSTIDs) induced by Perkins instability during the nighttime might contribute to the local time difference of SF in the summer.

4. The RSF mainly appeared in the frequency range of 1–7 Mhz with different magnitude extensions in different regions. In particular, the RSF covers a large frequency range of 1–13 Mhz in Sanya, and this phenomenon is most apparent in the equinox, indicating the frequently SSF occurrence in the equinox at Sanya. Moreover, the occurrence rate of mid-latitude SF reached its maximum in the midnight to post-midnight periods, confirming the previous results.

5. SF also appeared in the daytime in these stations, and AGW, TID, geomagnetic inclination/declination, etc. might contribute to the daytime SF formation. Due to the limitation of data, multi-instrument observations and simulations are needed to provide a statistical picture and to study the physical mechanisms of daytime SF in the future.

Author Contributions: Conceptualization, Y.Z. and J.F.; methodology, Z.Z.; investigation, C.Z. and Z.W.; validation, X.W. and B.C.; formal analysis, S.G. and Y.L.; resources, J.F. and S.G.; visualization, J.F. and Y, Z; funding acquisition, C.Z and X.W. All authors have read and agreed to the published version of the manuscript.

Funding: This work was supported by the National Key Research and Development Program of China (NO.2021YFC2802502), the Stable-Support Scientific Project of China Research Institute of Radio Wave Propagation (Grant No. A132102W06), the Hubei Natural Science Foundation (Grant No. 2022CFB651), the National Natural Science Foundation of China (NSFC Grant No. 42204161, 42074187), the Foundation of National Key Laboratory of Electromagnetic Environment (Grant No. 20200101), and the Excellent Youth Foundation of Hubei Provincial Natural Science Foundation (Grant No. 2019CFA054).

Data Availability Statement: Data sharing is not applicable to this article.

Acknowledgments: The authors would like to gratefully thank the anonymous reviewers for their insight and help. We appreciate the support by "The Young Top-notch Talent Cultivation Program of Hubei Province".

Conflicts of Interest: The authors declare no conflict of interest.

References

1. Booker, H.G. Turbulence in the ionosphere with applications to meteor trails, radio-star scintillation, auroral radar echoes, and other phenomena. *J. Geophys. Res.* **1956**, *61*, 673–705. [CrossRef]
2. Ratcliffe, J.A. *An Introduction to the Ionosphere and Magnetosphere*; Cambridge University Press: Cambridge, UK, 1972.
3. Booker, H.G.; Wells, H.W. Scattering of radio waves in the F region of ionosphere. *Terr. Magn. Atmos. Electr.* **1938**, *43*, 249. [CrossRef]

4. Farley, D.T.; Balsley, B.B.; Woodman, R.F.; McClure, J.P. Equatorial SF, implications of VHF radar observations. *J. Geophys. Res.* **1970**, *75*, 7199–7216. [CrossRef]
5. Sobral, J.H.A.; Abdu, M.A.; Zamlutti, C.J.; Batista, I.S. Association between plasma bubble irregularities and airglow disturbances over Brazilian low latitudes. *Geophys. Res. Letts.* **1980**, *1*, 980–982. [CrossRef]
6. Abdu, M.A.; Batista, I.S.; Sobral, J.H.A. A new aspects of magnetic declination control on equatorial spread F and F region dynamo. *J. Geophys. Res.* **1992**, *97*, 14897–14904. [CrossRef]
7. Abdu, M.A. Major phenomena of the equatorial ionosphere–thermosphere system under disturbed conditions. *J. Atmos. Sol. Terr. Phys.* **1997**, *13*, 1505–1519. [CrossRef]
8. Abdu, M.A. Outstanding problems in the equatorial ionosphere– thermosphere system relevant to spread F. *J. Atmos. Sol. Terr. Phys.* **2001**, *63*, 869–884. [CrossRef]
9. Abdu, M.A.; Kherani, E.A.; Batista, I.S.; Sobral, J.H.A. Equatorial evening prereversal vertical drift and spread F suppression by disturbance penetration electric fields. *Geophys. Res. Lett.* **2009**, *36*, L19103. [CrossRef]
10. Huang, C.-S.; Kelley, M.C. Nonlinear evolution of equatorial spread F:1 On the role of plasma instabilities and spatial resonance associated with gravity wave seeding. *J. Geophys. Res.* **1996**, *98*, 15631–15642. [CrossRef]
11. Huang, C.-S.; Foster, J.C.; Kelley, M.C. Long-duration penetration of the interplanetary electric field to the low-latitude ionosphere during the main phase of magnetic storms. *J. Geophys. Res.* **2005**, *110*, A11309. [CrossRef]
12. Fejer, B.G.; Scherliess, L.; de Paula, E.R. Effects of the vertical plasma drift velocity on the generation and evolution of equatorial spread F. *J. Geophys. Res.* **1999**, *104*, 19854–19869. [CrossRef]
13. Huba, J.D.; Joyce, G.; Krall, J. Three-dimensional equatorial spread F modeling. *Geophys. Res. Lett.* **2008**, *35*, L10102. [CrossRef]
14. Tsunoda, R.T. On equatorial spread F: Establishing a seeding hypothesis. *J. Geophys. Res.* **2010**, *115*, A12303. [CrossRef]
15. Yokoyama, T.; Shinagawa, H.; Jin, H. Nonlinear growth, bifurcation, and pinching of equatorial plasma bubble simulated by three-dimensional high-resolution bubble model. *J. Geophys. Res. Space Phys.* **2014**, *119*, 10474–10482. [CrossRef]
16. Hysell, D.L.; Milla, M.; Kuyeng, K. Radio Beacon and Radar Assessment and Forecasting of Equatorial F Region Ionospheric Stability. *J. Geophys. Res. Space Phys.* **2019**, *124*, 9511–9524. [CrossRef]
17. Wu, K.; Xu, J.Y.; Zhu, Y.J.; Yuan, W. Occurrence characteristics of branching structures in equatorial plasma bubbles: A statistical study based on all-sky imagers in China. *Earth Planet. Phys.* **2021**, *5*, 407–415. [CrossRef]
18. Mohandesi, A.; Knudsen, D.; Skone, S.; Langley, R.; Yau, A. Altitude Distribution of Large and Small-Scale Equatorial Ionospheric Irregularities Sampled from an Elliptical Low-Earth Orbit. *J. Geophys. Res. Space Phys.* **2022**, *127*, e2021JA030104. [CrossRef]
19. Dungey, J.W. Convective diffusion in the equatorial F region. *J. Atmos. Terr. Phys.* **1956**, *9*, 304–310. [CrossRef]
20. Fejer, B.G.; Kelley, M.C. Ionospheric irregularities. *Rev. Geophys.* **1980**, *18*, 401–454. [CrossRef]
21. Sultan, P.J. Linear theory and modeling of the Rayleigh-Taylor instability leading to the occurrence of equatorial spread F. *J. Geophys. Res.* **1996**, *101*, 26875. [CrossRef]
22. Huba, J. Generalized Rayleigh-Taylor Instability: Ion Inertia, Acceleration Forces, and E Region Drivers. *J. Geophys. Res. Space Phys.* **2022**, *127*, e2022JA030474. [CrossRef]
23. Woodman, R.F.; Hoz, C.L. Radar observations of F region equatorial irregularities. *J. Geophys. Res.* **1976**, *81*, 5447–5466. [CrossRef]
24. Singh, S.; Johnson, F.S.; Power, R.A. Gravity wave seeding of equatorial plasma bubbles. *J. Geophys. Res.* **1997**, *102*, 7399–7410. [CrossRef]
25. Tsunoda, R.T. Satellite traces: An ionogram signature for large-scale wave structure and a precursor for equatorial spread F. *Geophys. Res. Lett.* **2008**, *35*, L20110. [CrossRef]
26. Aveiro, H.C.; Hysell, D.L.; Park, J.; Lühr, H. Equatorial spread F related currents: Three-dimensional simulations and observations. *Geophys. Res. Lett.* **2011**, *38*, L21103. [CrossRef]
27. Perkins, F. Spread F and ionospheric currents. *J. Geophys. Res.* **1973**, *78*, 218–226. [CrossRef]
28. Bowman, G.G. A review of some recent work on mid-latitude spread-F occurrence as detected by ionosondes. *J. Geomagn. Geoelectr.* **1990**, *42*, 109–138. [CrossRef]
29. Yokoyama, T.; Yamamoto, M.; Fukao, S.; Cosgrove, R.B. Three dimensional simulation on generation of polarization electric field in the midlatitude E-region ionosphere. *J. Geophys. Res.* **2004**, *109*, A01309. [CrossRef]
30. Balan, N.; Otsuka, Y.; Nishioka, M.; Liu, J.Y.; Bailey, G.J. Physical mechanisms of the ionospheric storms at equatorial and higher latitudes during the recovery phase of geomagnetic storms. *J. Geophys. Res.* **2013**, *118*, 2660–2669. [CrossRef]
31. Zhou, C.; Tang, Q.; Huang, F.Q.; Liu, Y.; Gu, X.D.; Lei, J.H.; Ni, B.B.; Zhao, Z.Y. The simultaneous observations of nighttime ionospheric E region irregularities and F region medium-scale traveling ionospheric disturbances in midlatitude China. *J. Geophys. Res.* **2018**, *123*, 5195–5209. [CrossRef]
32. Liu, Y.; Zhou, C.; Xu, T.; Wang, Z.K.; Tang, Q.; Deng, Z.X.; Chen, G.Y. Investigation of midlatitude nighttime ionospheric E-F coupling and interhemispheric coupling by using COSMIC GPS radio occultation measurements. *J. Geophys. Res.* **2020**, *125*, e2019JA027625. [CrossRef]
33. Dabas, R.S.; Das, R.M.; Sharma, K.; Garg, S.C.; Devasia, C.V.; Subbarao, K.S.V.; Niranjan, K.; Rama Rao, P.V.S. Equatorial and low latitude spread-F irregularity characteristics over the Indian region and their prediction possibilities. *J. Atmos. Sol. Terr. Phys.* **2007**, *69*, 685–696. [CrossRef]
34. Hoang, T.L.; Abdu, M.A.; MacDougall, J.; Batista, I.S. Longitudinal differences in the equatorial spread F characteristics between Vietnam and Brazil. *Adv. Space Res.* **2010**, *45*, 351–360. [CrossRef]

35. Tulasi Ram, S.; Rama Rao, P.V.S.; Prasad, D.S.V.V.D.; Niranjan, K.; Gopi Krishna, S.; Sridharan, R.; Ravindran, S. Local time dependent response of postsunset ESF during geomagnetic storms. *J. Geophys. Res.* **2008**, *113*, A07310. [CrossRef]
36. Balan, N.; Liu, L.; Le, H. A brief review of equatorial ionization anomaly and ionospheric irregularities. *Earth Planet. Phys.* **2018**, *2*, 1–19. [CrossRef]
37. King, G.A. Spread-F on ionograms. *J. Atmos. Terr. Phys.* **1970**, *32*, 209–221. [CrossRef]
38. Su, S.Y.; Yeh, H.C.; Heelis, R.A. ROCSAT 1 ionospheric plasma and electrodynamics instrument observations of equatorial spread F: An early transitional scale result. *J. Geophys. Res.* **2001**, *106*, 29513–29519. [CrossRef]
39. Jin, H.; Zou, S.; Chen, G.; Yan, C.; Zhang, S.; Yang, G. Formation and Evolution of Low-Latitude F Region Field-Aligned Irregularities During the 7-8 September 2017 Storm: Hainan Coherent Scatter Phased Array Radar and Digisonde Observations. *Space Weather* **2018**, *16*, 648–659. [CrossRef]
40. Reinisch, B.W.; Huang, X. Automatic calculation of electron density profiles from digital ionograms: 3. Processing of bottomside ionograms. *Radio Sci.* **1983**, *18*, 477–492. [CrossRef]
41. Pezzopane, M.; Scotto, C. Automatic scaling of critical frequency foF2 and MUF(3000)F2: A comparison between Autoscala and ARTIST 4.5 on Rome data. *Radio Sci.* **2007**, *42*, RS4003. [CrossRef]
42. Jiang, C.; Yang, G.; Zhao, Z.; Zhang, Y.; Zhu, P.; Sun, H. An automatic scaling technique for obtaining F2parameters and F1critical frequency from vertical incidence ionograms. *Radio Sci.* **2013**, *48*, 739–751. [CrossRef]
43. Jiang, C.; Yang, G.; Zhou, Y.; Zhu, P.; Lan, T.; Zhao, Z.; Zhang, Y. Software for scaling and analysis of vertical incidence ionograms-ionoScaler. *Adv. Space Res.* **2017**, *59*, 968–979. [CrossRef]
44. Pillat, V.G.; Guimaraes, L.N.F.; Fagundes, P.R. A computational tool for ionosonde CADI's ionogram analysis. *Comput. Geosci.* **2013**, *52*, 372–378. [CrossRef]
45. Pezzopane, M.; Pillat, V.; Fagundes, P. Automatic scaling of critical frequency foF2 from ionograms recorded at São José dos Campos, Brazil: A comparison between Autoscala and UDIDA tools. *Acta Geophys.* **2017**, *65*, 173–187. [CrossRef]
46. Pillat, V.G.; Fagundes, P.R.; Guimarães, L.N.F. Automatically identification of Equatorial Spread-F occurrence on ionograms. *J. Atmos. Sol. Terr. Phys.* **2015**, *135*, 118–125. [CrossRef]
47. Lan, T.; Zhang, Y.; Jiang, C.; Yang, G.; Zhao, Z. Automatic identification of Spread F using decision trees. *J. Atmos Sol.-Terr. Phys.* **2018**, *179*, 389–395. [CrossRef]
48. Lan, T.; Hu, H.; Jiang, C.; Yang, G.; Zhao, Z. A Comparative Study of Decision Tree, Random Forest, and Convolutional Neural Network for Spread-F Identification. *Adv. Space Res.* **2020**, *65*, 2052–2061. [CrossRef]
49. Scotto, C.; Ippolito, A.; Sabbagh, D. A method for automatic detection of equatorial spread-F in Ionograms. *Adv. Space Res.* **2019**, *63*, 337–342. [CrossRef]
50. Rao, T.V.; Sridhar, M.; Ratnam, D.V. Auto-detection of sporadic E and spread F events from the digital ionograms. *Adv. Space Res.* **2022**, *70*, 1142–1152. [CrossRef]
51. Hajkowicz, L. Morphology of quantified ionospheric range spread-F over a wide range of midlatitudes in the Australian longitudinal sector. *Ann. Geophys.* **2007**, *25*, 1125–1130. [CrossRef]
52. Pezzopane, M.; Zuccheretti, E.; Abadi, P.; de Abreu, A.J.; de Jesus, R.; Fagundes, P.R.; Supnithi, P.; Rungraengwajiake, S.; Nagatsuma, T.; Tsugawa, T.; et al. Low-latitude equinoctial spread-F occurrence at different longitude sectors under low solar activity. *Ann. Geophys.* **2013**, *31*, 153–162. [CrossRef]
53. Wang, G.; Shi, J.; Wang, X.; Shang, S.P. Seasonal variation of spread-F observed in Hainan. *Adv. Space Res.* **2008**, *41*, 639–644. [CrossRef]
54. Lan, T.; Jiang, C.; Yang, G.; Zhang, Y.; Liu, J.; Zhao, Z. Statistical analysis of low-latitude spread F observed over Puer, China, during 2015–2016. *Earth Planets Space* **2019**, *71*, 138. [CrossRef]
55. Guo, B.; Xiao, S.; Shi, J.; Xiao, Z.; Wang, G.; Cheng, Z.; Shang, S.; Wang, Z.; Suo, Y. Comparative study on characteristics of spread-F at low-and mid-latitudes during high and low solar activities over East Asia. *Chin. J. Geophys.* **2017**, *60*, 3289–3300. [CrossRef]
56. Wang, N.; Guo, L.; Ding, Z.; Zhao, Z.; Xu, Z.; Xu, T.; Hu, Y. Longitudinal differences in the statistical characteristics of ionospheric spread-F occurrences at mid-latitude in Eastern Asia. *Earth Planets Space* **2019**, *71*, 47. [CrossRef]
57. Lan, T.; Jiang, C.; Yang, G.; Sun, F.; Xu, Z.; Liu, Z. Latitudinal Differences in Spread F Characteristics at Asian Longitude Sector during the Descending Phase of the 24th Solar Cycle. *Universe* **2022**, *8*, 485. [CrossRef]
58. Sun, W.; Xu, L.; Huang, X. Forecasting of ionospheric vertical total electron content (TEC) using LSTM networks. In Proceedings of the 2017 International Conference on Machine Learning and Cybernetics (ICMLC), Ningbo, China, 9–12 July 2017; pp. 340–344.
59. Chen, Z.; Jin, M.; Deng, Y. Improvement of a Deep Learning Algorithm for Total Electron Content Maps: Image Completion. *J. Geophys. Res. Space Phys.* **2019**, *124*, 790–800. [CrossRef]
60. Chen, X.; Lei, J.; Ren, D. A Deep Learning Model for the Thermospheric Nitric Oxide Emission. *Space Weather* **2021**, *19*, e2020SW002619. [CrossRef]
61. Li, X.; Zhou, C.; Tang, Q. Forecasting Ionospheric foF2 Based on Deep Learning Method. *Remote Sens.* **2021**, *13*, 3849. [CrossRef]
62. Xia, G.; Liu, M.; Zhang, F.; Zhou, C. CAiTST: Conv Attentional Image Time Sequence Transformer for Ionospheric TEC Maps Forecast. *Remote Sens.* **2022**, *14*, 4223. [CrossRef]

63. Jiang, C.; Zhang, Y.; Yang, G.; Zhu, P.; Sun, H.; Cui, X.; Song, H.; Zhao, Z. Automatic scaling of the sporadic E layer and removal of its multiple reflection and backscatter echoes for vertical incidence ionograms. *J. Atmos. Sol. Terr. Phys.* **2015**, *129*, 41–48. [CrossRef]

64. Kohavi, R. Scaling Up the Accuracy of Naïve-Bayes Classifiers: A Decision-Tree Hybrid. In Proceedings of the Second International Conference on Knowledge Discovery and Data Mining, Portland, OR, USA, 2–4 August 1996; AAAI Press: Palo Alto, CA, USA, 1996; pp. 202–207.

65. Piggot, W.R.; Rawer, K. URSI Handbook of Ionogram Interpretation and Reduction. In *International Union of Radio Science*; United States, Environmental Data Service: Washington, DC, USA, 1978.

66. Chen, W.S.; Lee, C.C.; Chu, F.D.; Su, S.Y. Spread F, GPS phase fluctuations, and medium-scale traveling ionospheric disturbances over Wuhan during solar maximum. *J. Atmos Sol. Terr. Phys.* **2011**, *73*, 528–533. [CrossRef]

67. Paul, K.S.; Haralambous, H.; Oikonomou, C.; Paul, A.; Belehaki, A.; Ioanna, T.; Kouba, D.; Buresova, D. Multi-station investigation of spread F over Europe during low to high solar activity. *J. Space Weather Space Clim.* **2018**, *8*, A27. [CrossRef]

68. Bhaneja, P.; Earle, G.D.; Bullett, T.W. Statistical analysis of midlatitude spread F using multi-station digisonde observations. *J. Atmos Sol.-Terr. Phys.* **2018**, *167*, 146–155. [CrossRef]

69. Igarashi, K.; Kato, H. Solar cycle variations and latitudinal dependence on the mid-latitude spread-F occurrence around Japan. In *The XXIV General Assembly*; International Union of Radio Science: Kyoto, Japan, 1993.

70. Huang, W.Q.; Xiao, Z.; Xiao, S.G.; Zhang, D.H.; Hao, Y.Q.; Suo, Y.C. Case study of apparent longitudinal differences of spread F occurrence for two midlatitude stations. *Radio Sci.* **2011**, *46*, RS1015. [CrossRef]

71. Devasia, C.V.; Jyoti, N.; Subbarao, K.S.V. On the plausible linkage of thermospheric meridional winds with equatorial spread F. *J. Atmos. Sol. Terr. Phys.* **2002**, *64*, 1–12. [CrossRef]

72. Liu, Y.; Deng, Z.; Xu, T.; Kong, J.; Zhou, C.; Yao, Y. Daytime F region echoes at equatorial ionization anomaly crest during geomagnetic quiet period: Observations from multi-instruments. *Space Weather* **2022**, *20*, e2021SW003002. [CrossRef]

73. Candido, C.M.N.; Batista, I.S.; Becker-Guedes, F.; Abdu, M.A.; Sobral, J.H.; Takahashi, H. Spread F occurrence over a southern anomaly crest location in Brazil during June solstice of solar minimum activity. *J. Geophys. Res.* **2011**, *116*, A06316. [CrossRef]

74. Shi, J.K.; Wang, G.J.; Reinisch, B.W.; Shang, S.P.; Wang, X.; Zherebotsov, G.; Potekhin, A. Relationship between strong range spread F and ionospheric scintillations observed in Hainan from 2003 to 2007. *J. Geophys. Res.* **2011**, *116*, A08306. [CrossRef]

75. Li, J.; Ma, G.; Maruyama, T.; Li, Z. Mid-latitude ionospheric irregularities persisting into late morning during the magnetic storm on 19 March 2001. *J. Geophys. Res.* **2012**, *117*, A08304. [CrossRef]

76. Kil, H.; Lee, W.K.; Paxton, L.J. Origin and distribution of daytime electron density irregularities in the low-latitude F region. *J. Geophys. Res. Space Phys.* **2020**, *125*, e2020JA028343. [CrossRef] [PubMed]

77. Jiang, C.; Yang, G.; Liu, J.; Yokoyama, T.; Komolmis, T.; Song, H. Ionosonde observations of daytime spread F at low latitudes. *J. Geophys. Res. Space Phys.* **2016**, *121*, 12093–12103. [CrossRef]

78. Yang, G.; Jiang, C.; Lan, T.; Huang, W.; Zhao, Z. Ionosonde observations of daytime spread F at middle latitudes during a geomagnetic storm. *J. Atmos. Terr. Phys.* **2018**, *179*, 174–180. [CrossRef]

79. Xiao, S.; Shi, J.; Zhang, D.; Hao, Y.; Huang, W. Observational study of daytime ionospheric irregularities associated with typhoon. *Sci. China Technol. Sci.* **2012**, *55*, 1302–1304. [CrossRef]

80. Huang, C.S.; de La Beaujardier, O.; Roddy, P.A.; Hunton, D.E.; Ballenthin, J.O.; Hairston, M.R. Long-lasting daytime equatorial plasma bubbles observed by the C/NOFS satellite. *J. Geophys. Res.* **2013**, *118*, 2398–2408. [CrossRef]

 *remote sensing*

Article

Solar Flare Effects Observed over Mexico during 30–31 March 2022

Maria A. Sergeeva [1,2,*], Olga A. Maltseva [3], Artem M. Vesnin [4], Donat V. Blagoveshchensky [5], Victor J. Gatica-Acevedo [1], J. Americo Gonzalez-Esparza [1], Aleksandr G. Chernov [6], Isaac D. Orrala-Legorreta [1,7], Angela Melgarejo-Morales [1], Luis Xavier Gonzalez [1,2], Mario Rodriguez-Martinez [8], Ernesto Aguilar-Rodriguez [1], Ernesto Andrade-Mascote [1] and Pablo Villanueva [1]

1   SCiESMEX, LANCE, Instituto de Geofisica, Unidad Michoacan, Universidad Nacional Autonoma de Mexico, Morelia C.P. 58089, Michoacan, Mexico
2   CONACYT, Instituto de Geofisica, Unidad Michoacan, Universidad Nacional Autonoma de Mexico, Morelia C.P. 58089, Michoacan, Mexico
3   Institute for Physics, Southern Federal University, 344090 Rostov-on-Don, Russia
4   Institute of Solar-Terrestrial Physics, Siberian Branch of Russian Academy of Sciences, 664033 Irkutsk, Russia
5   Institute of Radio Engineering and Information and Communications Technologies, Saint-Petersburg State University of Aerospace Instrumentation, 190000 Saint-Petersburg, Russia
6   Sitcomm LLC, 424031 Yoshkar-Ola, Russia
7   Facultad de Ciencias, Universidad Nacional Autonoma de Mexico, Coyoacán C.P. 04510, Mexico City, Mexico
8   Escuela Nacional de Estudios Superiores, Unidad Morelia, Universidad Nacional Autonoma de Mexico, Morelia C.P. 58190, Michoacan, Mexico
*   Correspondence: maria.a.sergeeva@gmail.com

Citation: Sergeeva, M.A.; Maltseva, O.A.; Vesnin, A.M.; Blagoveshchensky, D.V.; Gatica-Acevedo, V.J.; Gonzalez-Esparza, J.A.; Chernov, A.G.; Orrala-Legorreta, I.D.; Melgarejo-Morales, A.; Gonzalez, L.X.; et al. Solar Flare Effects Observed over Mexico during 30–31 March 2022. *Remote Sens.* **2023**, *15*, 397. https://doi.org/10.3390/rs15020397

Academic Editor: Fabio Giannattasio

Received: 28 November 2022
Revised: 27 December 2022
Accepted: 28 December 2022
Published: 9 January 2023

Abstract: Manifestations of two solar flares of March 2022 were studied over Mexico. The flare effects in the lower ionosphere had a ~3 min delay from the X1.3-flare onset and ~5 min from the M9.6-flare onset. The maximal impact on the HF signal amplitude was ~(14–15) min after the onset of both flares. The X1.3-flare provoked the shortwave fadeout during ~6 min. The effects in the lower ionosphere lasted longer than the flares and the effects at the F2 region and higher altitudes only during the flares. The interpretation of results showed the following. (1) Based on the absorption level estimated with minimum frequency and signal amplitude on ionograms, the major role of X-ray radiation in the electron concentration increase in the lower ionosphere was confirmed. At the same time, the EUV radiation impact on the lower ionosphere cannot be totally discarded. The lower ionosphere recovery began before and lasted after the X1.3-flare end, being more rapid at Eglin than in Mexico. During M9.6-flare, the responses at the two observation points were rather synchronized due to the more similar illumination conditions at the two meridians. (2) According to the dI variations characterizing the F2 region and higher, the M9.6-flare provoked medium-scale and the X1.3-flare provoked both medium- and small-scale ionospheric irregularities. The response duration corresponded to the dI series filtered with (10–20) min windows. The dI curve during the flares was characterized by the И-form and depended more on the active region position and the flare class than on the solar zenith angle. The available data do not allow us to unambiguously identify the reason for the negative dI: the applied filtering procedure or the physical effect. (3) During both flares, the major EUV impact on the lower ionosphere was by the flux at 133.5 nm and on the F2 region and higher altitudes at 25.6 nm. In addition, during the M9.6-flare, EUV 28.4, 30.4 and 121.6 nm spectral bands also played an important role in the F2 response. During the X1.3-flare, the EUV 25.6 nm flux and X-ray flux impacts on the F2 region were of the same level. The weakest impact was caused by the emission in the EUV 28.4 nm spectral band on the absorption in the lower ionosphere during both flares and on the electron density in the F2 region and higher during the X1.3-flare.

Keywords: solar flare; ionosphere; ionospheric sounding; ionosonde; GNSS; slant TEC; Mexico

1. Introduction

Ionospheric disturbances determine ionospheric parameter deviations from their quiet diurnal values with characteristic time scales from some minutes to several days. The ionosphere can be perturbed by different factors. Sometimes, the effects of different sources superpose.

Sudden Ionospheric Disturbances (SID) are the ionization increases of the dayside ionosphere caused by the bursts of ultraviolet and X-ray radiation from the Sun during solar flares. Sharp increases in solar X-rays ionizing radiation result in an increase of the ionization velocity and eventually in the corresponding electron concentration increases in the D and lower E regions of the ionosphere. SIDs in the upper E and F regions that are mostly ionized by ultraviolet radiation are usually less pronounced and of less duration [1]. Significant increases of radiation affect radio propagation conditions [2]. Each particular flare is a unique event, and the ionospheric response to it can vary. In the worst case, the total absorption of the energy of radio waves propagating by reflection in the ionosphere is observed. In general, the electron concentration increase caused by the intense flares is maximal within altitudes of 60–85 km. It can also reach 50–200% in the ionospheric E region and 10–30% in the F region [1,3].

In this work, we estimate the impacts of two intense flares (X1.3 and M9.6) that occurred on 30 and 31 March 2022 on the topside and the lower ionosphere over the Mexican region. Since the flares occurred close to the Mexican noon, its territory was most affected by the flares.

Data of different instruments may be used to estimate the flare impact on the ionosphere in general and the absorption level in particular, for example radars, e.g., [4,5], riometers (mostly at high-latitudes) e.g., [6], ionospheric sounding (ionosonde) data e.g., [7,8], satellite-to-ground radio methods [9,10], VLF signal data (including signal amplitudes) e.g., [11,12] and others.

The attenuation of HF radio waves is due to absorption in the lower ionosphere [2,3,13]. The fadeout events occur more often at high-latitudes [14]. Our focus is on the low- to mid-latitudes of the North American sector.

Intense solar flares can also perturb the ionosphere at higher altitudes, namely at the F2-layer heights and in the topside ionosphere [15–19] and references therein. One of the approaches to study the effects at these heights is the use of slant and/or vertical Total Tlectron Content (TEC) data (e.g., [19]). Habarulema et al. [20], citing the articles of 2021, pointed out that the flare effects can extend to geospace, which emphasizes the importance of TEC use for analysis. Similar to Habarulema et al. [20], we had the opportunity to involve both TEC and ionospheric sounding data. The peculiarity of the present study is the high temporal resolution of the ionosonde data (ionograms taken each 2 min) and the GNSS-derived filtered slant TEC (sTEC) data (time series resolution of 30 s).

There are many statistical and case studies on solar flare's influence on the ionosphere (we only mention some of them not to overload the text with numerous references that the reader can find in the review papers). Nonetheless, to date, the interest in studying flare effects, especially within the ionospheric regional context, has not faded. This is partially explained by the variability of the ionospheric responses to this type of Space Weather disturbance. As shown by Barta et al. and Habarulema et al. [7,20], the flare effects have been studied most frequently for the European and African regions. This study is focused on the low latitudes of the North American sector. Recently, Le et al. [21] revealed an interesting feature for this region. According to these authors, some flares resulted in TEC enhancements being the highest, not near the subsolar point, but rather far away from it. They found that the anomaly distributions of TEC deviations were observed mainly in the North American region during the flares that occurred in September–November. Though our study is focused on spring flares, such results indicate the overall importance of the studies for this region. During equinoxes, the ionospheric response to flares is more pronounced than during solstice periods because of the seasonal variation of neutral density [22]. This is our case, as the considered flares occurred 10 days after the spring equinox in 2022.

This paper is organized as follows. Section 2 provides a description of the two flares considered. Section 3 introduces the data used for the study. Section 4.1 discusses the flare effects on the lower ionosphere (D and lower E regions) revealed with the ionospheric sounding data. Section 4.2 presents the results of the flare effect detection at the higher ionospheric heights (from the F2 region to the topside ionosphere). Final remarks are given in Conclusions.

2. Description of the Considered Flares

Two intense solar flares occurred near the midday local time of the American sector on 30 and 31 March 2022. The flare parameters are given in Table 1. There are similarities and differences in the observation conditions of the considered events. Both were caused by the same active region (AR) on the solar disk (Figure 1), which was disappearing due to solar rotation (Figure 2). In addition to the close local time of occurrence (LT), both flares had comparable durations and occurred under undisturbed geomagnetic conditions. However, though the two events occurred on two consecutive days, meaning the same total daylight duration and seasonal conditions, there was a difference in the solar activity represented by the F10.7-index and some difference in the geomagnetic conditions. Note that no magnetic storm occurred on the flare days or before. Notwithstanding the F10.7 and Dst variations, it was possible to observe exactly the flare effects. The intensity of the flares was different.

Table 1. Parameters of the considered flares and conditions of their observation.

Date	Flare Onset, UT	Flare Max, UT	Flare End, UT	Duration (Phases), Min	LT = ~(UT-5)	X-Class	AR (Coordinates)	F10.7, s.f.u (at 20 UT)	Dst * (Daily max/min), nT
30 March 2022 DOY = 089	17:21	17:37	17:46	25 (16 + 9)	12:21 LT	X1.3	2975 (N13W31)	151.3	10/−9
31 March 2022 DOY = 090	18:17	18:35	18:45	28 (18 + 10)	13:17 LT	M9.6	2975 (N12W47)	239.5	22/−18

* Dst values are given only to characterize the overall background conditions.

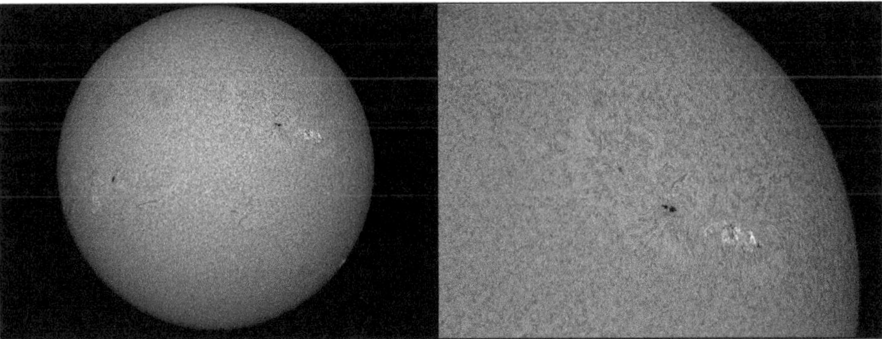

Figure 1. Images of the solar chromosphere in H-Alpha (6562.8 Å) taken in Morelia, Mexico: the whole solar disk (**left**) and the AR 2975 in focus (**right**) on 30 March 2022. The images are provided by the Laboratory of Geo-Spatial Sciences (LACIGE), Morelia, Mexico.

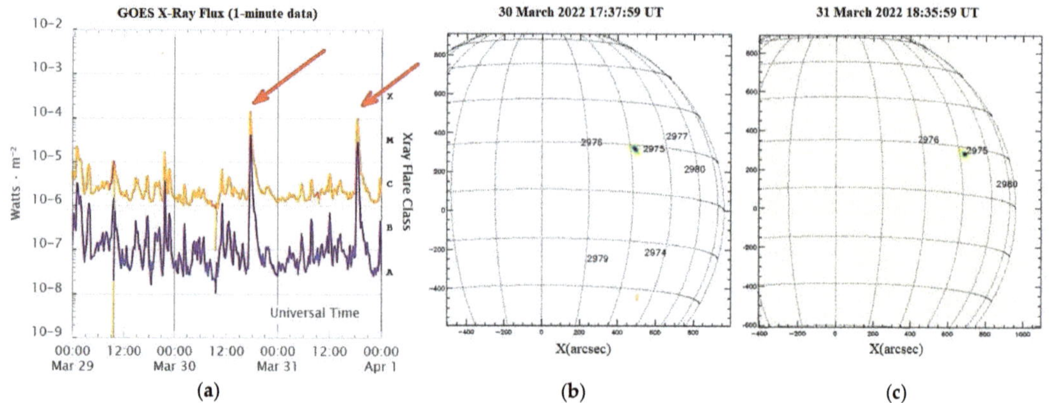

Figure 2. X-ray flux by GOES satellite data (flare moments marked with arrows) (**a**), and AR position on the solar disk on 30 March 2022 (**b**) and 31 March 2022 (**c**). Data supplied by courtesy of SolarMonitor.org and NOAA's Space Weather Prediction Center.

3. Data Used

The ionograms of Vertical Ionospheric Sounding (VIS) taken in the North American low- to mid-latitude sector were used for the analysis. Quasi-vertical ionospheric sounding experimental campaigns were performed in Mexico in 2021–2022. The transceiver and receiver stations were installed in the National Space Weather Laboratory of Mexico (LANCE) in Mexart, Mexico (MEXART, 19.8°N, 101.68°W), and in Morelia, Mexico (IGUM, 19.64°N, 101.22°W), respectively. The distance between the two stations is ~51 km. They operated in the range of 2–30 MHz with a frequency sweep rate of 150 kHz/s and a frequency turning step ≤ 1 Hz. The chirp sounding frequency sweep rate and frequency operation range were different in different time intervals and depended on the experiment tasks of the particular time period. During the last days of March 2022, the sounding signal parameters did not change, which allowed us to detect the HF propagation condition variations during the considered flares. We also studied VIS ionograms by the Eglin digisonde station located in the southern part of the US (EG931, 30.5°N, 86.5°W). Eglin ionograms were taken every 7.5 min [23]. Figure 3 shows a map of the experiment.

Figure 3. Locations of GNSS receiver stations (red and green circles), ionospheric stations of LANCE equipped with chirp ionosondes (black circles) and the digisonde station in Eglin (blue circle).

Data from 35 local GNSS stations were involved in the analysis (Figure 3). First, based on dual-frequency phase measurements, sTEC values were calculated along each line-of-site (LoS) between the visible GPS satellites and ground receivers. Then, to eliminate trends, the time series of sTEC for every continuous arc (interval when data exhibit no jumps, gaps or other artifacts) were detrended with splines [24]. Third, they were filtered with a centered moving average with three different sets of windows. As a result, we had time series filtered with 2 and 10 min, 10 and 20 min and 20 and 60 min windows. The purpose of filtration is to have variation with periods in the aforementioned ranges. Each of the three time series corresponded to different spatiotemporal scales of ionospheric disturbances. Finally, these three filtered sTEC variations were converted to equivalent vertical variations (dI) for each LoS using the mapping function [25] and references therein. The resulting dI curve is the deviation of the sTEC from its trend. dI is constructed in this way to have its mean values close to zero for every continuous arc. The time series of Rate of TEC index (ROTI) were involved to support the analysis [26,27].

The data on the X-ray and Extreme Ultraviolet (EUV) fluxes measured by the Extreme Ultraviolet Sensor and X-ray Irradiance Sensors at the Geostationary Operational Environmental Satellite (GOES-R) at the geostationary orbit were involved in the analysis [28]. Dst and F10.7 indices were used to estimate the geomagnetic field state and the overall level of solar activity, respectively.

4. Results of Observations

4.1. Flare Effects by VIS Data

4.1.1. Ionospheric Conditions by Ionograms

First, let us consider the ionospheric conditions at the low latitudes of the North American sector (geographic latitude ~20°N; geomagnetic latitude ~28°N). In March 2022, chirp ionosondes in the center of Mexico were operated in experimental mode. The ionogram background was contaminated with some noise due to radio interference near the receiver station. Nonetheless, the results of observations allowed us to clearly detect the flare effects. It is worth noting that at the end of March 2022, only the traces of reflection from the F2-layer were present on the ionograms during the time of interest (between 17 and 19 UT). We use the observations on 29 March and 1 April as a reference for comparison with the ionograms of 30 and 31 March. The examples of the ionograms for the first flare are shown in Figure 4.

On 30 March 2022, the flare's onset, peak and end were at 17:21, 17:37 and 17:46 UT, respectively. According to the sounding data (Figure 4), a signal amplitude decrease was observed on ionograms beginning from 17:24 UT. The shortwave fadeout (the absence of the reflected signal on the ionograms due to the growth of absorption in the lower ionosphere) was observed between 17:34 and 17:40 UT. Then, the appearance of the signal traces on the ionogram at higher frequencies of the considered frequency range began at 17:40–17:42 UT. The recovery lasted until approximately 18:48 UT, when the pre-flare amplitude level was reached.

To sum up, the lower ionosphere over Mexico (D region and lower part of E region) responded with a 3 min delay to the flare onset. The fadeout lasted about 6 min. The recovery of radio propagation conditions to their pre-flare level lasted about 1 h after the flare end.

Further, we estimate the ionospheric conditions at the middle latitudes of the considered sector (geographic latitude ~30°N; geomagnetic latitude ~40°N). Since the time step of taking ionograms at the Eglin station is 7.5 min, this complicates the identification of the exact time of the ionospheric responses to fast phenomena. According to the available data (Figure 5), no effect was detected at 17:22:30 UT (1.5 min after the flare onset). The observed signal was less intense and its frequency range shortened at 17:30:00 UT. The reflected signal was absent on the ionogram (fadeout) at 17:37:30 UT (30s after the flare peak) and began to recover at 17:45:00 UT (1 min before the flare end). In general, this timeline is in accord with the Mexican ionosonde measurements, though the lower temporal resolution of data by Eglin did not allow a detailed comparison

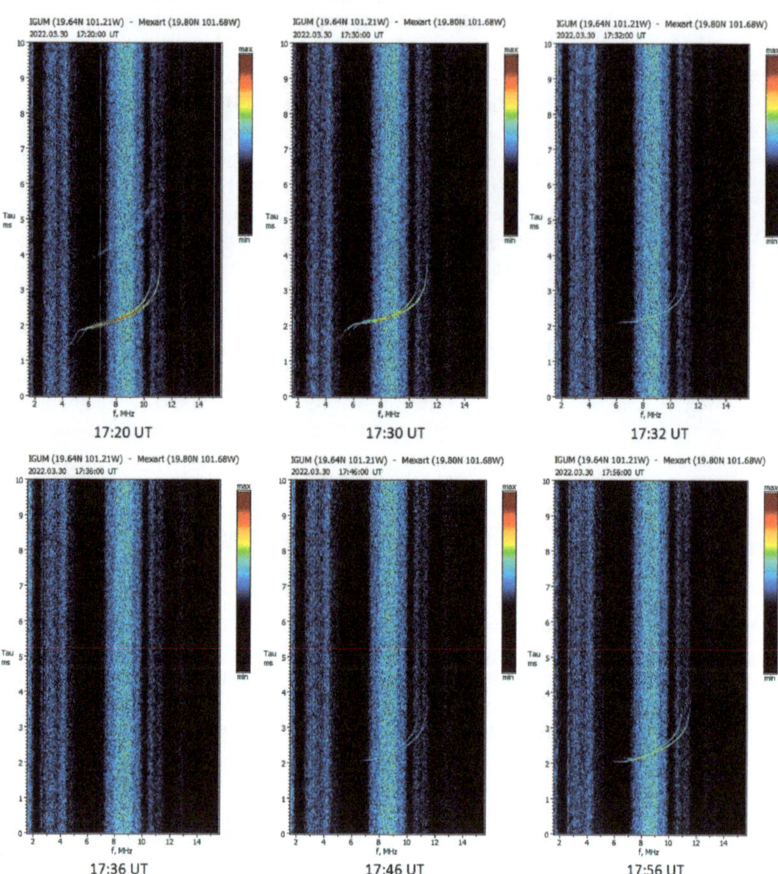

Figure 4. Ionograms taken in Mexico before, during and after the X1.3-flare on 30 March 2022.

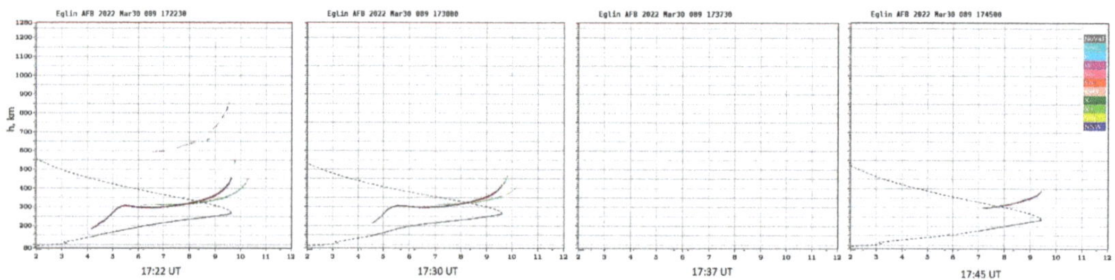

Figure 5. Ionograms taken at Eglin during and after the X1.3-flare on 30 March 2022.

At both stations, the ionospheric response was delayed some minutes after the flare onset. The total absorption of the signal was detected near the time of the flare peak. Here, we note that the sounding signal parameters (emitted power, type of the radio wave (continuous/pulse), antenna pattern, etc.) are different at two stations; therefore, we may compare the fadeout events only qualitatively. The recovery to the pre-flare conditions was longer at lower latitudes (before the flare end at Eglin and after the flare end in Mexico).

In regard to the less intense flare on 31 March 2022, the beginning of the impact on the lower ionosphere was detected at 18:22 UT (5 min after the flare onset) by the low-latitude VIS data (Figure 6). At 18:28 UT, the narrowing of the frequency range of the reflected signal began. The lowest signal amplitudes and the narrowest frequency range (18:32 UT) were detected 3 min before the flare peak. The recovery of the pre-flare conditions was slow. It began at 18:38 UT and lasted until approximately 19:12 UT. To note, the flare end was at 18:45 UT.

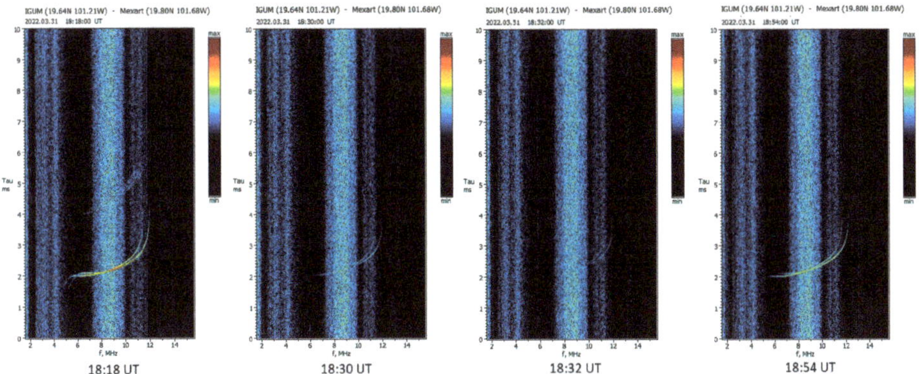

Figure 6. Ionograms taken before and during the M9.6-flare on 31 March 2022.

At the Eglin station (Figure 7), at 18:22:30 UT (5.5 min after the flare onset), still no effect was detected. At 18:30 UT (13 min after the onset), the reflected signal was observed in a very narrow frequency range. At 18:37:30 UT (2.5 min after the flare peak), the range was the narrowest. The recovery to the pre-flare conditions lasted between 18:45 and 19:15 UT (during 30 min after the flare end).

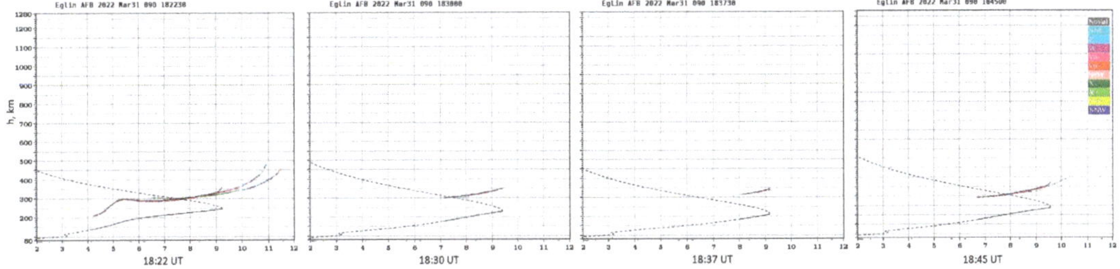

Figure 7. Ionograms taken at Eglin during and after the M9.6-flare on 31 March 2022.

The flare effects at the two stations began simultaneously. In general (as far as the temporal resolution allows estimating), the most pronounced effects and the recovery to the pre-flare HF propagation conditions were also in accord at two stations. In contrast, the day before, the responses in Mexico and Eglin to the more intense flare were not so synchronized (the effects in Mexico lasted longer). The illumination conditions (LT of the flare) at Mexican and Eglin meridians were more similar on 31 March than on 30 March. In this case, the longitudinal difference between the stations was more important than the latitudinal difference. The difference in the ionospheric response in Mexico and Eglin on 30 March is probably explained by the different Sun's zenith angle at the moment of the flare.

To sum up, according to the low-latitude ionosonde data of 2 min time resolution, the effects of the X1-flare lasted ~1 h 24 min. The effects of the M9-flare on the next day lasted ~40 min. In both cases, the effect duration was longer than the flare duration (in contrast to the effects in the F2 region and topside ionosphere by GNSS data, as shown further).

A detailed comparison of the flare impact processes with time is given in Figure 8. The more intense (X1.3) flare on 30 March provoked the more significant response (narrowing of the operation range and the signal amplitude decrease on ionograms) in the lower ionosphere of the low- to mid-latitude North American sector than the less intense (M9.6) flare on 31 March. Though, from first glance, this conclusion seems obvious, it is not obvious at all. There are important factors other than the X-class of the flare. For instance, Barta et al. [7] emphasized the role of the solar zenith angle.

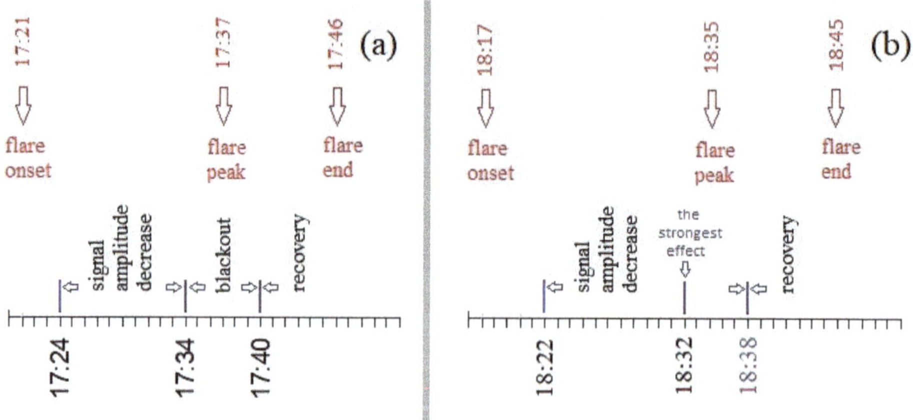

Figure 8. Timeline of the ionospheric response by the ionosonde data: X1.3-flare (**a**) and M9.6-flare (**b**).

4.1.2. Estimation of the Reflected Signal Attenuation in Mexico

Ionograms provide many ionospheric and radio propagation parameter values. One approach is to study the amplitude of the received ionospheric sounding signal (e.g., [8]). In our case, the relative amplitude of the reflected chirp signal is displayed on ionograms with color (color bar in the right part of the ionogram). When measurements are performed with the same transceiver settings (no changes of the sweep rate, emitted power, frequency and delay ranges, ionogram width, etc., during the period of interest), these amplitudes (dB) in a series of ionograms may be compared to each other. A qualitative assessment of the relative amplitudes of the received signal was obtained during the flares for the following cases.

(1) The assessment of the relative signal amplitude (A) at the critical frequency of the F2 layer for ordinary (foF2) or extraordinary (fxF2) components does not seem possible, as the trace of the signal at these frequencies was not observed well or was absent on several ionograms due to the increased absorption. Therefore, for each ionogram, we analyzed the maximal A without regard to the frequency at which this maximal A was observed. Since this frequency was different on each ionogram, our assessment is only of a qualitative character. Nevertheless, further analysis showed the validity of this approach.

(2–3) Estimation of the signal amplitude at the same particular frequency is not always possible, as the position of the traces on the ionograms changes with time (even under quiet conditions). We analyzed A (dB) at 7 and 11 MHz because in our case, the reflected signal was always present at these frequencies, with the exception of the fadeout interval.

For all three cases mentioned above, the relative amplitudes on the last ionogram taken before the flare onset were considered as the reference amplitudes. Then, the difference between the reference values and the values of amplitudes observed on the subsequent ionograms (taken after the flare onset) were calculated. Figure 9 shows the results for cases (1)–(3) described above for three time intervals—that is, during the X1.3-flare of 30 March and during the M9.6-flare of 31 March. As the time resolution of VIS data for 29 March does not allow us to perform comparison with this quiet day, the day after the second flare was chosen for comparison. The curves for 1 April are shown only during the time interval of the first flare (30 March). Variations on 1 April during the same time interval as for the second flare (31 March) are not shown for the economy of space. Essentially, they were rather similar to the variations shown in Figure 9c. We remark that 1 April may not be considered a geomagnetically quiet day, but it may be used as a reference day, first, because the time scale and the magnitude of the ionospheric response to flares and magnetic field disturbances are different and, second, because here, we compare the relative amplitudes of the signal that are affected by flares and not affected by geomagnetic disturbances.

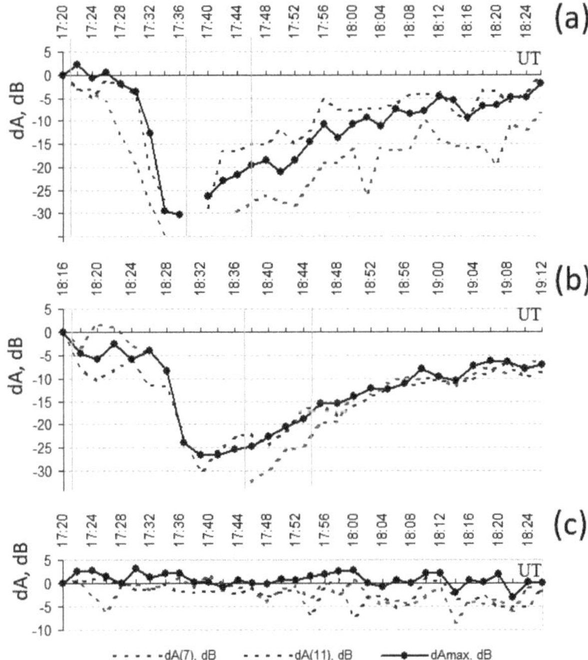

Figure 9. Deviation of the maximal relative amplitudes of the signal observed in ionograms during the intervals of the flares on 30 March (**a**) and 31 March (**b**) and for the reference day, 1 April (**c**). Thin gray vertical lines indicate the moments of the flare onset, peak and end.

Figure 9 illustrates the flare impacts: the decrease of the received signal amplitude by 27.28 dB (at 11 MHz) was detected 13 min after the solar flare onset on 30 March. This decrease was immediately followed by the fadeout event (~15 min after the flare onset). As for 31 March, the maximal amplitude decrease by 30.38 dB (at 11 MHz) was revealed 14 min after the flare onset. During the reference day of 1 April, the relative amplitude deviation did not exceed the value of 7 dB.

To conclude, the maximal impact of the flare on the reflected chirp signal amplitude in the center of Mexico was observed approximately 14–15 min after the flare beginning for both considered flares.

4.1.3. Correlation with Different Ranges of EUV and X-ray Radiation

The solar radiation spectral composition qualitatively changes during flares. The flare impact on the ionosphere depends on the amount of radiation emitted in different EUV and X-ray ranges. To estimate the change of the absorption level during the flares, we considered deviation of the relative amplitude of the reflected signal on the ionograms (dA) discussed in the previous section. In addition, the value of the minimal frequency at which the signal traces were observed on the ionogram (fmin) was considered. This parameter is mainly defined by the absorption level in the ionosphere.

The linear correlation coefficients (r) between the time series of dA and EUV/X-ray fluxes within different frequency bands were calculated. As the time step of ionograms was 2 min, the time series of EUV and X-ray fluxes were also reduced to this step. Please note that the series lengths were not the same as sometimes the signal trace was absent at some frequency or at all frequencies due to absorption. Nonetheless, this analysis allows us to obtain at least a qualitative assessment of the impacts of different radiation bands. The coefficient r was also calculated between the fmin value and the EUV/X-ray fluxes within different frequency bands. The results for the two flares are presented in Tables 2 and 3. Low r values are given in red and high r values in blue.

Table 2. The coefficient r during the X1.3 flare.

	EUV Bands, 30 March 2022						
	25.6 nm	**28.4 nm**	**30.4 nm**	**117.5 nm**	**121.6 nm**	**133.5 nm**	**140.5 nm**
dA(7), dB	−0.75	−0.65	−0.85	−0.70	−0.79	−0.76	−0.56
dA(11), dB	−0.82	−0.54	−0.86	−0.89	−0.74	−0.91	−0.78
dAmax, dB	−0.90	−0.73	−0.97	−0.86	−0.91	−0.90	−0.74
fmin	0.89	0.52	0.92	0.95	0.80	0.96	0.86

Table 3. The same as in Table 2, but for the M9.6-flare.

	EUV Bands, 31 March 2022						
	25.6 nm	**28.4 nm**	**30.4 nm**	**117.5 nm**	**121.6 nm**	**133.5 nm**	**140.5 nm**
dA(7), dB	−0.91	−0.54	−0.93	−0.67	−0.81	−0.82	−0.53
dA(11), dB	−0.76	−0.07	−0.82	−0.84	−0.67	−0.91	−0.75
dAmax, dB	−0.83	−0.19	−0.88	−0.83	−0.76	−0.91	−0.72
fmin	0.63	0.03	0.66	0.68	0.52	0.73	0.59

For the more intense flare, there is a rather strong correlation between the EUV radiation bursts and the change of the amplitude of the received sounding signal (Table 2). The effect was much less pronounced for the less intense flare (Table 3). In general, according to the data of both flares, the lowest or even absent correlation was observed between the absorption level (represented by the signal amplitude and fmin) and the emission in the 28.4 nm spectral band. The highest correlation was observed in the case of the 133.5 nm band, meaning the strongest impact on the lower ionosphere was caused by this EUV spectral band.

Figure 10 schematically shows the EUV flux variation for different bands: the values for each curve were multiplied by a coefficient k (different for each band) to adjust all the curves to a similar scale. This was done for illustrative purposes to compare the forms of the curves. Black curves with dots show the fmin values along the right Y-axis. It results that the increase of the emission in the 28.4 nm band (second curve from the top of the figure) was less abrupt in contrast to other spectral bands. This may explain why its effect on the absorption level was less pronounced.

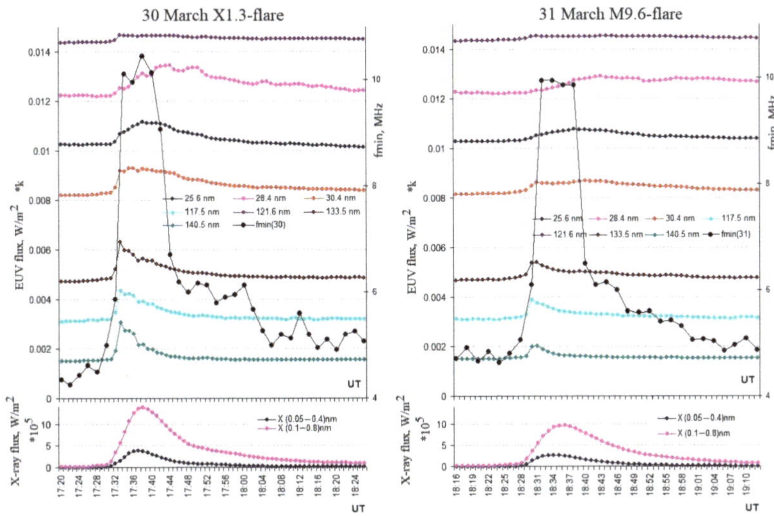

Figure 10. Schematic representation of the EUV and X-ray fluxes in different spectral bands (**left** *Y*-axis) and the corresponding fmin variation (**right** *Y*-axis) for the periods of X1.3-flare (**left**) and M9.6-flare (**right**). EUV flux value is given in the form of EUV × k, where k is the specific coefficient for each band (from top to bottom k values are 80; 130; 15; 30, 1.75; 20; 10). The X-ray flux value is given in the form of EUV × 10^5.

It was even more interesting to study the same correlation but with the X-ray spectral bands, as it is known that the physical source of strong increases of electron concentration in the ionospheric D region during flares is the X-ray radiation in the wavelength range shorter than 10 nm. The results are shown in Tables 4 and 5. X-ray variations are shown in Figure 10.

Table 4. The same as in Table 2, but for X-rays.

	X-ray Bands, 30 March 2022	
	(0.05–0.4) nm	(0.1–0.8) nm
dA(7), dB	−0.8	−0.84
dA(11), dB	−0.91	−0.88
dAmax, dB	−0.89	−0.93
fmin	0.96	0.92

Table 5. The same as in Table 3, but for X-rays.

	X-ray Bands, 31 March 2022	
	(0.05–0.4) nm	(0.1–0.8) nm
dA(7), dB	−0.93	−0.95
dA(11), dB	−0.9	−0.89
dAmax, dB	−0.93	−0.94
fmin	0.96	0.91

At first glance, it seems that the X-ray radiation within the shorter wavelengths (between 0.05 and 0.4 nm) had more effect on the fmin variations. At the same time, such a conclusion may be incorrect. Considering the qualitative character of our assessment and the fact that, in general, r in Tables 4 and 5 varies near the values of high correlation (0.8–0.95), we consider it impossible to make conclusions about which of the X-ray bands had a more pronounced impact on the absorption level in the lower ionosphere. The only conclusion from Figure 10 and Tables 4 and 5 is that the role of the X-ray radiation during the flares is the most important, which coincides with the well-known fact from the literature.

To conclude, the major role of the X-ray radiation in the electron concentration increase in the lower ionosphere of the considered region during the flares was confirmed. As for the EUV radiation impact on the lower ionosphere, it cannot be totally discarded. It was more pronounced for the more intense flare X1.3 than for M9.6. For the particular cases of the two considered flares, the EUV emission in the 28.4 nm spectral band provoked the weakest impact on the absorption conditions in the lower ionosphere (probably due to the smooth gradual increase of radiation in this band) and the 133.5 nm band had the most pronounced impact. Therefore, the parts of the spectrum that have the greatest influence in this case have been identified.

4.2. Flare Detection by GNSS Data

4.2.1. Features of dI Response

We considered sTEC variations that were filtered with a moving average of different windows (see Section 3). Due to the time scales of the ionospheric responses to the flares and the duration of these responses, the most informative data were the dI time series obtained with (2–10) min and (10–20) min windows. Some traces of responses were also detected in the dI(20–60) min series, but they were detected only at some LoSs and mostly did not exceed the background dI(20–60) fluctuations. To add, this filtering did not allow us to estimate adequately the time of the dI response because of its duration being less than 20 min. To discard the possible dI peaks caused by the multipath effects of GNSS signal reception, the dI data were analyzed not only during the flare day but also the day before (for comparison).

dI variations were studied at each LoS from different satellites. Figure 11 illustrates an example of dI(2–10) variations observed at the receiver station YOIG in the south of Mexico (Figure 3). A certain pattern of dI behavior was detected at all LoSs during the flare: the dI(2–10) decrease after the flare onset followed by the moderate increase and then, before or at the moment of the flare end, the return to the level of the background variations. The only different pattern is seen for the ray path YOIG-G07. Probably, it is explained by the local processes that were not overcome by the flare effect. The analysis of Figure 11 shows that the dI(2–10) variation caused by the flare was of a small amplitude compared with other effects, which, for example, are seen at LoSs YOIG-G09 and YOIG-G06 at 18:30 UT. These were provoked by the multipath effect due to some obstacle in the signal path. Another conclusion is the simultaneity of the dI response at all LoSs.

Let us describe the responses to each flare, as it is known that they can vary. For example, based on data for X-, M- and C-flares during 2014–2017, Syrovatskiy et al. [29] demonstrated that the response to a flare with a sufficiently strong X-ray flux may practically not manifest itself, and vice versa, the response to a weaker flare may turn out to be greater than to a stronger flare. Dmitriev et al. [15] also discussed the issue of why a weaker flare can produce a larger TEC enhancement and a very intense flare can produce effects comparable to those caused by moderate flares. These facts prove the presence of variability in the ionospheric response. The flare intensity defined in terms of the X-ray class of the flare may not be a measure of the ionospheric response in the F2-region [20,30,31].

Figure 12 shows examples of dI(2–10) and dI(10–20) variations for 30 March 2022 at LoSs from satellites G14 and G30 to different receivers. dI curves during the flare were characterized by the Π-form: the negative bay followed by the positive peak. The same dI behavior pattern was revealed at all LoSs for this day (not shown for the economy of space). Notwithstanding the small amplitudes of dI deviations, the simultaneity and uniformity of the response at all LoSs, as well as the moment of the ionospheric response occurrence, leave no doubt that it was caused exactly by the flare.

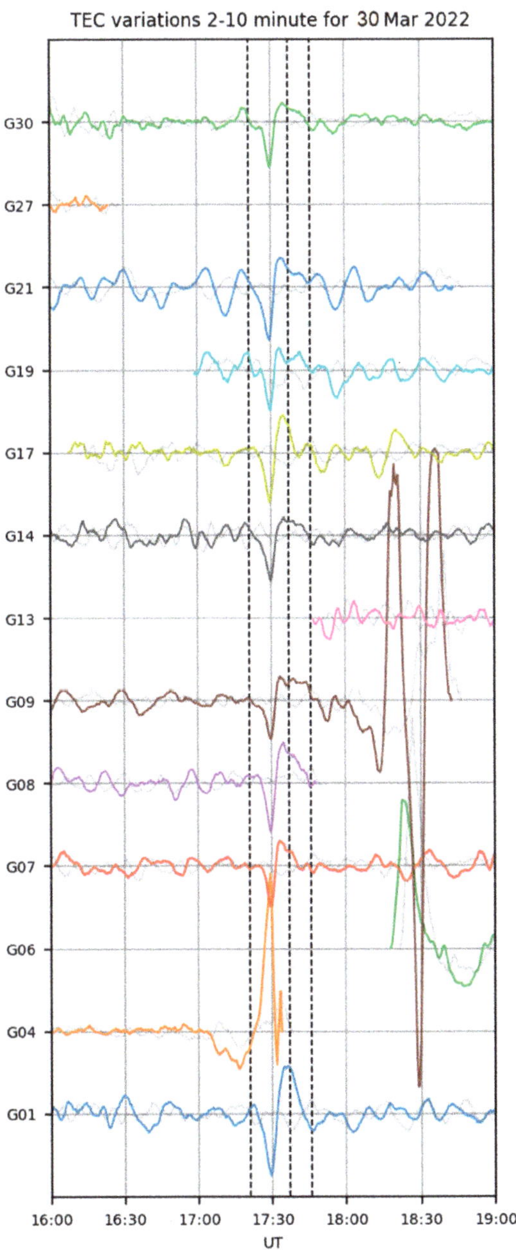

Figure 11. dI(2–10) variations registered at the YOIG station on 30 March 2022. Each curve represents the dI(2–10) for the corresponding GPS satellite. The Y-axis unit is 0.3 TECU. The dashed lines mark the flare onset, peak and end. Colored dI curves stand for the day of the flare and the gray curves stand for the day before (for comparison).

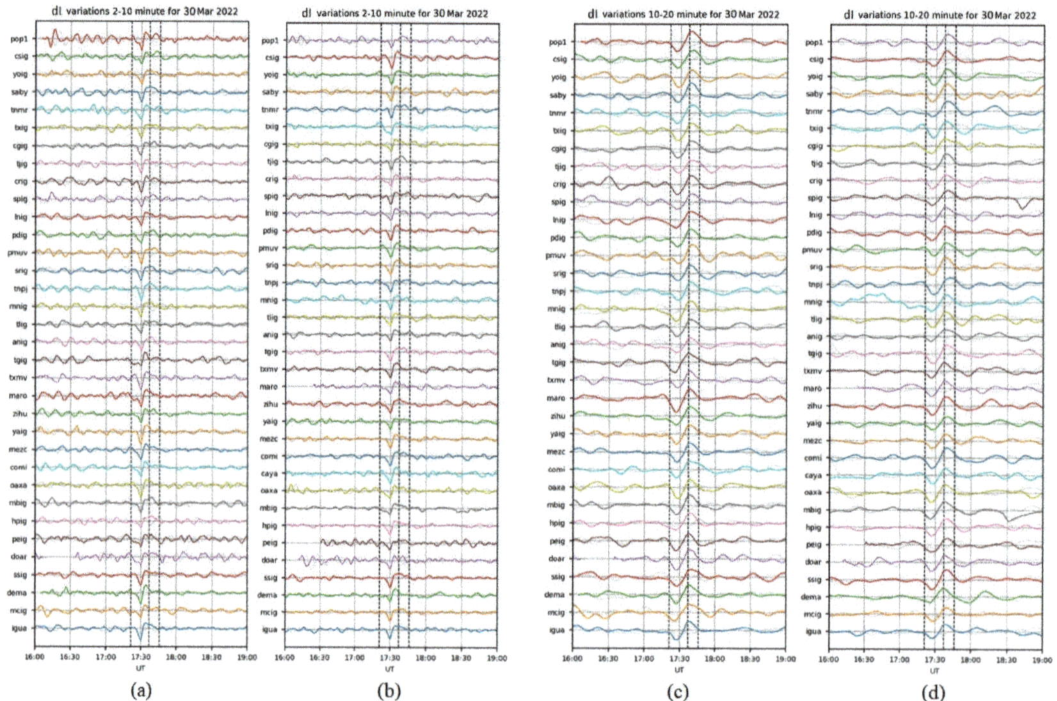

Figure 12. dI(2–10) variations at LoSs to satellites G14 (**a**) and (**b**) G30 and dI(10–20) variations at LoSs to satellites G14 (**c**) and (**d**) G30 for the case of X.3-flare on 30 March 2022. The Y-axis unit is 0.3 TECU. The dashed lines mark the flare onset, peak and end. Colored dI curves stand for the day of the flare and the gray curves stand for the day before for comparison.

In the case of the less intense flare of 31 March 2022 (Figure 13), the dI response at many LoSs did not exceed the level of background fluctuations. In general, the response (when detected) was less intense than in the previous case. It can be noted that sometimes dI(2–10) show the И-pattern but at the same time, the deviations are of the same amplitude as the background fluctuations. Only the simultaneity of the И-form occurrence indicates that it is a response. As for the dI(10–20) variations, the response was not present at some LoSs. When it was present, the parameter deviations exceeded the background fluctuations. At the majority of LoSs (please note that only the examples are shown in Figure 13), the negative bay in the temporal dI (10–20) variations was more pronounced than the following positive peak.

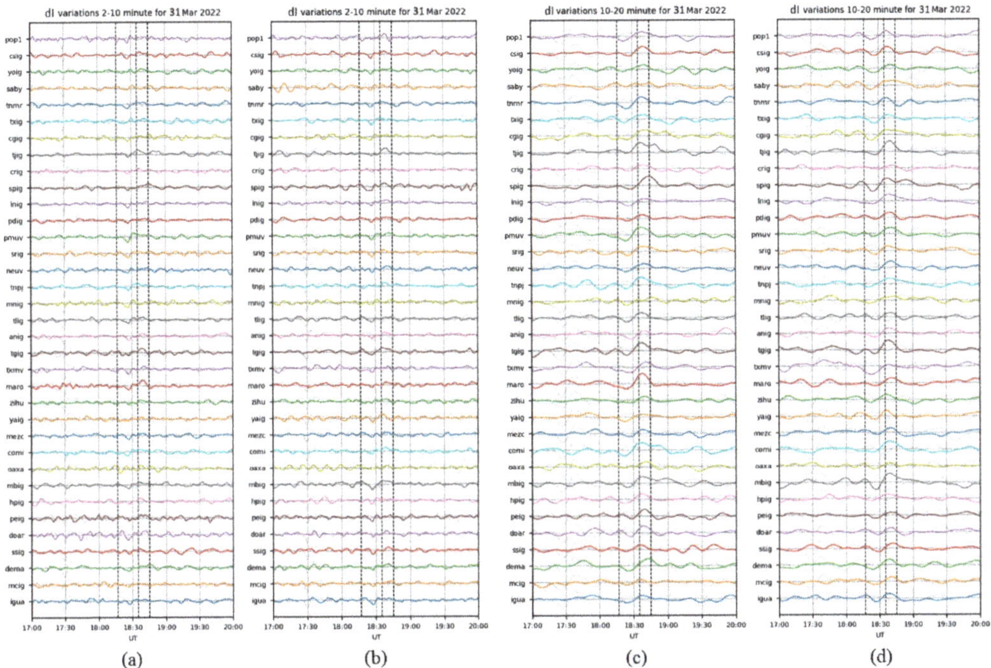

Figure 13. dI(2–10) variations at LoSs to satellites G14 (**a**) and (**b**) G30 and dI(10–20) variations at LoSs to satellites G14 (**c**) and (**d**) G30 for the case of X.3-flare on 31 March 2022. The Y-axis unit is 0.3 TECU. The dashed lines mark the flare onset, peak and end. Colored dI curves stand for the day of the flare and the gray curves stand for 30 March 2022.

Considering the general small amplitude of dI variations, we constructed the maps of the sub-ionospheric points at all LoSs before, during and after the flares as it results that such a representation of results allows us to better appreciate the flare effects on dI. The maps of the sub-ionospheric points were constructed with a time step of 30 s. The examples are given in Figure 14. Essentially, the green color of the points means no deviation of dI(10–20) over this point; the blue color means negative and the red color means positive deviation. The animation of the sequence of maps is available as Supplementary Materials to this article (https://zenodo.org/record/7363050, accessed on 25 December 2022).

There are different works dedicated to the roles of position of AR responsible for the flare, solar zenith angle at the moment of the flare and flare X-class in relation to the ionosphere response to the flare [18] and references therein. According to the literature, the impact of X-ray radiation during flares does not depend much on the flare's AR position on the solar disk. In contrast, EUV radiation during flares is more geo-effective if the corresponding AR is in the center of the solar disk [32,33]. Therefore, the smaller effects of the M9.6-flare on the lower ionosphere may be explained by the weaker X-class of the flare and on the F2 layer and topside ionosphere, at least partially, by the AR position, which was further from the solar disk center than during the X1.3-flare. The issue of the influence of the solar zenith angle on the ionospheric response was discussed by different authors, for example [2,10,30,31]. In our case, the solar zenith angle in the center of Mexico was $Z(sun) = {\sim}26°$ during the first flare and $Z(sun) = {\sim}17°$ during the second flare, meaning that in the second case the $Z(sun)$ conditions were more favorable to cause the major effects, as it was closer to the noon. However, smaller dI deviations were registered during the second flare. Considering the similarity of the other observation conditions (Table 1), this implies that the ionospheric response depended more on the AR position and the flare class than on the solar zenith angle. Nevertheless, this is the preliminary conclusion, as more statistics is needed.

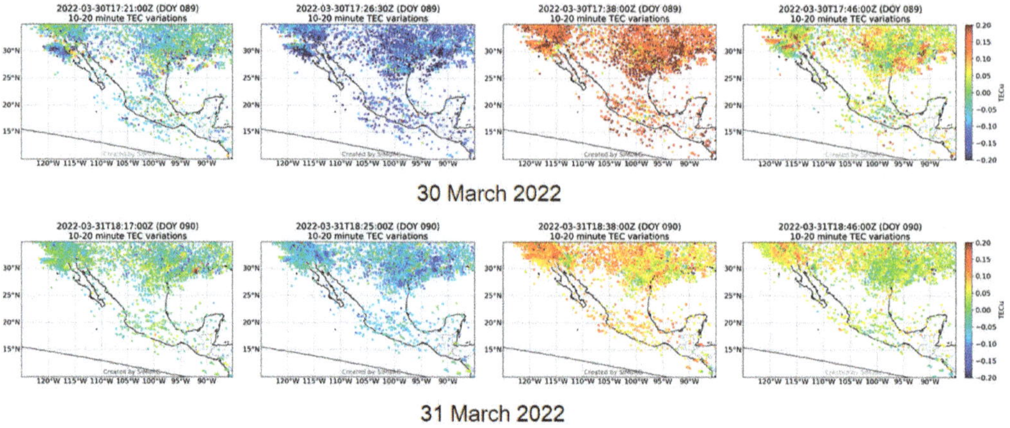

Figure 14. Maps of sub-ionospheric points over Mexico for 30 March 2022 (**upper panels**) and 31 March 2022 (**lower panels**). Deviations of dI(10–20) are represented by the color of each point. The maps are shown for the following moments from left to right: flare onset, major negative deviation, major positive deviation and flare end.

The presence of the negative bay in the dI variation (И-form) can be a result of either the sTEC filtering procedure or the physical effect. There are works that evidenced negative TEC deviations during flares. For example, Thome and Wagner [34] reported negative disturbances (3–10%) within heights of 280–600 km as a response to two flares. According to the simulation results by Leonovich and Taschilin [35], the electron concentration decrease in the outer ionosphere can be caused by O+ ion outflow towards the plasmasphere. After the "switching off" of a flare, the plasma pressure in the F2 region decreases rapidly and the difference in pressure between the upper and lower ionosphere cannot maintain O+ ion flow in the upward direction. The ionosphere returns to its undisturbed state [35]. There are more works on this issue. Mendillo and Evans [5] reported such an upward outflow caused by flares. Mendillo et al. [16] showed that an enhancement of upward plasma flux occurs for large flares observed on Earth. To provide a more recent example, Liu et al. [19] studied the ion upflow near the X9.3-flare peak on 6 September 2017 and found that ambipolar diffusion enhancement is the main driver for it. The last mentioned work also provides a list of references for different mechanisms proposed to explain the ionospheric upflow in the literature. In addition, the reader can find a list of publications on the dynamics resulting in the ionospheric outflow of O+ ions in work by Lin and Ilie [36]. From the other hand, Yasyukevich et al. [22] applied the same methodology as in the present work to study solar flare effects during September 2017. According to these authors, as the filtering results in a zero mean value, any dI increase is shown as a decrease with the minimum at the moment of the event onset. In our case, the dI minimum was between the flare onset and the rapid increase of radiation to its peak during the flare. Additional studies are needed, especially since the processing methods used by different authors are different. To verify the "physical" version in our case and to study the ion/electron behavior exactly in the F2 ionospheric layer and/or higher, low-orbit data are needed. Unfortunately, no satellite with the necessary data passed over the North American low-latitude region at the moments of the flares. Therefore, we cannot draw a definitive conclusion on this issue. According to the ionograms, foF2 did not show (short-term) rapid large deviations, which means there was no much change in the electron concentration during the flares at the F2 layer heights. Although foF2 had some decrease after the first flare onset, it was very small (~1% of the value). In addition, it should be noted that the foF2 values showed significant day-to-day variability. Basically, short-term foF2 variations before, during and after the flares on 20 and 31 March 2022 did not seem different from the background fluctuations during the same period on the reference day of 1 April. Therefore, it can be inferred that the detected dI response was a result of changes at the higher altitudes, as it is known that the ionospheric regions above the F2 layer can contribute, by various estimates, up to 25–30% of the TEC value (e.g., [37]). Mendillo et al. [10] studied the TEC response to the great flare with data from North America, Europe and Africa and also emphasized the role of the changes above the F2-layer peak height.

Furthermore, due to the application of moving average as a filter to the dI time series, it is difficult to estimate the real duration of the effects in the ionosphere. In this regard, for both flares, the characteristic coincidence of the phases of dI(10–20) variation and flare phases was noted. First, at the moment of flare onset, dI(10–20) began to decrease and further to increase. The increase was maximal at the moment of the flare peak. Then, the dI(10–20) value decreased to its pre-flare value. This recovery ended at the moment of the flare end (examples given in Figures 11 and 12). The described pattern seems interesting and may not be ignored. We will proceed from the assumption that dI(10–20) are closer by their time scale to the duration of the sTEC response to the flares.

4.2.2. Correlation with Different Ranges of EUV and X-ray Radiation

The assumption that the response of the ionosphere in the F2 region and above depends on the particular level of radiation in different EUV spectral bands was made previously in several classical [1,13] and more recent works [16]. It would be interesting to estimate, at least qualitatively, their impacts on the dI change. Similar to Section 4.1.3, the coefficient r was calculated between the time series of dI and EUV and the X-ray fluxes of different spectral bands. The 30 s dI series were adjusted to 1 min GOES data. It was shown that the dI response at different LoSs was of the same pattern; consequently, it was sufficient to calculate r for the parameters only at one LoS. For this purpose, the time series of dI(2–10), dI(10–20) and ROTI at the ray path between the GPS satellite G17 and YOIG station (considered before) were used for the analysis. We recall that dI(2–10) corresponds to irregularities of the small scale and dI(10–20) to those of the medium scale. ROTI was involved as an additional parameter to better understand dI variations because it is always positive, which may be useful to identify the beginning of the ionospheric response. The length of the time series was limited by the flare duration because (as shown above) the dI variations are not exclusively affected by flares. They can present other features over a longer period. The results are presented in Tables 6 and 7. Considering the overall strong dependence of the ionosphere behavior on solar radiation as well as the fact that the fluxes of UV and soft X-rays at wavelengths correlate with each other, for our particular case we consider correlation to be strong if $r \geq 0.9$ and moderate if $0.7 \leq r \leq 0.9$.

Table 6. Correlation between sTEC derivatives and fluxes of different spectral bands during X1.3-flare.

r (17:20–17:46 UT)	EUV 25.6 nm	EUV 28.4 nm	EUV 30.4 nm	EUV 117.5 nm	EUV 121.6 nm	EUV 133.5 nm	EUV 140.5 nm	X (0.05–0.4) nm	X (0.1–0.8) nm
dI(2-10)	0.60	0.39	0.63	0.65	0.59	0.60	0.58	0.64	0.58
dI(10-20)	0.93	0.74	0.87	0.71	0.84	0.70	0.51	0.93	0.93
ROTI	0.51	0.17	0.68	0.92	0.63	0.90	0.95	0.54	0.43

Table 7. The same as in Table 6, but for M9.6-flare.

r (18:16–18:45 UT)	EUV 25.6 nm	EUV 28.4 nm	EUV 30.4 nm	EUV 117.5 nm	EUV 121.6 nm	EUV 133.5 nm	EUV 140.5 nm	X (0.05–0.4) nm	X (0.1–0.8) nm
dI(2-10)	0.54	0.59	0.65	0.51	0.66	0.58	0.46	0.34	0.40
dI(10-20)	0.97	0.91	0.95	0.50	0.91	0.64	0.31	0.73	0.87
ROTI	−0.44	−0.44	0.51	−0.54	−0.59	−0.58	−0.50	−0.34	−0.35

In general, during the more intense X1.3-flare correlation between the dI series filtered with the (10–20) min window and the fluxes of all spectral bands was rather high. Radiation at 25.6 nm had the major impact on dI(10–20). Surprisingly (as EUV is known to play the main role in the higher ionosphere), X-ray flux of both shorter and longer wavelengths showed a strong correlation with dI(10–20) (of the same level as EUV at 25.6 nm). Correlations for dI(2–10) were mostly weak. Correlations for ROTI were strong to 117.5, 133.5 and 140.5 nm and almost absent for other bands. In general, r for the 28.4 nm band (see Figure 14 upper panel) was rather low (similarly to the lower ionosphere). This was probably due to the more gradual increase at this wavelength.

Figure 15 (upper panel) illustrates the sTEC derivative variations from Table 6 and the bursts of radiation in the chosen bands. Similar to Figure 10, EUV and X-ray flux values are multiplied by the specific coefficient for each band (coefficient values are not listed as they have no physical meaning) for illustrative purposes. Both dI(2–10) and dI(10–20) increases to their peak values occurred with the flux intensification at the mentioned EUV bands and the corresponding X-ray burst. The beginning ROTI increase corresponded to an increase of X-rays as well. At that, we should note that due to the rules of ROTI calculation, there is an uncertainty of its peak beginning within a 5 min interval. This probably explains why ROTI and dI(2–10) response beginnings are not synchronized in the upper plot of Figure 14. If the ROTI curve is "moved" 4 min to the left on the *X*-axis, its response beginning will correspond to a dI(2–10) decrease.

Positive dI(2–10) and dI(10–20) manifested themselves with the EUV flux increase at particular wavelengths (dotted curves). Negative dI bay is seen after the onset of the flare at 17:21 UT before the peaks of radiation at different bands. It is worth noting that the flare onset is defined as the first minute in a sequence of 4 min of the monotonic increase in (0.1–0.8) nm flux. The fact that the transition from negative to positive dI occurs exactly at the moment of the solar radiation's rapid increase to its peak calls attention. Maybe, it suggests that during the slow increase of radiation, a certain process is responsible for the dI decrease and then, during the rapid growth of radiation, another process (ionization) begins to dominate in the dI behavior. Still, as if there is no certainty that the dI transition from negative to positive values has a physical meaning.

As for the less intense flare on 31 March, it was shown that it provoked mostly medium-scale irregularities. This is confirmed by the results in Table 7: only correlations for dI(10–20) are worth noting. ROTI did not show any response in this case. In contrast to the more mixed influence during the more intense flare, EUV bands that impacted more can be clearly revealed. In particular, the 25.6, 28.4, 30.4 and 121.6 nm spectral bands showed strong correlation with the ionospheric response. Radiation at 25.6 nm played a major role (very strong correlation, r = 0.97). Similar to the previous case, negative dI are observed after the flare onset and the transition from negative to positive dI occurred exactly at the moment of the beginning of the radiation bursts. In addition, we recall that during M9.6, the positive dI(10–20) were rather weak at some LoSs (probably due to the less intense radiation), but the negative dI(10–20) bays were well detected.

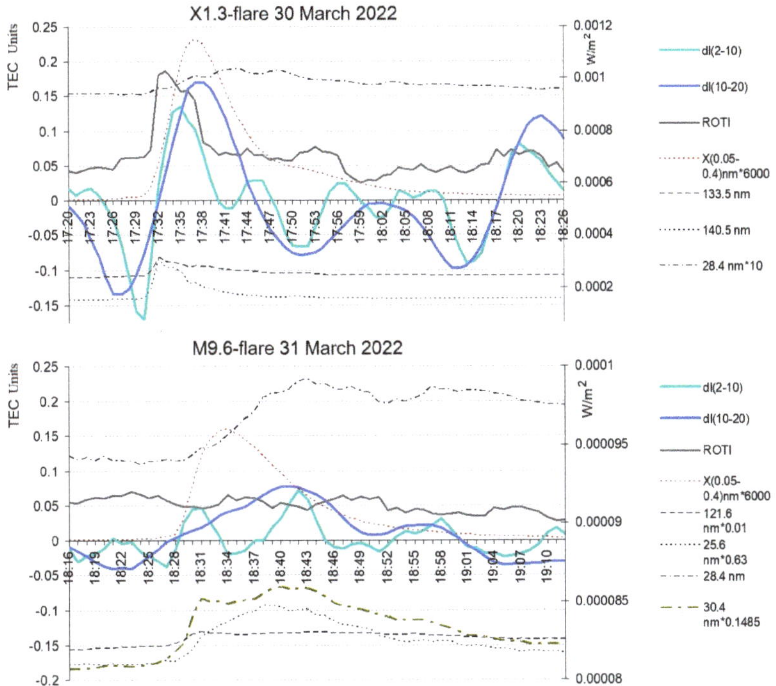

Figure 15. dI variations filtered with the windows (2–10) and (10–20) and ROTI variations (continuous curves) during the X1.3 flare on 30 March 2022 (**upper panel**) and the M9.6 flare on 31 March 2022 (**lower panel**). Dotted lines schematically show the bursts of solar radiation in the chosen spectral bands.

It was shown previously that the flare impact on the ionosphere depends on the amount of radiation emitted in different EUV and X-ray bands [11,19,35,38]. Figure 5 in [37] shows an interesting picture of how increases in different solar radiation spectral intervals influence the contribution of different ionospheric regions (heights) to the TEC value. According to these authors, the electron content increase in the topside ionosphere is related to EUV spectral bands of 55–65 and 85–95 nm. Berdermann et al. [11] stated that the EUV component around 30 nm can seriously affect GNSS positioning services. According to Hernández-Pajares et al. [38], vertical TEC is highly correlated with EUV photon flux at the 26–34 nm spectral band, which is geo-effective in the ionization of mono-atomic oxygen in the atmosphere. Regarding the possible electron density decrease in the topside ionosphere, Leonovich and Taschilin [35] concluded that it can be related to changes in the 15–20 nm, 30–35 nm and 35–40 nm bands. Liu et al. [19] affirmed that the changes in the vertical plasma density gradient are mainly due to EUV 15.5–79.8 nm.

In our case during both flares, EUV radiation at 25.6 nm played the major role, which is in accord with the results of some authors mentioned above. Surprisingly, during the X1.3-flare X-ray flux, the impact was of the same level as the EUV 25.6 nm in the F2 region. During the M9.6-flare, EUV spectral bands of 28.4, 30.4 and 121.6 nm also showed strong correlation with the ionospheric F2 response. Basically, the comparison of our results with other works implies that the dI decrease was a physical effect.

To sum up, a coherent dI response was observed at all LoSs during the first flare and almost at all LoSs during the second flare. X1.3 flare provoked small- and medium-scale ionospheric irregularities. The M9.6-flare provoked mostly medium-scale irregularities. These disturbances can be characterized by the electron concentration change of 1-5% of the background variation (small-scale) and 5-30% of the background variation (medium-scale).

Such deviations correspond to the dI amplitudes. According to the form of the dI(10–20) curves and the results of Tables 6 and 7, it seems that the duration of the ionospheric response corresponds to dI variation filtered with the windows (10–20) min. During both flares, the response consisted of the negative dI bay observed after the onset of the flare, which was followed by the positive dI peak observed between the rapid increase of radiation at particular EUV and X-ray wavelengths and the end of the flare. During the less intense flare, the positive dI deviation was not pronounced at some LoSs. In general, the amplitudes of dI(10–20) deviations were lower than in the case of the more intense flare. To note, in both cases, the response was not intense, as other effects (e.g., multipath effect) manifested themselves in the dI variations stronger. The presence of the negative dI may be explained by the filtering procedure or have the physical meaning. There are some indirect indications in favor of the latter. However, without low-orbit satellite data, it is impossible to make a definitive conclusion on this issue.

5. Conclusions

Manifestations of two intense flares were studied in the ionosphere of the low- to mid-latitude North American sector (mostly Mexican region). The end of March 2022 was characterized by the notable change in the background solar activity represented by the F10.7-index. However, this circumstance did not have much effect on the ionospheric response to the flares on 30–31 March 2022. This is because during the flares, not only the intensification of solar radiation but also the significant change in the spectrum of solar radiation (radiation at different bands) plays an important role. Studying the impact of these processes on such regions as Mexico, combining several methods, allows us to compare the results with previous studies and obtain new data. This is especially important because of the increase in the number of solar flares expected due to the solar activity growth at the ascending part of the solar cycle 25. The peculiarity of the present study is the high temporal resolution of the ionosonde data (2 min) and the GNSS-derived data (30 s). The following results were obtained:

(1) The more intense X1.3-flare on 30 March 2022 provoked a more significant response in the lower ionosphere (including the HF fadeout event) than the less intense M9.6-flare on 31 March 2022.

(2) The narrowing of the frequency operation range and the signal amplitude decrease on ionograms manifested themselves with a ~3 min delay from the X1.3-flare onset and ~5 min delay from the M9.6-flare onset.

(3) X1.3-flare caused the shortwave fadeout near the flare peak moment (maximal radiation), which lasted ~6 min in Mexico and ~(1–8) min in Eglin. No fadeout event was caused by the M9.6-flare. For both considered flares, the maximal flare impact on the amplitude of the received signal in Mexico was observed ~(14–15) min after the flare beginning.

(4) The recovery of the lower ionosphere conditions (absorption) began before the X1.3 flare end and lasted after it, being more rapid at Eglin than in Mexico. As for M9.6 flare, the responses at two observation points were rather synchronized due to the more similar illumination conditions (LT of the flare) at two meridians than the day before, when X1.3-flare occurred.

(5) The overall X1.3-flare effects in the lower ionosphere lasted ~1 h 24 min and those of the M9.6-flare ~40 min. In both cases, the effects in the lower ionosphere lasted longer than the flare duration, in contrast to the effects revealed by GNSS data (F2 layer and topside ionosphere).

(6) The analysis of different spectral bands of solar radiation confirmed the major role of X-rays in the electron concentration increase in the lower ionosphere during the flares. At the same time, the analysis showed that the EUV impact on the lower ionosphere cannot be totally discarded. It was more pronounced for the more intense flare.

(7) For the particular cases of the two considered flares, EUV emission in the 28.4 nm spectral band provoked the weakest impact on the absorption conditions in the lower

ionosphere (probably due to the smooth gradual increase of radiation in this band) and the 133.5 nm band had the most pronounced impact.

(8) A coherent (simultaneous) dI response (implying the integral response in the F2-layer and topside ionosphere) was observed at all LoSs during the X1.3-flare and almost at all LoSs during the M9.6-flare. Both consisted of the negative dI bay after the flare onset, followed by the positive dI peak observed between the rapid increase of radiation at the particular EUV and X-ray wavelengths and the end of the flare. During the less intense flare, the positive dI were not pronounced at some LoSs.

(9) In both cases, the dI response was not as intense as the other effects (e.g., multipath effect). Notwithstanding the small amplitudes of dI deviations, the simultaneity and uniformity of the response at all LoSs, as well as the moment of the ionospheric response occurrence, leave no doubt that it was caused exactly by the flare. The ionospheric response duration corresponded to dI filtered with the centered moving average with (10–20) min windows. The dI(10–20) deviation amplitudes were smaller during the less intense flare.

(10) The preliminary conclusion is that the ionospheric response by dI depended more on the AR position on the Sun and the flare class than on the solar zenith angle. However, more statistics is needed.

(11) It was revealed that the X1.3-flare provoked small- and medium-scale ionospheric irregularities, and the M9.6-flare provoked mostly medium-scale irregularities in the F-layer.

(12) The presence of the negative dI may be explained by the filtering procedure or have the physical meaning. There are some indirect indications in favor of the last mentioned. However, in the absence of low-orbit satellite data, it was impossible to make a definitive conclusion on this issue.

(13) EUV is known to play the main role in the F2 region of the ionosphere. During the X1.3-flare, EUV radiation at 25.6 nm had the major impact on dI(10–20). Surprisingly, X-ray flux of both shorter and longer wavelengths showed a strong correlation with dI(10–20) which was of the same level as EUV at 25.6 nm. ROTI was affected by 117.5, 133.5 and 140.5 nm bands. Similar to the lower ionosphere, the 28.4 nm band impact on the F2-layer was rather low but only during the X1.3-flare. During the M9.6-flare, 25.6, 28.4, 30.4 and 121.6 nm EUV spectral bands showed strong correlation with the ionospheric response. Radiation at 25.6 nm again played a major role (very strong correlation, r = 0.97).

Supplementary Materials: Maps of sub-ionospheric points over Mexico for the flare periods during 30 March 2022 and 31 March 2022 are available in the Zenodo repository at https://doi.org/10.5281/zenodo.7363050 (accessed on 27 November 2022).

Author Contributions: Conceptualization, M.A.S.; methodology, M.A.S., O.A.M. and A.M.V.; experiment design, M.A.S., A.M.V., A.G.C. and D.V.B.; experiment performance and data acquisition, E.A.-M., P.V., I.D.O.-L., V.J.G.-A., A.M.-M., M.R.-M., E.A.-R. and L.X.G.; data processing and validation, A.M.V., I.D.O.-L., A.G.C., V.J.G.-A., A.M.-M. and L.X.G.; data analysis, M.A.S., D.V.B., O.A.M. and J.A.G.-E.; visualization, M.A.S., A.M.V. and M.R.-M.; resources, J.A.G.-E.; writing—original draft preparation, M.A.S. and O.A.M.; writing—review and editing, M.A.S. and O.A.M. All authors have read and agreed to the published version of the manuscript.

Funding: LANCE acknowledges partial support from CONACyT-AEM, Grant 2017-01-292684 and CONACyT LN-315829. L.X. Gonzalez was supported by the CONACyT-AEM Grant AEM-2018-01-A3-S-63804. E. Aguilar-Rodriguez was supported by DGAPA/PAPIIT project IN103821. M. Rodriguez-Martinez was supported by the DGAPA PAPIME project PE103419 and the CONACyT grant INFR:253691. O.A. Maltseva was supported by Ministry of Science and Higher Education of the Russian Federation (State task in the field of scientific activity 2023). A. Vesnin was supported by the Ministry of Education and Science (Basic Research Program II.16). V.J. Gatica-Acevedo and A. Melgarejo-Morales express their gratitude to CONACyT.

Data Availability Statement: The OMNI data (Dst and F10.7 indices) were obtained from the GSFC/SPDF OMNIWeb interface at https://omniweb.gsfc.nasa.gov (accessed on 25 December 2022). The authors also thank NOAA's National Centers for Environmental Information and NOAA's Space Weather Prediction Center for the opportunity to use GOES satellite data available at https://data.ngdc.noaa.gov/platforms/solar-space-observing-satellites/goes/ (accessed on 25 December 2022) and at www.swpc.noaa.gov/products/goes-x-ray-flux (accessed on 25 December 2022), correspondingly. The data shown in Figure 2b,c are supplied by courtesy of SolarMonitor.org. The sTEC values used to obtain dI series were derived from GNSS data from stations of (a) TLALOCNet and SSN-TLALOCNet networks, whose data is available at both TLALOCNet (http://tlalocnet.udg.mx, accessed on 25 December 2022) and UNAVCO (http://unavco.org, accessed on 25 December 2022) data archives and (b) SSN GPS network, whose data is available upon request to SSN (www.ssn.unam.mx, accessed on 25 December 2022). The authors express their gratitude to the services of DIDBase (https://ulcar.uml.edu/DIDBase, accessed on 25 December 2022) for the ionosonde data available for educational purposes. Other data used in this study are available in the Zenodo repository at doi:10.5281/zenodo.7363174 (accessed on 25 December 2022).

Acknowledgments: The authors also thank NOAA's National Centers for Environmental Information for the opportunity to download GOES satellite data. This work is based on GPS data provided by TLALOCNet at Servicio de Geodesia Satelital [39] and the Servicio Sismológico Nacional [40], both at the Instituto de Geofísica, Universidad Nacional Autónoma de México (UNAM) with support from UNAM-PAPIIT project IN107321, and the National Science Foundation grant 2025104 to the College of New Jersey, and also the GAGE facility operated by UNAVCO, Inc. with support from the National Science Foundation, the National Aeronautics and Space Administration, and the U.S. Geological Survey under NSF Cooperative Agreement EAR-1724794. We gratefully acknowledge the personnel from SGS, especially Luis Salazar-Tlaczani, as well as all the personnel from SGS and UNAVCO, for station maintenance, data acquisition, IT support and data distribution. GNSS station locations are indicated in Figure 3. The maps of sub-ionospheric points and animations based on these maps were constructed using the SIMuRG online tool [25]. The authors express their gratitude to this service for the opportunity of map construction and to GNSS data providers whose data are used by SIMuRG, in particular the following networks: International GNSS Service (IGS), UNAVCO, Northern California Earthquake Data Center, of Continuously Operating Reference Stations (CORS) managed by the National Geodetic Survey (NGS), an office of NOAA's National Ocean Service. We thank the anonymous reviewers for their comments and suggestions.

References

1.	Bryunelli, B.E.; Namgaladze, A.A. *Physics of the Ionosphere*; Nauka: Moscow, Russia, 1988; 528p, ISBN 5-02-000716-1.
2.	Davies, K. *Ionospheric Radio Propagation*; Monograph 80; National Bureau of Standards: Gaithersburg, MD, USA, 1965; 487p.
3.	Mitra, A.P. *Ionospheric Effect of Solar Flares*; Reidel: Norwell, MA, USA, 1974.
4.	Berngard, O.I.; Ruohoniemi, J.M.; Nishitani, N.; Shepherd, S.G.; Bristow, W.A.; Miller, E.S. Attenuation of decameter wavelength sky noise during X-ray solar flares in 2013–2017 based on the observations of midlatitude HF radars. *J. Atmos. Sol.-Terr. Phys.* **2018**, *173*, 1–13. [CrossRef]
5.	Mendillo, M.; Evans, J.V. Incoherent scatter observations of the ionospheric response to a large solar flare. *Radio Sci.* **1974**, *9*, 197–203. [CrossRef]
6.	Blagoveschensky, D.V.; Sergeeva, M.; Raita, T. Riometer absorption during four similar storms. *Adv. Space Res.* **2022**, *69*, 1. [CrossRef]
7.	Barta, V.; Sátori, G.; Berény, K.A.; Kis, A.; Williams, E. Effects of solar flares on the ionosphere as shown by the dynamics of ionograms recorded in Europe and South Africa. *Ann. Geophys.* **2019**, *37*, 747–761. [CrossRef]
8.	Zaalov, N.Y.; Moskaleva, E.V.; Rogov, D.D.; Zernov, N.N. Influence of X-ray and polar cap absorptions on vertical and oblique sounding ionograms on different latitudes. *Adv. Space Res.* **2015**, *56*, 2527–2541. [CrossRef]
9.	Liu, L.B.; Wan, W.X.; Chen, Y.D.; Le, H.J. Solar activity effects of the ionosphere: A brief review. *Chin. Sci. Bull.* **2011**, *56*, 1202–1211. [CrossRef]
10.	Mendillo, M.; Klobuchar, J.A.; Fritz, R.B.; da Rosa, A.V.; Kersle, L.; Yeh, K.C.; Flaherty, B.J.; Rangaswamy, S.; Schmid, P.E.; Evans, J.V.; et al. Behavior of the Ionospheric F Region during the Great Solar Flare of 7 August 1972. *J. Geophys. Res.* **1974**, *79*, 4.

11. Berdermann, J.; Kriegel, M.; Banyrs, D.; Heymann, F.; Hoque, M.M.; Wilken, V.; Borries, C.; HeBelbarth, A.; Jakowski, N. Ionospheric response to the X9.3 Flare on 6 September 2017 and its implication for navigation services over Europe. *Space Weather* **2018**, *16*, 1604–1615. [CrossRef]

12. Rathore, V.S.; Kumar, S.; Singh, A.K.; Singh, A.K. Ionospheric response to an intense solar flare in equatorial and low latitude region. *Indian J. Phys.* **2018**, *92*, 1213–1222. [CrossRef]

13. Kelley, M.C. *The Earth's Ionosphere: Plasma Physics and Electrodynamics*, 2nd ed.; International Geophysics Series 96; Elsevier: Amsterdam, The Netherlands, 2009; ISBN 978-0-12-088425-4.

14. Hunsucker, R.D.; Hargreaves, J.K. *The High-Latitude Ionosphere and Its Effects on Radio Propagation*; Cambridge University Press: Cambridge, UK, 2003.

15. Dmitriev, A.V.; Yeh, H.-C.; Chao, J.-K.; Veselovsky, I.S.; Su, S.-Y.; Fu, C.C. Top-side ionosphere response to extreme solar events. *Ann. Geophys.* **2006**, *24*, 1469–1477. [CrossRef]

16. Mendillo, M.; Erickson, P.J.; Zhang, S.-R.; Mayyasi, M.; Narvaez, C.; Thiemann, E.; Chamberlain, P.; Andersson, L.; Peterson, W. Flares at Earth and Mars: An ionospheric escape mechanism? *Space Weather* **2018**, *16*, 1042–1056. [CrossRef]

17. Leonovich, L.A.; Tashchilin, A.V. Disturbances in the Topside Ionosphere during Solar Flares. *Geomagn. Aeron.* **2008**, *48*, 759–767. [CrossRef]

18. Afraimovich, E.L.; Astafyeva, E.I.; Demyanov, V.V.; Edemskiy, I.K.; Gavrilyuk, N.S.; Ishin, A.B.; Kosogorov, E.A.; Leonovich, L.A.; Lesyuta, O.S.; Palamartchouk, K.S.; et al. A review of GPS/GLONASS studies of the ionospheric response to natural and anthropogenic processes and phenomena. *J. Space Weather Space Clim.* **2013**, *3*, A27. [CrossRef]

19. Liu, X.; Liu, J.; Wang, W.; Zhang, S.-R.; Zhang, K.; Lei, J.; Liu, L.; Chen, X.; Li, S.; Zhang, Q.-H.; et al. Explaining solar flare-induced ionospheric ion upflow at Millstone Hill (42.6 N). *J. Geophys. Res.-Space* **2022**, *127*, e2021JA030185. [CrossRef]

20. Habarulema, J.B.; Tshisaphungo, M.; Katamzi-Joseph, Z.T.; Matamba, T.M.; Nndanganeni, R. Ionospheric response to the M- and X-class solar flares of 28 October 2021 over the African sector. *Space Weather* **2022**, *20*, e2022SW003104. [CrossRef]

21. Le, H.J.; Liu, L.B.; Chen, Y.D.; Zhang, H. Anomaly distribution of ionospheric total electron content responses to some solar flares. *Earth Planet. Phys.* **2019**, *3*, 481–488. [CrossRef]

22. Yasyukevich, Y.; Astafyeva, E.; Padokhin, A.; Ivanova, V.; Syrovatskii, S.; Podlesnyi, A. The 6 September 2017 X-class solar flares and their impacts on the ionosphere, GNSS, and HF radio wave propagation. *Space Weather* **2018**, *16*, 1013–1027. [CrossRef] [PubMed]

23. Reinisch, B.W.; Galkin, I.A. Global ionospheric radio observatory (GIRO). *Earth Planets Space* **2011**, *63*, 377–381. [CrossRef]

24. Maletckii, B.; Yasyukevich, Y.; Vesnin, A. Wave Signatures in Total Electron Content Variations: Filtering Problems. *Remote Sens.* **2020**, *12*, 1340. [CrossRef]

25. Yasyukevich, Y.V.; Kiselev, A.V.; Zhivetiev, I.V.; Edemskiy, I.K.; Syrovatskii, S.V.; Maletckii, B.M.; Vesnin, A.M. SIMuRG: System for Ionosphere Monitoring and Research from GNSS. *GPS Solut.* **2020**, *24*, 69. [CrossRef]

26. Pi, X.; Mannucci, A.J.; Lindqwister, U.J.; Ho, C.M. Monitoring of global ionospheric irregularities using the worldwide GPS-network. *Geophys. Res. Lett.* **1997**, *24*, 2283–2286. [CrossRef]

27. Sergeeva, M.A. Space Weather General Concepts. In *Book Space Weather Impact on GNSS Performance*; Springer: Cham, Switzerland, 2022; pp. 89–150. [CrossRef]

28. Machol, J.L.; Eparvier, F.G.; Viereck, R.A.; Woodraska, D.L.; Snow, M.; Thiemann, E.; Woods, T.N.; McClintock, W.E.; Mueller, S.; Eden, T.D.; et al. GOES-R Series Solar X-ray and Ultraviolet Irradiance. In *Book GOES-R Series*; Elsevier: Amsterdam, The Netherlands, 2020; pp. 233–242. [CrossRef]

29. Syrovatskiy, S.V.; Yasyukevich, Y.V.; Vesnin, A.M.; Edemskiy, I.K.; Voeykov, S.V.; Zhivetiev, I.V. The effect of solar flares on the ionosphere of the Earth during 24th cycle of solar activity. *Mem. Fac. Phys. Mosc. Univ.* **2018**, *4*, 1840403.

30. Kunitsyn, V.E.; Nazarenko, M.O.; Nesterov, I.A.; Padokhin, A.M. Solar Flare Forcing on Ionization of Upper Atmosphere. Comparative Study of Several Major X_Class Events of 23rd and 24th Solar Cycles, Physics of Earth, Atmosphere and Hydrosphere. *Mosc. Univ. Phys. B* **2015**, *70*, 312–318. [CrossRef]

31. Le, H.; Liu, L.; Chen, Y.; Wan, W. Statistical analysis of ionospheric responses to solar flares in the solar cycle 23. *J. Geophys. Res.-Space* **2013**, *118*, 576–582. [CrossRef]

32. Donnelly, R.F. Empirical models of solar flare X ray and EUV emission for use in studying their E and F region effects. *J. Geophys. Res.* **1976**, *81*, 4745–4753. [CrossRef]

33. Qian, L.; Burns, A.G.; Chamberlin, P.C.; Solomon, S.C. Flare location on the solar disk: Modeling the thermosphere and ionosphere response. *J. Geophys. Res.* **2010**, *115*, e2009JA015225. [CrossRef]

34. Thome, G.D.; Wagner, L.S. Electron density enhancements in the E and F regions of the ionosphere during solar flares. *J. Geophys. Res.* **1981**, *76*, 6883–6895. [CrossRef]

35. Leonovich, L.A.; Taschilin, A.V. Aeronomic effects of the solar flares in the topside ionosphere. *Earth Planets Space* **2009**, *61*, 643–648. [CrossRef]

36. Lin, M.-Y.; Ilie, R. A Review of Observations of Molecular Ions in the Earth's Magnetosphere-Ionosphere System. *Front. Astron. Space Sci.* **2022**, *8*, 745357. [CrossRef]

37. Leonovich, L.A.; Tashchilin, A.V.; Portnyagina, O.Y. Dependence of the Ionospheric Response on the Solar Flare Parameters Based on the Theoretical Modeling and GPS Data. *Geomagn. Aeron.* **2010**, *50*, 201–210. [CrossRef]

38. Hernández-Pajares, M.; García-Rigo, A.; Juan, J.M.; Sanz, J.; Monte, E.; Aragón-Àngel, A. GNSS measurement of EUV photons flux rate during strong and mid solar flares. *Space Weather* **2012**, *10*, e2012SW000826. [CrossRef]
39. Cabral-Cano, E.; Pérez-Campos, X.; Márquez-Azúa, B.; Sergeeva, M.A.; Salazar-Tlaczan, L.; DeMets, C.; Adams, D.; Galetzka, J.; Hodgkinson, K.; Feaux, K.; et al. TLALOCNet: A Continuous GPS-Met Backbone in Mexico for Seismotectonic, and Atmospheric Research. *Seismol. Res. Lett.* **2018**, *89*, 373–381. [CrossRef]
40. Pérez-Campos, X.; Espíndola, V.H.; Pérez, J.; Estrada, J.A.; Monroy, C.C.; Bello, D.; González-López, A.; Gonzalez Avila, D.; Contreras Ruiz Esparza, M.G.; Maldonado, R.; et al. The Mexican National Seismological Service: An Overview. *Seismol. Res. Lett.* **2018**, *89*, 318–323. [CrossRef]

Technical Note

High-Resolution Observation of Ionospheric E-Layer Irregularities Using Multi-Frequency Range Imaging Technology

Bo Chen [1], Yi Liu [1], Jian Feng [2], Yuqiang Zhang [1,*], Yufeng Zhou [3], Chen Zhou [1] and Zhengyu Zhao [1]

1   Department of Space Physics, School of Electronic Information, Wuhan University, Wuhan 430072, China
2   China Research Institute of Radiowave Propagation (CRIRP), Qingdao 266107, China
3   Beijing Institute of Applied Meteorology, Beijing 100029, China
*   Correspondence: yqzhang_3@stu.xidian.edu.cn

Abstract: E-region field-aligned irregularities (FAIs) are a hot topic in space research, since electromagnetic signal propagation through ionospheric irregularities can undergo sporadic enhancements and fading known as ionospheric scintillation, which could severely affect communication, navigation, and radar systems. However, the range resolution of very-high-frequency (VHF) radars, which is widely used to observe E-region FAIs, is limited due to its bandwidth. As a technology that is widely used in atmosphere radars to improve the range resolution of pulsed radars by transmitting multiple frequencies, this paper employed the multifrequency radar imaging (RIM) technique in a Wuhan VHF radar. The results showed that the range resolution of E-region FAIs greatly improved when compared with the results in traditional single-frequency mode, and that finer structures of E-region FAIs can be obtained. Specifically, the imaging results in multifrequency mode show that E-region FAIs demonstrate an overall descending trend at night, and it could be related to the tides or gravity waves due to their downward phase velocities or even driven by downwind shear. In addition, typical quasi-periodic (QP) echoes with a time period of around 10 min could be clearly seen using the RIM technique, and the features of the echoes suggest that they could be modulated by gravity waves. Furthermore, the RIM technique can be used to obtain the fine structure of irregularities within a short time period, and the hierarchical structure of E-region FAIs can be easily found. Therefore, the multifrequency imaging RIM technique is suitable for observing E-region FAIs and their evolution, as well as for identifying the different layers of E-region FAIs. Combined with the RIM technique, a VHF radar provides an effective and promising way to observe the structure of E-region FAIs in more detail to study the physical mechanism behind the formation and evolution of ionospheric E-region irregularities.

Keywords: field-aligned irregularities (FAIs); range imaging (RIM); very-high-frequency (VHF) radar

Citation: Chen, B.; Liu, Y.; Feng, J.; Zhang, Y.; Zhou, Y.; Zhou, C.; Zhao, Z. High-Resolution Observation of Ionospheric E-Layer Irregularities Using Multi-Frequency Range Imaging Technology. *Remote Sens.* **2023**, *15*, 285. https://doi.org/10.3390/rs15010285

Academic Editor: Fabio Giannattasio

Received: 2 December 2022
Revised: 30 December 2022
Accepted: 30 December 2022
Published: 3 January 2023

1. Introduction

Ionospheric field-aligned irregularities (FAIs) are ionized "clumps" or "wave-like" structures of various scales floating in a normal ionospheric structure, and they are of critical importance for electromagnetic signal propagation [1–3]. Therefore, FAIs are one of the most important sources of disturbance for navigation, communication, and radar systems, and a remarkable set of comprehensive statistical studies have been conducted on ionospheric FAIs [4–17]. In addition, modeling and case studies have been carried out to reproduce ionospheric irregularities and to understand their mechanism [18–28]. Among them, midlatitude ionospheric E-region FAIs have been investigated for more than four decades since the first observation was made with a VHF radar over Arecibo [4–6,14,15,17–31].

In order to detect and characterize ionospheric irregularities, different instruments have been used, such as ionosondes, an airglow imager, coherent/incoherent scatter radars, and a global positioning satellite receiver [14,32,33]. Very-high-frequency (VHF) radars are one of the most widely used instruments due to their capability to detect small-scale irregularities, and these radars include the Japan middle and upper (MU) atmosphere radar, the standard frequency-agile radar (FAR), the Gadanki MST radar, Crete Island, the Chung-Li VHF radar, the Sanya VHF radar, and the Wuhan VHF radar. By using the radar observations, Yamamoto et al. [30] was the first to find that midlatitude E-region FAIs can be classified into two types: daytime continuous echoes at 90–100 km and nighttime quasi-periodic (QP) echoes at 100–130 km. Since then, considerable research has been devoted to the investigation of E-region FAIs, and several mechanisms have been proposed to explain the generation and evolution of FAIs. These mechanisms include atmosphere gravity waves (AGW), gradient-drift instability (GDI), Kelvin–Holmoholtz instability (KHI), and sporadic E (Es) layer instability [18–24]. Notably, the formation of midlatitude E-region FAIs is generally thought to be closely related to the Es layer since the seasonal variation in E-region FAI occurrence is consistent with the local Es occurrence [4,5]. Early and recent studies have shown that there is a positive correlation between the appearance of E-region FAIs and the difference in the critical frequency of a sporadic E layer (foEs) and the blanketing frequency of a sporadic E layer (fbEs) [14,34]. The simulation results also showed that the plasma density gradient in the Es layer represents a profile of the formation of E-region FAIs through gravity-wave modulation, Kelvin–Helmholtz instability generated by wind shear, and polarization electric fields associated with high-density plasma cloud embedded within sporadic E (Es) layers [18,19,21,23]. However, some questions such as the ionospheric E–F coupling phenomena and the effect of geomagnetic activity on E-region irregularity still remain to be thoroughly explained due to observational and modeling limitations.

Range resolution is one of the most important parameters in radar detection. As the most widely used instrument for E-region FAI measurements, the range resolution of a VHF radar is generally several hundred meters due to the bandwidth limitation; therefore, resolving the small-scale structure in a radar-illuminated volume is impossible [35]. To overcome this restriction, multifrequency radar imaging (RIM) technology, which has been used successively in atmospheric radars, was introduced into observations of E-region FAIs. Recently, Chen et al. [36–38] applied the RIM technique to the MU radar, and they found that the range resolution greatly improved such that finer structures could be obtained when compared with previous observations. However, no other specific studies have used RIM technology in VHF radars to observe irregularities. The Wuhan VHF radar is located at midlatitude for East Asia, a latitude lower than that of the MU radar. Previous studies have suggested that E-region FAIs have different features at different latitudes [3,15]. Therefore, in this paper, the RIM technique is employed to develop high-resolution imaging of ionospheric E-region FAIs based on the Wuhan VHF radar to expand upon the study of the fine structure of ionospheric irregularities at mid- and low latitudes. This paper is organized as follows: Section 2 presents the RIM technique based on Capon's multifrequency interferometric imaging and the experimental setup of the VHF radar. Section 3 provides the results, which include the results of a standard range resolution after single-frequency acquisition and the results after the RIM technique processing. A discussion about the possible sources and mechanism behind the formation of E-region FAIs is given in Section 4. Section 5 presents a summary.

2. Method

2.1. Multifrequency Range Imaging Algorithm

In order to eliminate the limitations due to assumptions in the single Gaussian layer in the frequency-domain interference algorithm (FDI) and to further improve the resolution, Palmer et al. [39] developed a technique called range imaging, which uses the best combination of multiple frequency signals to reconstruct a fine-scale atmospheric structure within the resolution range. In this section, we use the Wuhan VHF radar to perform multifrequency range imaging to obtain fine-scale structures in the FAI echoing volume.

The RIM technique uses the Fourier method, the Capon method, the maximum entropy method, and other inversion algorithms to obtain an estimation of the so-called power density or brightness function of spatial refractive index fluctuations [40]. Among the inversion algorithms used with RIM, the Capon method [41] is robust [39,42]. Without considering Doppler-frequency sorting of the echoes, the Capon equations can be simply expressed as follows:

$$P(r) = \frac{1}{e^H R^{-1} e} \tag{1}$$

$$R = \begin{bmatrix} R_{11} & R_{12} & \cdots & R_{1n} \\ R_{21} & R_{22} & \cdots & R_{2n} \\ \vdots & \vdots & \cdots & \vdots \\ R_{n1} & R_{n2} & \cdots & R_{nn} \end{bmatrix} \tag{2}$$

$$e = \left[e^{j2k_1 r}, e^{j2k_2 r}, \dots, e^{j2k_n r} \right]^T \tag{3}$$

where $P(r)$ is the imaged power at the range r after range-scanning processing. The superscripts H and -1 in Equation (1) and T in Equation (3) represent, respectively, the Hermitian, inverse, and transposition operators; k_n is the wave number of the nth carrier frequency; and R_{mn} is the non-normalized cross-correlation function of the signals calculated at zero-time lag for a pair of frequencies. R_{mn} can be simply written as

$$\begin{aligned} R_{mn} &= \langle R_m R_n{}^* \rangle \\ &= \langle A_m \exp[j(-2k_m r_m + \varphi_m)] \cdot A_n \exp[j(2k_n r_n - \varphi_n)] \rangle \\ &= \langle A_m A_n \exp[j2((k_n r_n - k_m r_m) + (\varphi_m - \varphi_n))] \rangle \end{aligned} \tag{4}$$

wherein $\langle \cdot \rangle$ represents the overall average, r_m and r_n represent the detection range of the scatterer, and φ_m and φ_n represent the phase terms related to the response of the system to different transmission frequencies.

2.2. Equipment and Experimental Setup

The Wuhan VHF radar, primarily operated at a central frequency of 48.2 MHz, can now generate carrier frequencies from 48.0 MHz to 48.4 MHz, with the smallest frequency step being 1 kHz. The carrier frequencies are generated with the same oscillator and can be changed from pulse to pulse to meet the basic requirement of RIM, that is, the echoes with various carrier frequencies are received almost simultaneously. The radar pulse shape can be Gaussian or rectangular, and complementary codes are usually employed to increase the signal-to-noise ratio (SNR). The VHF radar system used in the experiment is shown in Table 1.

Table 1. VHF radar system introduction.

System	VHF Pulse Doppler Radar
Center frequency	48.2 MHz
Antenna	1 sending and 24 receiving
Transmitter	Peak power: 20 Kw
Receiver	6 Digital channel receiving system
Peripherals	24 antennas for layer E FAI observation

Previous studies have found that the appearance of E-region irregularities in Wuhan is highest in summer [4,14,15]. Therefore, the experiment in this paper focuses on the period from June to August 2022 and uses RIM technology to carry out the radar frequency scanning test of the ionospheric E layer. In the current RIM test, 11 frequency points between 48.15 MHz and 48.25 MHz were used. The frequency interval was set as 10 KHz, and the bandwidth of the transmitted signal was 320 kHz. The selection and use of the above radar parameters were subject to the limitations in the capacity of the Wuhan VHF radar system. Table 2 lists some important radar parameters used in the test. The pulse-repetition frequency of these test-transmitted signals was 525 Hz, the sampling time was about 2 ms, and the number of coherent accumulations was 8. In total, 256 data points were used for each carrier frequency to estimate the non-normalized value in Equation (4), so the time resolution of the signal was about 1 min.

Table 2. Wuhan VHF radar parameters in the experiment.

Parameter	Numerical Value
Observation site	(114.2°E, 30.3°N)
Observation time	From 15 July 2022 to now
Distance range	130 km–250 km
Original range resolution	500 m
Time resolution	1 min
Pulse repetition rate	550 Hz
Coherent accumulation times	8
Coding system	16 bit complementary code
Transmission signal bandwidth	320 kHz
Carrier frequency (MHz)	48.15, 48.16, 48.17, 48.18, 48.19 48.20, 48.21, 48.22, 48.23, 48.24, 48.25

In this paper, a long-term experimental observation was carried out with this set of equipment. The specific detection method is shown in Figure 1. Several groups of pulse signals with similar carrier-frequency intervals were transmitted through the transmitter, corresponding to different beams kn and km, as shown in Figure 1. These pulse signals generated scattering echo after encountering ionospheric irregularities. Then, the echo signal was transmitted to the receiver through the 12×2 counter circle receiving an antenna array, and finally, the original data from the experimental analysis were obtained through digital signal processing of the receiver.

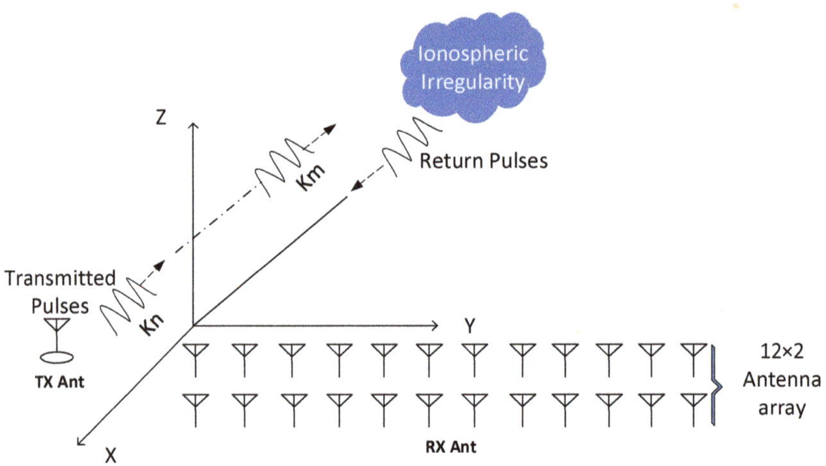

Figure 1. VHF radar configuration of multifrequency range imaging technology for irregular body detection.

Taking the observation data within one minute as an example, the observation data in this experiment were processed as follows: first, we recorded 241×256 data corresponding to each frequency of each channel and processed 11 groups of data using Formula (2) for the baseband IQ data of a single channel. Thereafter, we obtained an 11×11 matrix R within each threshold value and calculated its inverse matrix R^{-1}. Then, the wave number k_n was calculated according to the frequency, and each threshold scanning range gate ($\Delta r = 500$ m) was subdivided into 50 layers, that is, the spacing r between adjacent layers after subdivision was 10 m. Finally, we used formula (3) to calculate the threshold scanning vector e and to conjugate transpose matrix e^H, and substituted them into Formula (1) to obtain the scattering imaging power $P(r)$, which represents the "brightness density function" at each time period and detection range. As a result, the refined spatial distribution of the E-region FAIs can be seen from the figure.

3. Result

In order to determine the common height distribution of E-region irregularities, Figure 2 takes the normalized amplitude data of echo signals at different detection ranges within the period 14:01–14:03 UT on 10 July 2022 as an example. The figure clearly shows fewer data points, and that the amplitude distribution is sparse in most of the range gates, with amplitude values generally around 0. Specifically speaking, most of the normalized amplitude values are lower than 0.1, and the normalized amplitude around 155–165 km is larger than 0.2 at 14:01 UT. At 14:02 UT, most of the normalized amplitude is lower than 0.2, and the amplitude around 155–170 km is larger, with values more than 0.2. The normalized amplitude is lower than 0.2, and the amplitude around 160–175 km is larger than 0.2 at 14:03 UT. Therefore, the normalized amplitude values around 155–175 km vary widely and are significantly higher than those in the other detection ranges, which indicates the existence of irregularities in this distance range. The data at other time periods are similar to the above-demonstrated situation after extensive statistical analysis. Note that the relationship between the actual altitude H and the detection distance D is $H = \sqrt{2}/2 \times D$, since the elevation angle of our antenna array is $45°$. Therefore, the altitude of the common distribution of FAIs is between 95 and 123 km, consistent with that from previous studies [14,30]. In the following, we mainly focus on analyzing the morphology of irregularities within this altitude.

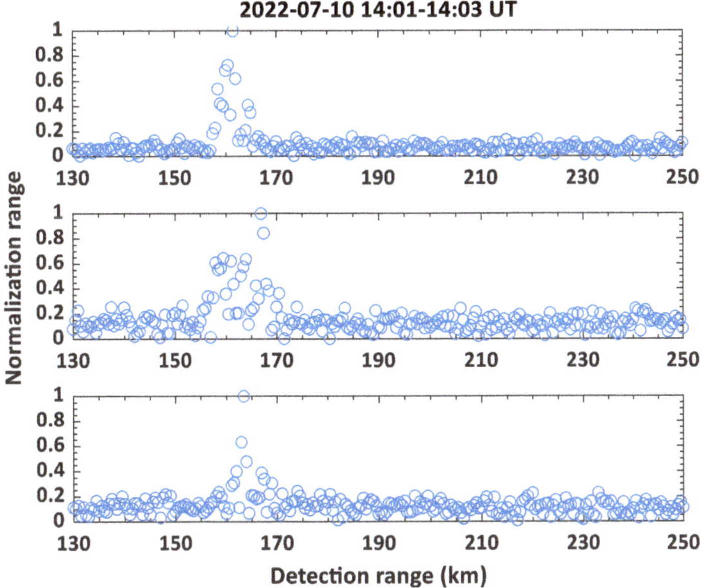

Figure 2. Normalized amplitude distribution of VHF radar original detection data of the E layer measured in Wuhan within the period 14:01–14:03 UT on 10 July 2022 (from top to bottom is 14:01–14:03 UT).

In order to verify the effectiveness and feasibility of the RIM algorithm, this paper uses single-frequency and multifrequency range imaging techniques to process the data on 26 July 2022. By using the data corresponding to a frequency of 48.2 MHz in the radar frequency scan test, the echo signal intensity is calculated to obtain the distribution of scattered power in the single-frequency mode, as shown in Figure 3. Note that the x axis represents UT hour, and the y axis represents detection ranges. In addition, the color represents intensity of the echo scattering power in Figure 3. The results show more obvious FAI echoes at the detection range of 155–175 km (corresponding to an altitude of 110–123 km) during 12:00–21:00 UT, which is in line with the common distribution of FAIs at the detection range given in Figure 2. However, due to the low range resolution, the echo pattern is more blurred in Figure 3. The FAI clumps show an obviously diffused state, and the spatial–temporal distribution features are difficult to accurately extract for the period 15:00–19:00 UT, especially for the time period of 17:00–19:00 UT.

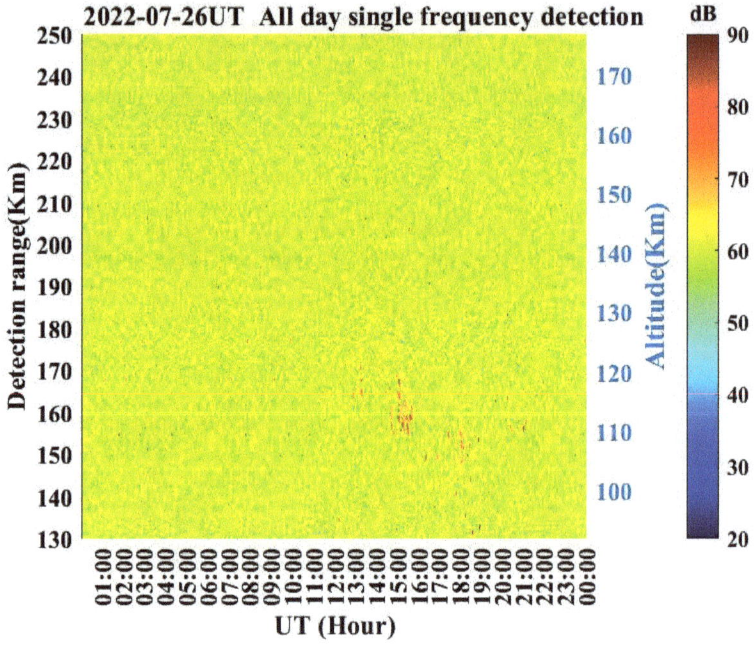

Figure 3. Echo scattering power under single-frequency mode of E layer measured in Wuhan on 26 July 2022. Note that the x axis represents UT hour, and the y axis represents detection range. In addition, the color represents intensity of the echo scattering power.

Figure 4 displays the distribution of scattered power in multifrequency scanning mode during the same time period. It is of great importance that the x axis represents UT hour, and the y axis represents detection ranges. In addition, the color represents intensity of the echo scattering power in Figure 4. From the figure, the echo morphology is basically consistent with those in Figure 3; however, the echo morphology is more complete and the structure is clearer in multifrequency mode. Specifically, FAIs first appeared at 13:00 UT, with sporadic distribution in the detection range of 160–170 km. From 15:00 UT, a large range of irregularities appeared in the detection range of 150–170 km, lasting about one hour. At around 16:00 UT, the irregularities were found to be scattered at the detection range of 150 km. A large range of irregularities appeared again at the detection range of 140–160 km at 18:00 UT, and the overall altitude decreased by 7 km, on average, compared with the FAI two hours previously. Then, the FAI gradually disappeared within one hour. During the 20:30–21:30 UT period, the FAIs occurred again around the detection range of 155 km. Compared with the results of the single-frequency mode, the images processed by RIM technology are refined and complete in terms of both the morphological structure and echo distribution of the FAI, which is conducive to further research on its physical mechanism and evolution.

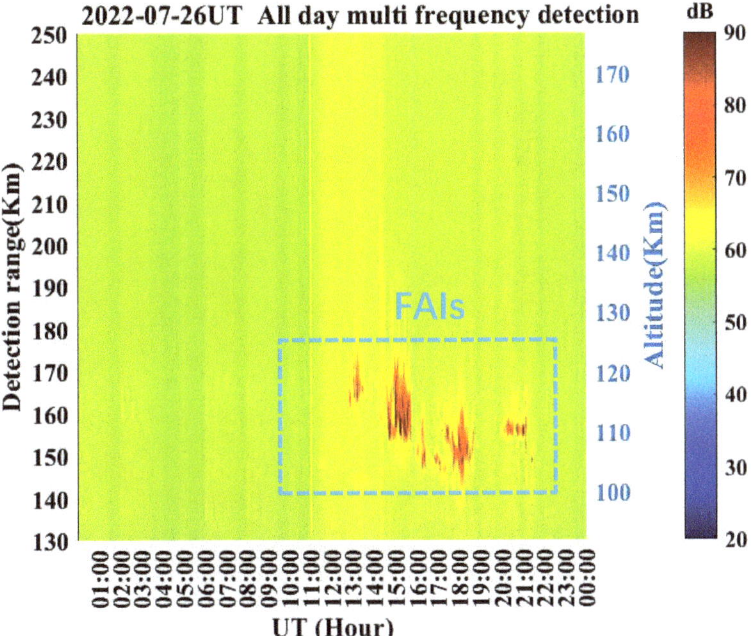

Figure 4. Same as Figure 3 but for the echo scattering power in multifrequency scanning mode.

As the FAI for a whole day, shown in Figures 3 and 4 above, is concentrated in some specific time periods and distance ranges, in order to further compare echoes under two modes. In this paper, data from the time range 15:00–16:00 UT and the detection range of 135–175 km were selected for analysis. The echo signal in this time period is strong and the echo quality is excellent, which is more conducive to observing the FAI structure after RIM processing and to conducting a comparative analysis with the single-frequency imaging results. The results are shown in Figures 5 and 6. Note that the x axis represents UT hour, and the y axis represents detection ranges. In addition, the color represents intensity of the echo scattering power in Figures 5 and 6.

Figure 5 shows that the echo distribution is accompanied by an obvious amplitude mutation on each range gate during 15:00–15:40 UT, due to the low initial range resolution in single-frequency mode. This directly leads to two problems: First, it is impossible to analyze the distribution of FAIs inside the original range gates. Second, the amplitude of FAI echoes between each range gate changed suddenly, which makes it impossible to accurately determine the position of the FAI and to observe its hierarchical structure.

Compared with the results of the single-frequency mode, Figure 6 shows that the echo in the multifrequency mode lasts longer and covers a wider area and that the echo signal is more continuous during the same time period. The FAI processed by RIM technology can not only obtain more precise echo-amplitude distribution inside each range gate, but also solve the amplitude mutation problem between each range gate. Specifically, the FAI structures in the P1–P4 period circled in Figure 6 can also accurately determine the evolution of FAIs. Specifically, the overall altitude of the FAI decreased by 4 km in the period of P1 to P2, and the FAI in P2 remained stable for about 10 min within the detection range of 155–160 km. The FAI in the P3 period obviously drifted downward, from the initial range of 165 km (altitude of 116 km) to 155 km (altitude of 110 km). Finally, it is stably distributed within the range of 157–163 km in the P4 period, lasting for 5 min.

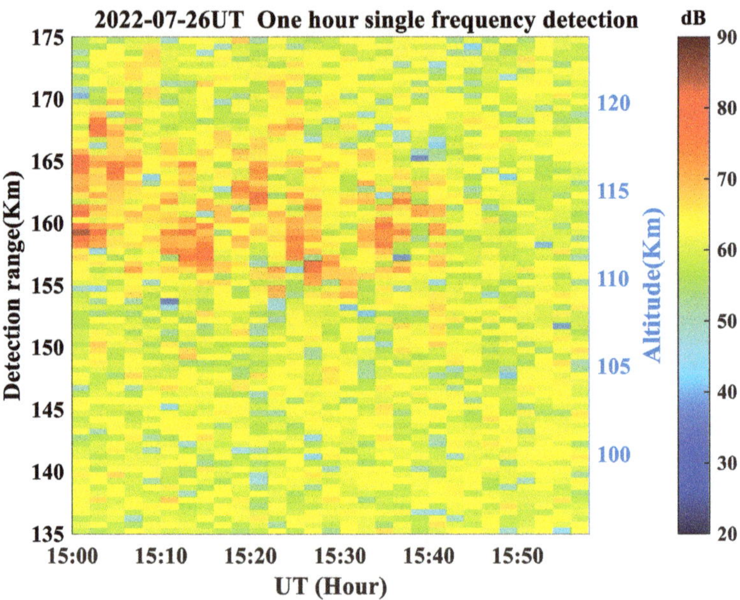

Figure 5. Echo scattering power under single-frequency mode of E layer measured in Wuhan at 15:00–16:00 UT on 26 July 2022. Note that the *x* axis represents UT hour, and the *y* axis represents detection range. In addition, the color represents intensity of the echo scattering power.

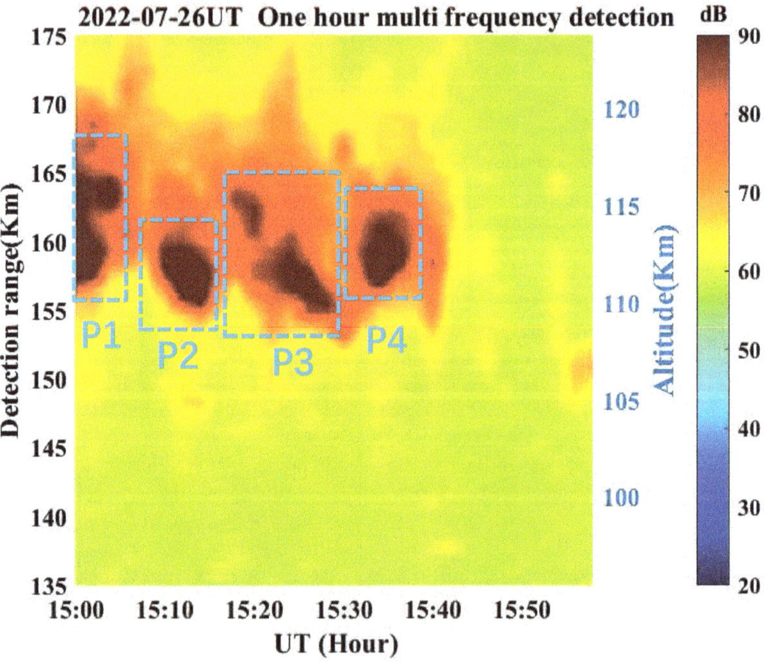

Figure 6. Same as Figure 5 but for the echo scattering power in multifrequency scanning mode.

Further research shows that the multifrequency range imaging technique can also be used to obtain the fine structure of irregularities within a short time period. Taking the results of 15:01–15:04 UT on 15 August 2022 as an example, Figures 7 and 8 show the comparison between the imaging of single-frequency mode and those processed by RIM technology. It is important to note that the *x* axis represents UT hour, and the *y* axis represents detection ranges. In addition, the color represents intensity of the echo scattering power in Figures 7 and 8. Figure 7 shows that the imaging results are relatively fuzzy due to the low range resolution in the single-frequency mode. In terms of temporal and spatial distribution, the continuity of the scattering echo power density function is poor in time and range, resulting in the imaging results being unable to reflect the specific altitude and duration of the FAI. By contrast, from Figure 8, the image results are more refined after RIM processing, and the hierarchical structure of FAIs is clearer. For example, the FAI layering structure is obvious at 15:03 UT, and they can be divided into four layers, located at the detection ranges of 187 km, 173–180 km, 169–171 km, and 164 km (corresponding to altitudes of 132 km, 122 km–127 km, 119 km–121 km, and 116 km, respectively). Therefore, the height information and inter-layer spacing of each layer structure can be effectively extracted. Specifically, FAIs were concentrated in the range of 165–170 km in the first two minutes, and a small amount of FAIs were found in the range of 175 km according to the temporal–spatial distribution of FAIs in these four minutes. At 15:03 UT, the FAIs were mainly distributed in the range of 173–177 km, with obvious multilayer structures. Moreover, FAIs were concentrated within the range of 170–180 km at 15:04 UT. In summary, the FAI was relatively stable in terms of spatial–temporal distribution within a short time period but was also accompanied by changes in altitude and stratification. Therefore, the temporal distribution of irregularities within a short time period can be observed through the results of RIM technology processing, which would provide observational support for research on the physical properties and evolution process of FAIs within a short time period.

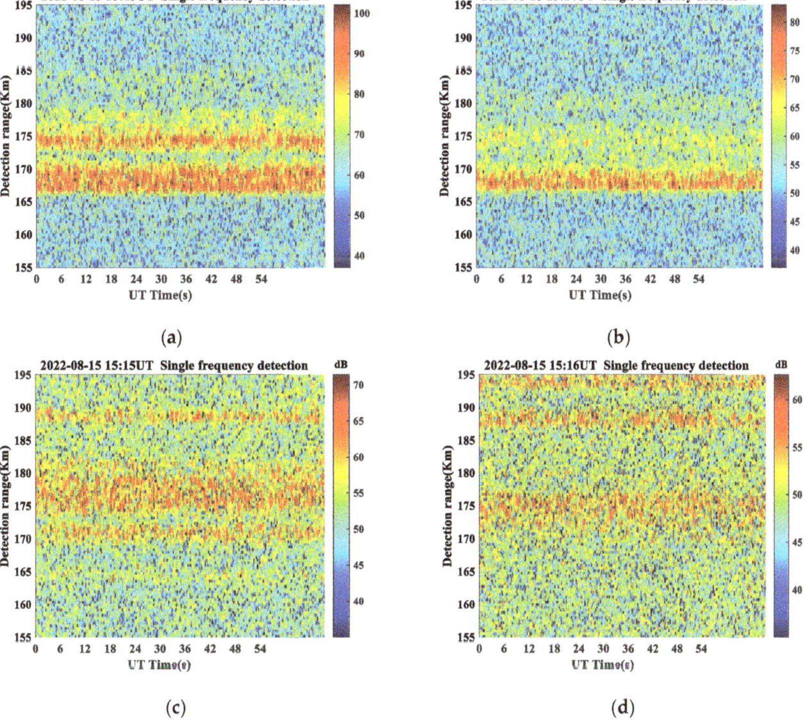

Figure 7. Echo scattering power under single frequency mode of E layer measured at 15:12–15:16 UT on 15 August 2022 in Wuhan. (**a**) Echo scattering power measured at 15:12–15:13 UT. (**b**) Echo scattering

power measured at 15:13–15:14 UT. (**c**) Echo scattering power measured at 15:14–15:15 UT on 15 August 2022. (**d**) Echo scattering power measured at 15:15–15:16 UT. Note that the *x*-axis represents UT hours and the *y*-axis represents the detection range. In addition, the color indicates the intensity of the echo scattering power.

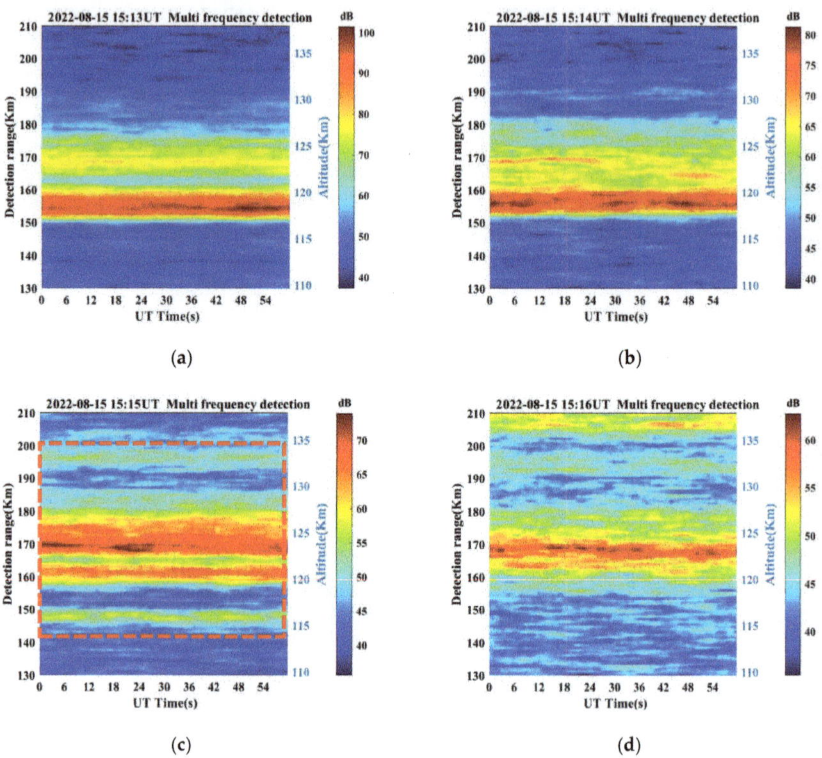

Figure 8. Same as Figure 7 but for the echo scattering power under the multifrequency scanning mode of E layer measured at 15:12–15:16 UT on 15 August 2022 in Wuhan. (**a**) Echo scattering power measured at 15:12–15:13 UT. (**b**) Echo scattering power measured at 15:13–15:14UT. (**c**) Echo scattering power measured at 15:14–15:15 UT. (**d**) Echo scattering power measured at 15:15–15:16 UT.

4. Discussion

RIM is a technology that is widely used in atmosphere radars to improve the range resolution of pulsed radars by transmitting multiple frequencies. Chen et al. [36–38] have verified its feasibility and effectiveness in the detection of E-region FAIs using the MU radar. To expand this technique into the observation of E-region FAIs using other VHF radars, in this paper, the RIM technique was employed using the Wuhan VHF radar. The results show that the range resolution of E-region FAIs significantly improved when compared with original single-frequency imaging. It is of great importance to note that the improvement of range resolution depends on two aspects. On the one hand, it depends on the wavelength of the carrier frequency. The central frequency of our experiment is 48.2 MHz, and the corresponding wavelength is about 6.2 m. Therefore, our imaging limit of range resolution cannot exceed 3.1 m. On the other hand, it depends on the characteristics of the detection target. For irregularities, its horizontal and vertical distribution has a large spatial scale, and the distance between layers is also large. We have carried out many experiments, and the results show that the range resolution of 10 m produces the ideal result to further improve the range resolution of imaging and the computational complexity of the algorithm; however, it cannot improve the imaging result. Therefore, this paper uses 10 m as the final range resolution for multifrequency modes. In addition, the results show the details of some typical features of FAI, such as QP echoes and hierarchical structures; they can, therefore, be used in an investigation of the formation and evolution processes of the E-region FAIs discussed below.

Processed with multifrequency RIM technology, Figure 4 clearly shows that the E-region FAIs demonstrate an overall descending trend when compared with Figure 3, and their height distributions decreased from 160–170 km at 13:00 UT to 140–160 km at 18:00 UT. Previous studies have also found a downward trend to be one of E-region FAIs' characteristics and proposed that it may be caused by tides or gravity waves with downward phase velocities or even driven by downwind shear [43,44]. The Es layer, which is closely related to the E-region FAI, also shows a downward trend, which was confirmed in previous studies [45,46]. Therefore, the descending E-region FAIs might be related to the downward trend of the Es layer. However, the physical mechanisms behind the downward trend of E-region FAIs still needs to be explored with more observations and modeling.

In Figure 6, the echoes of E-region FAIs show a quasi-period feature, which belongs to the QP echoes (type 2 echoes). From the figure, the period of the echoes is almost 10 min during the P1-P4 period, which is a typical period of a gravity wave. In addition, the echo power displays a periodic variation in a wave-like pattern. Therefore, we preliminarily conclude that the generation of QP echoes is modulated by a gravity waves. Woodman [18] first suggested that the existing Es layers might be modulated by passing atmospheric gravity waves (AGW), and Tsunoda et al. [19] further proposed that the polarization process will be excited in the presence of background ionospheric winds and wind-driven dynamo electric fields during AGW modulation. Moreover, neutral dynamic instability (gradient-drift instability and Kelvin–Helmholtz instability) and plasma instability (Es-layer instability) were suggested to be responsible for the formation of QP echoes [21–26]. Although it remains unclear which mechanism is the dominant mechanism, the high-resolution observation being processed with the RIM technique provides a convenient and promising way to investigate this question.

In addition to the results of a period, the RIM technique can also be applied to observe irregularities within a short time period, as Figure 8 clearly shows the hierarchical structure of E-region FAIs. Early and recent studies have found that the hierarchical structure of E-region FAIs could also be seen at other observation sites, suggesting that it is one of the frequent phenomena of E-region FAIs. Using an Arecibo incoherent scatter radar (ISR) observation, Early et al. [45] found that the maximum of an electron density gradient can appear at different heights (two to three, or more) in the E-region, that is, a multiple Es-layer structure. Szuszczewicz et al. [47] also found the appearance of a multiple Es layer from a Townsville ionosonde observation, and this structure could be found at other ionosonde observations [6,43]. Therefore, the hierarchical structure of E-region FAIs could be closely related to the multiple E-layer structure. The mechanisms behind the formation of E-region FAIs described above could be used to explain the hierarchical structure of E-region FAIs. The multilayered FAI structure at night may be caused by the gradient-drift instability acting on multiple ionized density-gradient layers formed by tides or gravity waves. Specifically, the multiple Es layer indicates the development of an electron density gradient and strong spatial structuring in the electron density, and the ambient electric field in the Es layers could excite the stronger FAIs around the local Es altitude [48]. This example demonstrates that the spatial resolution of the E-region FAIs is improved by the RIM technique and is suitable for the observation of multiple layers of E-region FAIs.

5. Conclusions

In this study, we applied the multifrequency range imaging RIM technique to improve the range resolution of a VHF radar for observing E-region FAIs. The results are generally consistent with those of previous studies for E-region FAIs; however, the spatial resolution of E-region FAIs greatly improved when compared with the single-frequency imaging technique, and finer structures such as QP echoes can be obtained through the RIM technique. The quasi-period features of the echoes indicate that the generation of QP echoes are modulated by gravity waves. Moreover, the RIM technique was found to also be applicable to the observation of irregularities within a short time period, and is particularly suitable for the observation of multiple layers of E-region FAIs. The multilayered FAI structure at night might be caused by a gradient-drift instability acting on multiple ionized density-gradient layers formed by tides or gravity waves. By using the RIM technique, the VHF radar provides convenient and promising ways for more refined observations of E-region FAIs, as well as the physical mechanisms involved in the generation and evolution of the FAIs. More observations and simulations will be combined with these high-resolution imaging results in our future work to study E-region FAIs.

Author Contributions: Conceptualization, Y.Z. (Yuqiang Zhang), J.F. and Y.L.; methodology, Z.Z.; investigation, C.Z. and Y.Z. (Yufeng Zhou); validation, J.F. and B.C.; formal analysis, B.C. and Y.L.; resources, Y.Z. (Yuqiang Zhang) and Z.Z.; visualization, J.F. and Y.Z. (Yufeng Zhou); funding acquisition, Y.L. and B.C. All authors have read and agreed to the published version of the manuscript.

Funding: This work was supported by the Stable-Support Scientific Project of the China Research Institute of Radio Wave Propagation (grant No. A132102W06), the Hubei Natural Science Foundation (grant No. 2022CFB651), the National Natural Science Foundation of China (NSFC grant No. 42204161, 42074187), the Foundation of National Key Laboratory of Electromagnetic Environment (grant No. 20200101), and the Excellent Youth Foundation of Hubei Provincial Natural Science Foundation (grant No. 2019CFA054).

Data Availability Statement: Not applicable.

Acknowledgments: The authors gratefully thank the anonymous reviewers for their insight and help. The observation data on the VHF radar are available by contacting Yuqiang Zhang (yqzhang_3@stu.xidian.edu.cn).

Conflicts of Interest: The authors declare no conflict of interest.

References

1. Booker, H.G. Turbulence in the ionosphere with applications to meteor trails. radio-star scintillation, auroral radar echoes, and other phenomena. *J. Gephys. Res.* **1956**, *61*, 673–705. [CrossRef]
2. Ratcliffe, J.A. *An Introduction to the Ionosphere and Magnetosphere*; Cambridge University Press: Cambridge, UK, 1972.
3. Balan, N.; Liu, L.; Le, H. A brief review of equatorial ionization anomaly and ionospheric irregularities. *Earth Planet. Phys.* **2018**, *2*, 257–275. [CrossRef]
4. Yamamoto, M.; Fukao, S.; Ogawa, T.; Tsuda, K.; Kato, S. A morphological study on mid-latitude E-region field-aligned irregularities observed with the MU radar. *J. Atmos. Sol. Terr. Phys.* **1992**, *54*, 769–777. [CrossRef]
5. Haldoupis, C.; Schlegel, K. Characteristic of midlatitude coherent backscatter from the ionospheric E region obtained with Sporadic E scatter experiment. *J. Geophys. Res.* **1996**, *101*, 13387–13397. [CrossRef]
6. Fukao, S.; Yamamoto, M.; Tsunoda, R.T.; Hayakawa, H.; Mukai, T. The SEEK (Sporadic-E Experiment over Kyushu) Campaign. *Geophys. Res. Lett.* **1998**, *25*, 1761–1764. [CrossRef]
7. Woodman, R.F.; Chau, J.L.; Aquino, F.; Rodriguez, R.R.; Flores, L.A. Low-latitude field-aligned irregularities observed in the E region with the Piura VHF radar, First results. *Radio Sci.* **1999**, *34*, 983–990. [CrossRef]
8. Chau, J.L.; Woodman, R.F.; Flores, L.A. Statistical characteristics of low-latitude ionospheric field-aligned irregularities obtained with the Piura VHF radar. *Ann. Geophys.* **2002**, *20*, 1203–1212. [CrossRef]
9. Patra, A.K.; Sripathi, S.; Sivakumar, V.; Rao, P.B. Statistical characteristics of VHF radar observations of low latitude E-region field-aligned irregularities over Gadanki. *J. Atmos. Sol.-Terr. Phys.* **2004**, *66*, 1615–1626. [CrossRef]
10. Ning, B.; Hu, L.; Li, G.; Liu, L.; Wan, W. The first time observations of low-latitude ionospheric irregularities by VHF radar in Hainan. *Sci. China Technol. Sci.* **2012**, *55*, 1189–1197. [CrossRef]
11. Li, G.Z.; Ning, B.Q.; Hu, L.; Li, M. Observations on the field-aligned irregularities using Sanya VHF radar, 2. Low latitude Ionospheric E-region quasi-periodic echoes in the East Asian sector. *Chin. J. Geophys.* **2013**, *56*, 2141–2151. (In Chinese)
12. Li, G.Z.; Ning, B.Q.; Hu, L. Interferometry observations of low latitude E-region irregularity patches using the Sanya VHF radar. *Sci. China Technol. Sci.* **2014**, *57*, 1552–1561. [CrossRef]
13. Wang, C.; Chu, Y.-H.; Su, C.; Kuong, R.; Chen, H.-C.; Yang, K. Statistical investigations of layer-type and clump-type plasma structures of 3-m field-aligned irregularities in nighttime sporadic E region made with Chung-Li VHF radar. *J. Geophys. Res.* **2011**, *116*, A12311. [CrossRef]

14. Zhou, C.; Liu, Y.; Tang, Q.; Gu, X.; Nin, B.; Zhao, Z. Investigation on the occurrence of mid-latitude E-region irregularity by Wuhan VHF radar and its relationship with sporadic E layer. *IEEE Trans. Geosci. Remote Sens.* **2018**, *99*, 7207–7216. [CrossRef]

15. Tsunoda, R.T.; Buoncore, J.J.; Saito, A.; Kishimoto, T.; Fukao, S.; Yamamoto, M. First observations of quasi-periodic radar echoes from Stanford, California. *Geophys. Res. Lett.* **1999**, *26*, 995–998. [CrossRef]

16. Aa, E.; Zou, S.; Liu, S. Statistical Analysis of Equatorial Plasma Irregularities Retrieved From Swarm 2013–2019 Observations. *J.Geophys. Res.* **2020**, *125*, e27022. [CrossRef]

17. Liu, Y.; Zhou, C.; Xu, T.; Deng, Z.; Du, Z.; Lan, T.; Tang, Q.; Zhu, Y.; Wang, Z.; Zhao, Z. Geomagnetic and solar dependencies of midlatitude E-region irregularity occurrence rate, A climatology based on Wuhan VHF radar observations. *J. Geophys. Res. Space Phys.* **2022**, *127*, e2021JA029597. [CrossRef]

18. Woodman, R.F.; Yamamoto, M.; Fukao, S. Gravity wave modulation of gradient drift instabilities in mid-latitude sporadic E irregularities. *J. Geophys. Res.* **1991**, *18*, 1197–1200.

19. Tsunoda, R.T.; Fukao, S.; Yamamoto, M. On the origin of quasiperiodic radar backscatter from midlatitude sporadic E. *Radio Sci.* **1994**, *29*, 349–365. [CrossRef]

20. Yokoyama, T.; Horinouchi, T.; Yamamoto, M.; Fukao, S. Modulation of the midlatitude ionospheric E region by atmospheric gravity waves through polarization electric field. *J. Geophys. Res.* **2004**, *109*, A12307. [CrossRef]

21. Larsen, M.F. A shear instability seeding mechanism for quasiperiodic radar echoes. *J. Geophys. Res.* **2000**, *105*, 24931–24940. [CrossRef]

22. Maruyama, T.; Fukao, S.; Yamamoto, M. A possible mechanism for echo striation generation of radar backscatter from midlatitude sporadic E. *Radio Sci.* **2000**, *35*, 1155–1164. [CrossRef]

23. Bernhardt, P.A. The modulation of sporadic-E layers by Kelvin–Helmholtz billows in the neutral atmosphere. *J. Atmos. Sol.-Terr. Phys.* **2002**, *64*, 1487–1504. [CrossRef]

24. Cosgrove, R.B.; Tsunoda, R.T. Simulation of the nonlinear evolution of the sporadic-E layer instability in the nighttime midlatitude ionosphere. *J. Geophys. Res.* **2003**, *108*, 1283. [CrossRef]

25. Cosgrove, R.B.; Tsunoda, R.T. Instability of the E-F coupled nighttime midlatitude ionosphere. *J. Geophys. Res.* **2004**, *109*, A04305.

26. Cosgrove, R.B.; Tsunoda, R.T.; Fukao, S.; Yamamoto, M. Coupling of the Perkins instability and the sporadic E layer instability derived rom physical arguments. *J. Geophys. Res.* **2004**, *109*, A06301.

27. Liu, Y.; Zhou, C.; Tang, Q.; Kong, J.; Gu, X.D.; Ni, B.B.; Yao, Y.B.; Zhao, Z.Y. Evidence of mid- and low-latitude nighttime ionospheric E-F coupling, coordinated observations of sporadic E layers, F-region field aligned irregularities, and medium-scale traveling ionospheric disturbances. *IEEE Trans. Geosci. Remote Sens.* **2019**, *57*, 7547–7557. [CrossRef]

28. Wu, J.; Zhou, C.; Wang, G.; Liu, Y.; Jiang, C.; Zhao, Z. Simulation of Es Layer Modulated by Nonlinear Kelvin Helmholtz Instability. *J. Geophys. Res. Space Phys.* **2021**, *127*, e2021JA030065. [CrossRef]

29. Ecklund, W.L.; Carter, D.A.; Balsley, B. Gradient drift irregularities in mid-latitude sporadic E. *J. Geophys. Res.* **1981**, *86*, 858–862. [CrossRef]

30. Yamamoto, M.; Fukao, S.; Woodman, R.F.; Ogawa, T.; Tsuda, T.; Kato, S. Midlatitude E region field-aligned irregularities observed with the MU radar. *J. Geophys. Res.* **1991**, *96*, 15943–15949. [CrossRef]

31. Haldoupis, C.; Schlegel, K.; Farley, D.T. An explanation for type 1 radar echoes from the midlatitude E-region ionosphere. *Geophys. Res. Lett.* **1996**, *23*, 97–100. [CrossRef]

32. Jin, H.; Zou, S.; Chen, G.; Yan, C.; Zhang, S.; Yang, G. Formation and Evolution of Low-Latitude F Region Field-Aligned Irregularities During the 7–8 September 2017 Storm, Hainan Coherent Scatter Phased Array Radar and Digisonde Observations. *Space Weather* **2018**, *16*, 648–659. [CrossRef]

33. Jin, S.; Wang, Q.; Dardanelli, G. A Review on Multi-GNSS for Earth Observation and Emerging Applications. *Remote Sens.* **2022**, *14*, 3930. [CrossRef]

34. Maruyama, T.; Saito, S.; Yamamoto, M.; Fukao, S. Simultaneous observation of sporadic E with a rapid-run ionosonde and VHF coherent backscatter radar. *Ann. Geophys.* **2006**, *24*, 153–162. [CrossRef]

35. Chen, J.-S.; Zecha, M. Multiple-frequency range imaging using the OSWIN VHF radar, Phase calibration and first results. *Radio Sci.* **2009**, *44*, RS1010. [CrossRef]

36. Chen, J.-S.; Chu, Y.-H.; Su, C.-L.; Hashiguchi, H.; Li, Y. Range imaging of E-region field-aligned irregularities by using a multifrequency technique, Validation and initial results. *IEEE Trans. Geosci. Remote Sens.* **2016**, *54*, 3739–3749. [CrossRef]

37. Chen, J.-S.; Wang, C.-Y.; Chu, Y.-H.; Su, C.-L.; Hashiguchi, H. 3-D Radar Imaging of E-Region Field-Aligned Plasma Irregularities by Using Multireceiver and Multifrequency Techniques. *IEEE Trans. Geosci. Remote Sens.* **2018**, *56*, 5591–5599. [CrossRef]

38. Chen, J.-S.; Wang, C.-Y.; Chu, Y.-H. Measurement of Aspect Angle of Field-Aligned Plasma Irregularities in Mid-Latitude E Region Using VHF Atmospheric Radar Imaging and Interferometry Techniques. *Remote Sens.* **2022**, *14*, 611. [CrossRef]

39. Palmer, R.D.; Yu, T.Y.; Chilson, P.B. Range imaging using frequency diversity. *Radio Sci.* **1999**, *34*, 1485–1496. [CrossRef]

40. Yu, T.Y.; Palmer, R.D. Atmospheric radar imaging using multiple-receiver and multiple-frequency techniques. *Radio Sci.* **2001**, *36*, 1493–1503. [CrossRef]

41. Capon, J. High-resolution frequency-wavenumber spectrum analysis. *Proc. IEEE* **1969**, *57*, 1408–1418. [CrossRef]

42. Luce, H.; Hassenpflug, G.; Yamamoto, M.; Fukao, S. High-resolution vertical imaging of the troposphere and lower stratosphere using the new MU radar system. *Ann. Geophys.* **2006**, *24*, 791–804. [CrossRef]

43. Ning, B.; Li, Z.; Hu, L.L.M. Observations on the field-aligned irregularities using Sanya VHF radar, 1. Ionospheric E-region continuous echoes. *Chin. J. Geophys.* **2013**, *56*, 719–730. (In Chinese)
44. Zhang, S.; Zhao, Z.; Zhou, C.; Ni, B.G.X. Wuhan VHF coherent scattering radar and its initial observation results. *Mod. Electron. Tech.* **2016**, *39*, 1–5.
45. Earle, G.D.; Bishop, R.L.; Collins, S.C.; Gonzalez, S.A.; Sulzer, M.P. Descending layer variability over Arecibo. *J. Geophys. Res.* **2000**, *105*, 24951–24961. [CrossRef]
46. Mathews, J.D. Sporadic E, Current views and recent progress. *J. Atmos. Sol. Terr. Phys.* **1998**, *60*, 413–435. [CrossRef]
47. Szuszczewicz, E.P.; Roble, R.G.; Wilkinson, P.J.; Hanbaba, R. Coupling mechanisms in the lower ionospheric-thermospheric system and manifestations in the formation and dynamics of intermediate and descending layers. *J. Atmos. Terr. Phys.* **1995**, *57*, 1483–1496. [CrossRef]
48. Zhu, Y.; Tang, Q.; Deng, Z.; Zhou, C.; Liu, Y.; Xu, T.; Wang, Z.; Zhao, Z.; Wei, F.; Feng, X. The study of daytime ionospheric E-region radar echoes simultaneously observed by Qujing VHF radar and multi-ionosondes. *Space Weather* **2022**, *20*, e2021SW002998. [CrossRef]

remote sensing

Article

Modeling and Forecasting Ionospheric foF2 Variation in the Low Latitude Region during Low and High Solar Activity Years

Cheng Bi [1], Peng Ren [1], Ting Yin [2,*], Zheng Xiang [1] and Yang Zhang [1]

1 School of Telecommunication Engineering, Xidian University, Xi'an 710071, China
2 Beijing Electronic Science & Technology Institute, Beijing 100070, China
* Correspondence: yinting@ntsc.ac.cn

Abstract: Prediction of ionospheric parameters, such as ionospheric F2 layer critical frequency (foF2) at low latitude regions is of significant interest in understanding ionospheric variation effects on high-frequency communication and global navigation satellite system. Currently, deep learning algorithms have made a striking accomplishment in capturing ionospheric variability. In this paper, we use the state-of-the-art hybrid neural network combined with a quantile mechanism to predict foF2 parameter variations under low and high solar activity years (solar cycle-24) and space weather events. The hybrid neural network is composed of a convolutional neural network (CNN) and bidirectional long short-term memory (BiLSTM), in which CNN and BiLSTM networks extracted spatial and temporal features of ionospheric variation, respectively. The proposed method was trained and tested on 5 years (2009–2014) of ionospheric foF2 observation data from Advanced Digital Ionosonde located in Brisbane, Australia (27°53′S, 152°92′E). It is evident from the results that the proposed model performs better than International Reference Ionosphere 2016 (IRI-2016), long short-term memory (LSTM), and BiLSTM ionospheric prediction models. The proposed model extensively captured the variation in ionospheric foF2 feature, and better predicted it under two significant space weather events (29 September 2011 and 22 July 2012).

Keywords: ionosphere; F2 layer; ionosonde; solar cycle-24; space weather; deep learning

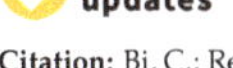

Citation: Bi, C.; Ren, P.; Yin, T.; Xiang, Z.; Zhang, Y. Modeling and Forecasting Ionospheric foF2 Variation in the Low Latitude Region during Low and High Solar Activity Years. *Remote Sens.* **2022**, *14*, 5418. https://doi.org/10.3390/rs14215418

Academic Editor: Fabio Giannattasio

Received: 4 October 2022
Accepted: 25 October 2022
Published: 28 October 2022

Publisher's Note: MDPI stays neutral with regard to jurisdictional claims in published maps and institutional affiliations.

1. Introduction

The high-frequency signal is reflected by the ionospheric F2 layer in the case of long-distance communication, and the ionospheric variation directly affects the communication quality [1,2]. Predicting and analyzing the variation of the ionospheric F2 layer is necessary to indicate adverse space weather for initiating necessary measures in high-frequency communication and global positioning system. The foF2 is a significant feature of the F2 layer, and it is utilized to indicate the ionospheric variation. To this end, significant research works were carried out to explore variations of ionospheric foF2 in different latitude regions.

The International Reference Ionosphere (IRI) project sponsored by the committee on space research and the international union of radio science is an empirical global ionospheric model. With the continuous update of prediction algorithms, three versions are currently released in 2007, 2012, and 2016. IRI models predicted ionospheric parameters and adopted real-time assimilative mapping to improve the capability of ionospheric responses during storm events. With the development of artificial intelligence technology, ionospheric prediction models based on machine learning concepts have been widely used and developed rapidly in the last one or two decades, and it performs better than traditional ionospheric prediction models. Currently, several neural networks and machine learning algorithms have been applied to predict the ionospheric variation, such as support vector machine, Elman, and extreme learning machine models. Bai et al. [3] combined the entropy weight method to develop the foF2 prediction model and provided reliable long-term predictions. A support vector machine model was utilized to establish an empirical local ionospheric forecasting model to predict foF2 in Lanzhou, China [4]. Similarly, Olga Maltseva [5] utilized the total electron content (TEC) to estimate foF2 and demonstrated the results for three stations in the southern hemisphere.

Deep learning models have been developed to establish the ionospheric prediction model in the last decades, and it has made a striking accomplishment in predicting ionospheric variation. Deep learning-based ionospheric prediction models have improved performance compared with conventional neural networks. To this end, Sun et al. [6] utilized the LSTM model and 8-year global positioning system total electronic content data set (from Bangladesh Station) to successfully predict the TEC. Suin et al. [7] developed the LSTM-foF2 model and tested it at the Jeju station. Similarly, Li et al. [8] utilized the LSTM model to predict ionospheric foF2 variation from 2006 to 2019 years. Kim et al. [9] developed an ionospheric prediction model-based LSTM algorithm and analyzed variation during geomagnetic storm periods. In addition, Venkateswara et al. [10] utilized the BiLSTM model to develop an ionospheric foF2 prediction model during geomagnetic storm events and quiet periods. However, unique phenomena affect ionospheric variation in the low latitude regions, such as the equatorial plasma bubbles and equatorial ionization anomaly, only using a single deep learning model, such as LSTM and BiLSTM cannot learn enough features of ionospheric variation. As a result, it is hard to forecast foF2 accurately, especially during space weather events.

In this paper, we proposed a hybrid neural network combined with quantile mechanism to model and forecast foF2 parameter variations at low latitude regions under solar cycle-24 and space weather events. The proposed ionosphere foF2 prediction model performed well compared with previous similar investigation models, such as LSTM-based, BiLSTM-based, and IRI-2016. The rest of this paper is organized as follows. In Section 2, we describe the data that are used as well as the proposed methodology. To validate the effectiveness of the proposed method, we conduct simulations with quiet and stormy ionosphere in Section 3. Then, we discuss the obtained results in Section 4. Finally, Section 5 concludes the paper.

2. Materials and Methods

2.1. Ionospheric Data

The foF2 observations were acquired from the Australian Bureau of Meteorology, Space Weather Services. The spatial location of the observation station is located at $27°53'S$, $152°92'E$. The solar cycle and geomagnetic information were downloaded from Goddard Space Flight Center, NASA (https://omniweb.gsfc.nasa.gov/ow.html (accessed on 31 December 2014)) and Helmholtz Centre Potsdam GFZ German Research Centre for Geosciences (https://www.gfz-potsdam.de/en (accessed on 31 December 2014)). The sunspot number (SSN) was provided by the Royal Observatory of Belgium, Brussels (http://sidc.oma.be/silso (accessed on 31 December 2014)).

2.2. Solar, Geomagnetic, Season, and Daily Cycle Index

Lee-Anne et al. [11] pointed out that foF2 is related to the maximum ionospheric electron density, and it is a function of latitude, longitude, local time, season, solar activity, and magnetic activity. The season is quantified by the day number (DAY) as suggested by Lee-Anne et al. However, DAY suffers from the problem that the temporally adjacent data for 31 December and 1 January are numerically far apart, making it difficult for the neural network to view them as being adjacent. Therefore, the DAY is split into two inputs [12,13].

$$DAYs = sin(\frac{2\pi \times DAY}{365}) \ and \ DAYc = cos(\frac{2\pi \times DAY}{365}), \tag{1}$$

Similarly, universal time (UT) is used to describe diurnal variation and cope with the same scheme regarding seasonal variation.

$$UTs = sin(\frac{2\pi \times UT}{24}) \ and \ UTc = cos(\frac{2\pi \times UT}{24}), \tag{2}$$

Ideally, ionospheric models should rely on the index that reflects the solar cycle variation. In the absence of long-term records for extreme ultraviolet-specific indices, ionospheric modelers have relied on two solar indices from a different wavelength range, such as the sunspot number and solar radio flux of 10.7 cm wavelength (F10.7). Although SSN is strongly correlated with F10.7, both parameters are used for modeling [14]. E.O. Joshua [15] proposed that the variations of foF2 are complex and cannot be described, leaving one to consider only SSN. In particular, for the phenomenon of hysteresis during the high SSN, E.O. Joshua and Kane [16] considered that F10.7 cm or dF10.7 should also be used for modeling foF2. In addition, Hongmei Bai. et al. [17] utilized the mutual information method to discuss the nonlinear dependence of foF2 with respect to the solar activity indices and proposed that foF2 is more dependent on SSN rather than F10.7 in the high and low solar activity. In moderate solar activity, the foF2 is more dependent on F10.7 rather than SSN. Therefore, the model inputs should include F10.7 and SSN, and the hybrid neural network can better learn foF2 variation under different solar activities.

Similarly, a variety of indices can quantify magnetic activity, in which the ap and interplanetary magnetic field Bz component (IMF-Bz) index plays a significant role in foF2 prediction. In addition, the disturbance storm time (Dst) index is utilized to define the even more refined quiet conditions and geomagnetic storms [18].

Similarly, we use solar activity parameters F10.7 and SSN along with geomagnetic index, ap, Dst, IMF-Bz, season, and universal time as model input parameters. Figure 1 plots F10.7, ap, Dst index, and foF2 sequence for the 2009–2014 period.

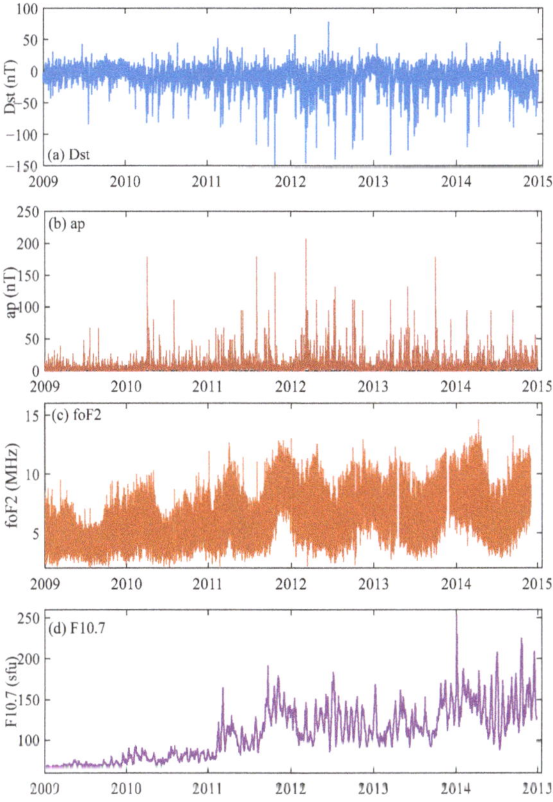

Figure 1. Variation of (**a**) geomagnetic index, Dst; (**b**) geomagnetic index, ap; (**c**) foF2 measurement samples; (**d**) solar activity index, F10.7.

2.3. Hybrid Neural Network for Ionospheric Prediction

The hybrid neural network-based ionospheric foF2 prediction model is composed of a convolutional neural network, bidirectional long short-term memory network, and quantile regression mechanism. In contrast to traditional neural networks, CNN has two major characteristics in terms of local connection and weight sharing. The CNN network generally consists of the convolutional layer, and the core is the convolutional operation. With the advantages of convolutional operation, CNN can abstract and express the features of ionospheric variation.

In contrast to the traditional neural network, the recurrent neural network (RNN) network adds unit loops in the hidden layer to record historical information. The continuous partial derivative multiplications are produced when gradient descent is utilized for training the neural network. However, the long-term information indicates multiple iterations of the loops process. Therefore, the RNN network loses the ability to deal with time sequences due to its computation complexity. The LSTM is a special case of the RNN network. Compared with the neurons in RNN, the long short-term memory network's neuron has several particular purpose nodes, such as input gate i, a forget gate f, an output gate o, and an internal memory unit c. The LSTM reduces the computation complexity encountered in RNN [19,20] via its unique structure. The obtained results are better than the time recurrent neural network and hidden Markov model. However, the LSTM is unable to encode back-to-front information encountered in time series. Therefore, the model is unable to obtain enough feature information [21,22], which causes the reduction in prediction accuracy. On the contrary, BiLSTM was proposed to solve this problem, indeed, it greatly reduced the prediction error.

The architecture of BiLSTM is implemented using MATLAB 2022a version deep learning toolboxes. Each of the unidirectional LSTMs is composed of forwarding and backward LSTM units. Similarly, the LSTM units are defined as:

$$f_t = \sigma(W_{xf}x_t + W_{hf}h_{t-1} + b_f), \tag{3}$$

$$i_t = \sigma(W_{xi}x_t + W_{hi}h_{t-1} + b_i), \tag{4}$$

$$c_t = f_t c_{t-1} + i_t \cdot tanh(W_{xc}x_t + W_{hc}h_{t-1} + b_c), \tag{5}$$

$$o_t = \sigma(W_{xo}x_t + W_{ho}h_{t-1} + b_o), \tag{6}$$

where x_t is input vector to the LSTM unit; σ represents sigmoid function; W_{xf}, W_{xi}, W_{xc}, W_{xo}, W_{hf}, W_{hi}, W_{hc}, and W_{ho} are input and hidden weight matrices, respectively; b_f, b_i, b_c, and b_o are bias vector parameters which need to be learned during training. The LSTM hidden layer h_t output is a function of output gate and memory cell with an activation function tanh, the hidden layer is defined as:

$$h_t = o_t \cdot \tanh(c_t), \tag{7}$$

where tanh is the activation function. Detailed description of the LSTM architecture is provided in [23,24].

The asymmetric distribution was worked out by the quantile regression [25–27], and it meticulously reflects the mapping information between inputs and response variables in a diverse range. Generally, the measurement samples of foF2 demonstrate characteristics of dynamic time-sequential and asymmetric distribution. Therefore, to enhance the prediction ability of the hybrid neural network on ionospheric foF2 sequence, we embed the QR mechanism in it. The τ quantile regression is defined as:

$$Q_\tau(\tau) = \min_{\gamma \in R}\left\{\sum_{i:Y_i \geq \gamma} \tau|Y_i - \gamma| + \sum_{i:Y_i \leq \gamma}(1-\tau)|Y_i - \gamma|\right\}, \tag{8}$$

The MSE function is generally utilized as the loss function in deep neural network regression layer. Here, we embed the QR and update it in the hybrid neural network architecture. According to Equation (8), the loss function is provided by:

$$Loss = \sum_{i:Y_i \geq \gamma} \tau |Y_i - \gamma| + \sum_{i:Y_i \leq \gamma} (1 - \tau)|Y_i - \gamma|, \tag{9}$$

where Y_i and γ represent observed and predicted values, respectively.

Figure 2 depicts the architecture of the proposed hybrid neural network with quantile regression mechanism, consisting of the input layer, convolution layers, sequence unfolding layer, BiLSTM layer, quantile regression layer, fully connected layer, and output layer. Among them, the BiLSTM layer is composed of forwarding and backward LSTM units. The parameters of ionospheric foF2, UTs, UTc, DAYs, DAYc, F10.7, Dst, IMF-Bz, ap, and SSN are applied to the input layer. The process can be expressed as:

$$X^{foF2} = (UTs,\ UTc, DAYs, DAYc, Dst, IMF - Bz,\ ap, SSN, F10.7, foF2), \tag{10}$$

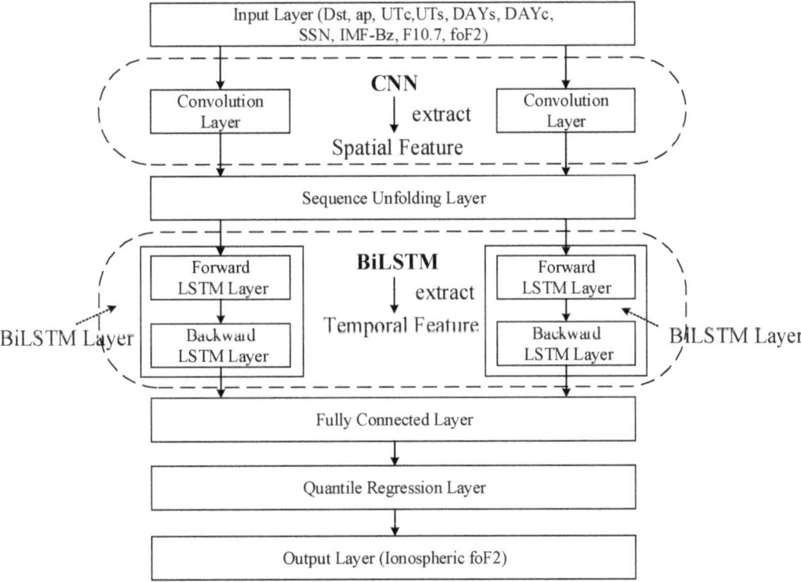

Figure 2. The hybrid neural network foF2 prediction architecture.

The output layer Y_t^{foF2} combines the two layers, BiLSTM and CNN, with the quantile regression as:

$$Y_t^{foF2} = Concat\left(BiLSTM_t^{output}, CNN_t^{output}\right), \tag{11}$$

It is noted that the model performance is assessed through the root-mean-square error (RMSE), mean absolute error (MAE), and mean absolute percentage error (MAPE).

$$RMSE = \sqrt{\frac{\sum_{i=1}^{N}\left(U_i^{obs} - U_i^{model}\right)^2}{N}}, \tag{12}$$

$$MAE = \frac{\sum_{i=1}^{N}\left|U_i^{obs} - U_i^{model}\right|}{N}, \tag{13}$$

$$\text{MAPE} = \frac{\sum_{i=1}^{N} \left| \frac{U_i^{obs} - U_i^{model}}{U_i^{obs}} \right|}{N}, \tag{14}$$

where U_i^{obs} indicates the foF2 measurement data, and U_i^{model} is the model output.

3. Results

The program deploys that foF2 observations were acquired from 2009 to 2014. The total number of foF2 samples is split as 75% for training, and 12.5% for validation and testing/prediction, respectively. The proposed model performance is evaluated in four scenarios composed of low solar activity year (2010), high solar activity year (2014), and two geomagnetic storm events, as shown in Table 1. The configurations of the models are presented in Table 2 during scenario 1, and the rest of the scenarios are similar, as well. In addition, the weight of the quantile regression (τ) is set to 0.75. For IRI-2016 model parameters, we used Sunspot number-R12, F-peak storm model (on), and the rest are maintained as default settings.

Table 1. The variation scenario on simulation conditions.

	Scenario 1	Scenario 2	Scenario 3	Scenario 4
Time	January–December 2010	January–December 2014	26 September 2011	15 July 2012
Geomagnetic Activity	–	–	Storm (<-30 nT)	Storm (<-30 nT)
Season	–	–	Spring	Winter
Solar Activity	Low	High	High	High

Table 2. Model configurations of LSTM-foF2 and BiLSTM-foF2 ionospheric prediction model.

Model Configuration	LSTM-foF2	BiLSTM-foF2
Learning Method	Deep Learning	Deep Learning
Numbers of Hidden Unit 1	250	250
Numbers of Hidden Unit 2	250	250
Epoch	200	200
Minimum Batch Size	512	512

The performance of the proposed foF2 prediction model is evaluated under four scenarios, which contained quiet period and two space weather events. Here, we consider the seasons of spring (August, September, and October), summer (January, November, and December), autumn (February, March, and April), and winter (May, June, and July). In addition, we select one month of each season to evaluate model performance, and the detailed configuration is shown in Table 3.

Table 3. Model configurations for 2010 and 2014.

Model Configuration		Summer	Autumn	Winter	Spring
Season	2010	January	March	June	September
	2014	January	March	June	September
Smooth monthly values of SSN	2010	14	18.5	24.6	29.5
	2014	109.3	114.3	114.1	101.9

3.1. Scenario 1: Analysis of Ionospheric foF2 Variation during Low Solar Activity Year (2010)

The solar cycle-24 began in December 2008 and ended in December 2019. The solar activity was minimal until 2010, but reached its maximum in 2014 with a smooth SSN monthly value of 81.8. The detailed variation of solar radiation flux F10.7 index in scenario 2 was depicted in Figure 3, and the measurement ionospheric foF2 sequence is utilized to evaluate models' performance under low solar activity year. Here, we selected hourly foF2 for evaluation, and the measured data are compared with the results of the proposed method, the BiLSTM-based, the LSTM-based, and the IRI-2016 foF2 prediction model.

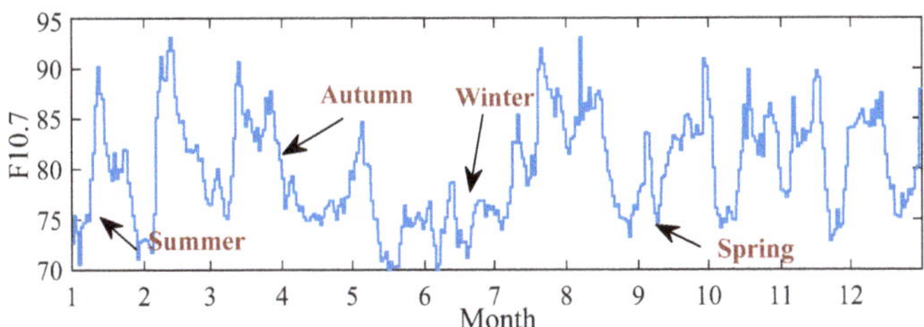

Figure 3. The variation of F10.7 index during 2010.

Figure 4 demonstrates the performance of test sample in different models. The LSTM network exhibits the highest prediction error, while the estimates of the proposed method are closer to the measured data.

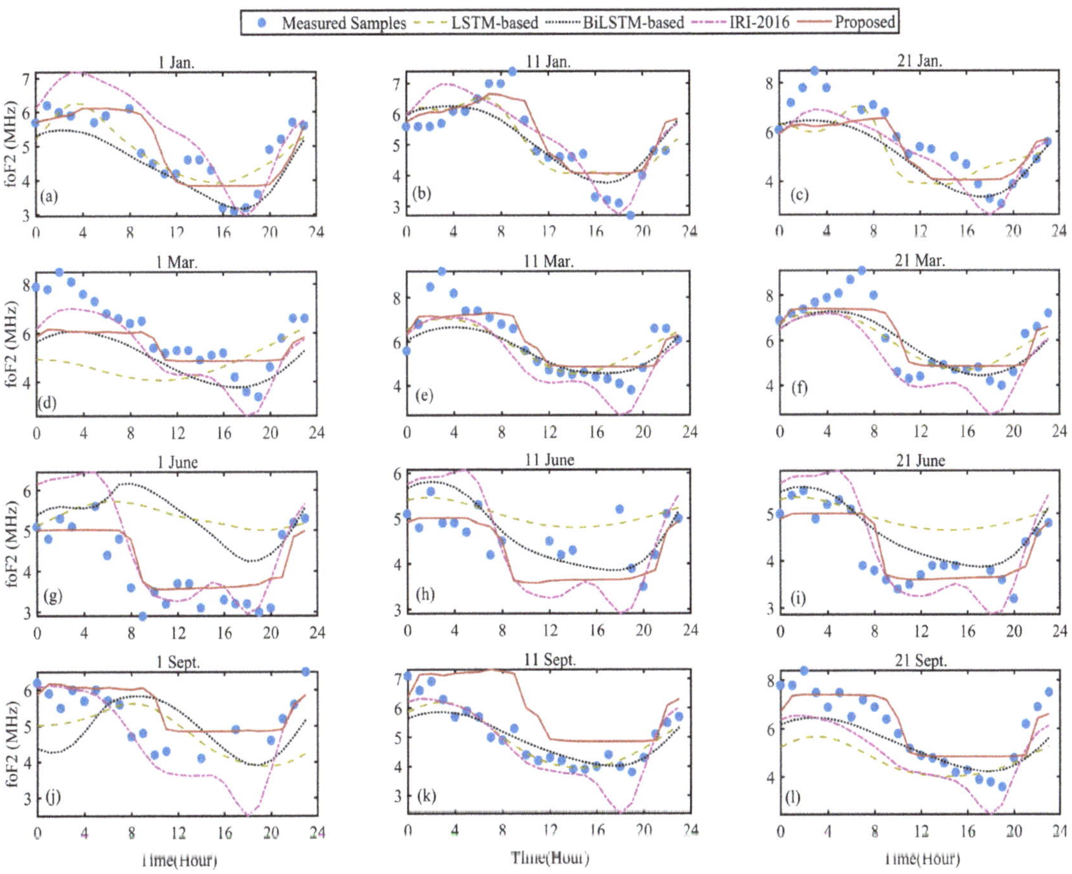

Figure 4. (**a–l**) The proposed foF2 prediction model results comparison during 2010.

The proposed model prediction results are compared with measured foF2 data and other models' output under the different months presented in Figure 4. Figure 4a–l indicates the prediction foF2 results in one day under January (summer), March (autumn), June (winter), and September (spring), respectively. Considering the comparison of RMSE, MAE, and MAPE of whole months, the proposed method and other foF2 prediction models are summarized in Tables 4–7.

Table 4. Performance evaluation of prediction model for foF2 during January 2010.

Model Configuration (January)	RMSE (MHz)	MAE (MHz)	MAPE
IRI-2016	0.801	0.643	12.286%
LSTM-foF2	0.904	0.692	13.193%
BiLSTM-foF2	0.871	0.678	12.618%
Proposed method	0.77	0.60	11.802%

Table 5. Performance evaluation of prediction model for foF2 during March 2010.

Model Configuration (March)	RMSE (MHz)	MAE (MHz)	MAPE
IRI-2016	1.123	0.935	15.478%
LSTM-foF2	1.033	0.788	12.971%
BiLSTM-foF2	0.986	0.782	12.671%
Proposed method	0.878	0.701	12.468%

Table 6. Performance evaluation of prediction model for foF2 during June 2010.

Model Configuration (June)	RMSE (MHz)	MAE (MHz)	MAPE
IRI-2016	0.648	0.527	11.99%
LSTM-foF2	0.947	0.745	19.29%
BiLSTM-foF2	0.754	0.592	14.529%
Proposed method	0.583	0.46	10.24%

Table 7. Performance evaluation of prediction model for foF2 during September 2010.

Model Configuration (September)	RMSE (MHz)	MAE (MHz)	MAPE
IRI-2016	0.811	0.654	12.482%
LSTM-foF2	0.979	0.777	13.832%
BiLSTM-foF2	0.8	0.627	12.005%
Proposed method	0.688	0.542	10.608%

The results of the proposed method are well correlated with the measured data during the considered scenario, with RMSE values of 0.77, 0.878, 0.583, and 0.688 MHz. The proposed model has a noteworthy improvement compared with the LSTM-foF2 method during autumn (March) and spring (September). Similarly, the performance of proposed method has reduced with RMSE of 11.6%, 10.95%, 22.68%, and 14% during each month, respectively, compared with the BiLSTM-based foF2 prediction model. The response in MAPE is comparatively low with the proposed method compared with other foF2 prediction models. Similarly, the IRI-2016 and BiLSTM-foF2 models are fairly followed by the observed values but with further RMSE compared with the proposed method. The MAE and RMSE of the proposed method are lower than the other models, representing that the proposed method performed well during quiet periods.

3.2. Scenario 2: Analysis of Ionospheric foF2 Variation during High Solar Activity Year (2014)

In contrast to 2010, 2014 was one of the most active years in solar cycle-24, and the annual average sunspot number is counted at 113.3. Both the sunspot number and solar radiation flux F10.7 indicators can reflect solar activity, and the detailed variation of daily SSN was present in Figure 5. The solar flare is an important sign of activity, and it is a powerful burst of radiation. The sun welcomed 2014 with two mid-level flares on 31 December 2013 and 1 January 2014, and the severe solar flare (level-X1.6) event occurred on 17 September. To evaluate the ionospheric foF2 prediction models, we select the ionospheric foF2 measurement sequence during January, March, June, and September.

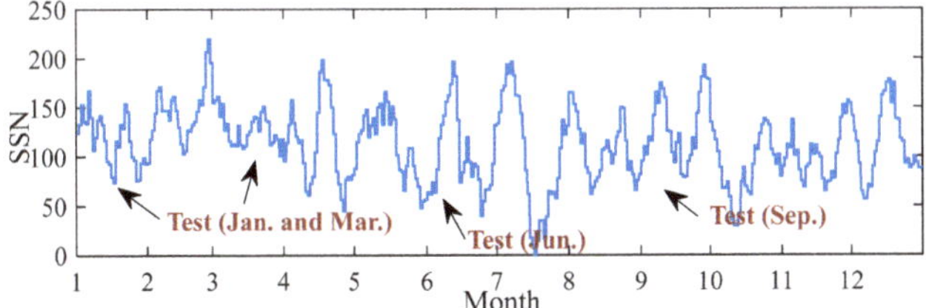

Figure 5. The variation of SSN during 2014.

The proposed model prediction results compared with measured foF2 data and other models' output under the different month are presented in Figure 6, and Figure 6a to Figure 6l depict one day for each month under spring (September), summer (January), autumn (March), and winter (June), respectively. Considering the comparison of RMSE, MAE, and MAPE of whole months, the proposed method and other foF2 prediction models are summarized in Tables 8–11.

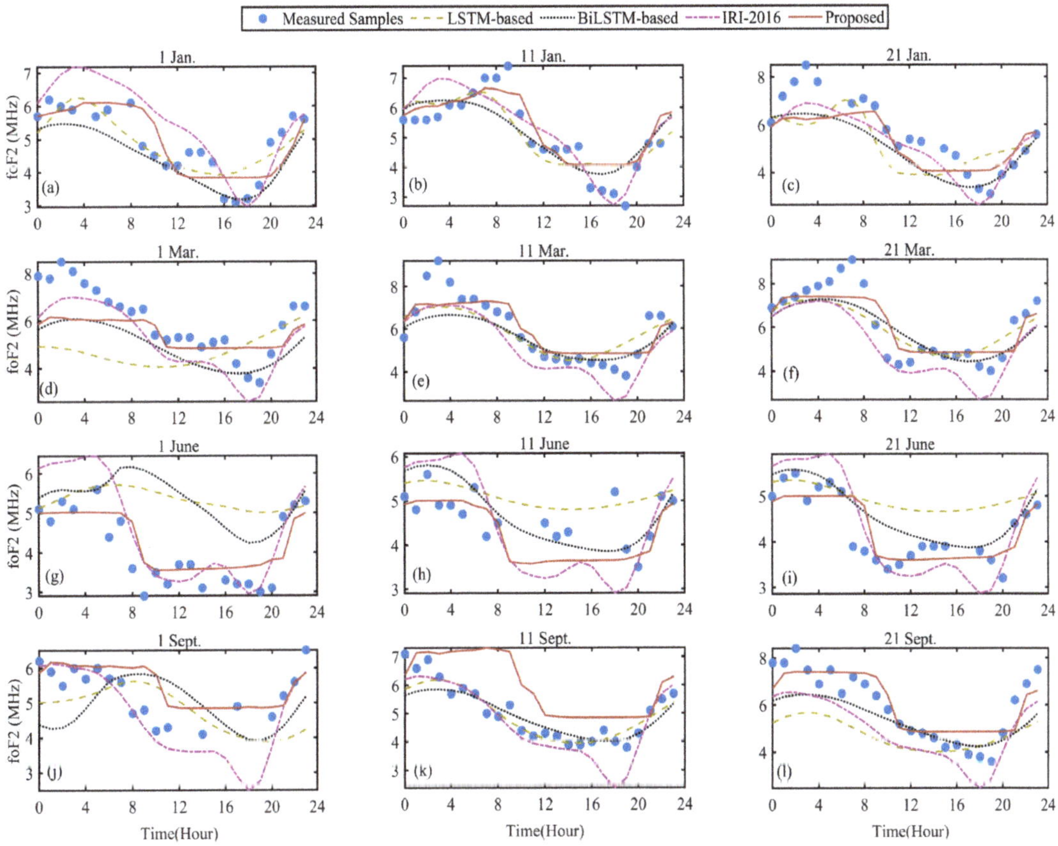

Figure 6. (a–l) The proposed foF2 prediction model results comparison during 2014.

Table 8. Performance evaluation of prediction model for foF2 during January 2014.

Model Configuration (January)	RMSE (MHz)	MAE (MHz)	MAPE
IRI-2016	1.05	0.851	10.26%
LSTM-foF2	1.03	0.812	9.763%
BiLSTM-foF2	0.959	0.767	9.52%
Proposed method	0.837	0.669	8.183%

Table 9. Performance evaluation of prediction model for foF2 during March 2014.

Model Configuration (March)	RMSE (MHz)	MAE (MHz)	MAPE
IRI-2016	1.582	1.451	15.442%
LSTM-foF2	1.816	1.527	15.812%
BiLSTM-foF2	1.808	1.577	16.106%
Proposed method	1.262	0.998	10.356%

Table 10. Performance evaluation of prediction model for foF2 during June 2014.

Model Configuration (June)	RMSE (MHz)	MAE (MHz)	MAPE
IRI-2016	1.163	0.96	16.99%
LSTM-foF2	1.42	1.104	22.656%
BiLSTM-foF2	1.16	0.945	17.643%
Proposed method	0.778	0.626	11.684%

Table 11. Performance evaluation of prediction model for foF2 during October 2014.

Model Configuration (October)	RMSE (MHz)	MAE (MHz)	MAPE
IRI-2016	1.281	1.054	14.667%
LSTM-foF2	1.134	0.911	12.847%
BiLSTM-foF2	1.06	0.844	11.961%
Proposed method	0.837	0.644	8.969%

The proposed method exhibits the lowest prediction error with an RMSE value of 0.778 MHz in June. On the contrary, the LSTM-foF2 model has poor results, with an RMSE value of 1.42 MHz. The response is relatively low with the LSTM-foF2 model compared with the measured samples, IRI-2016, and BiLSTM-based outputs. Similarly, the BiLSTM-foF2 model responds better when compared with IRI-2016 in winter (January) and spring (September), but their RMSE values are more when compared with the proposed method. The RMSE, MAE, and MAPE results indicate that the proposed method is well correlated with the measured samples on high solar activity year.

3.3. Prediction Analysis Results of Space Weather Events

Here, we evaluate the performance between the proposed method and other foF2 prediction models under two geomagnetic storm events that occurred in the period between 2011 and 2012. Scenario 3 to scenario 4, respectively, correspond to the storm events under different seasons.

3.3.1. Scenario 3: The foF2 Prediction Analysis for the 26 September 2011 Geomagnetic Storm

Several geomagnetic storm events occurred in 2011, among the unique intense storm developed on 26 September 2011. Figure 7 depicts scenario 3 of the proposed model and other model results with measured data during the storm event.

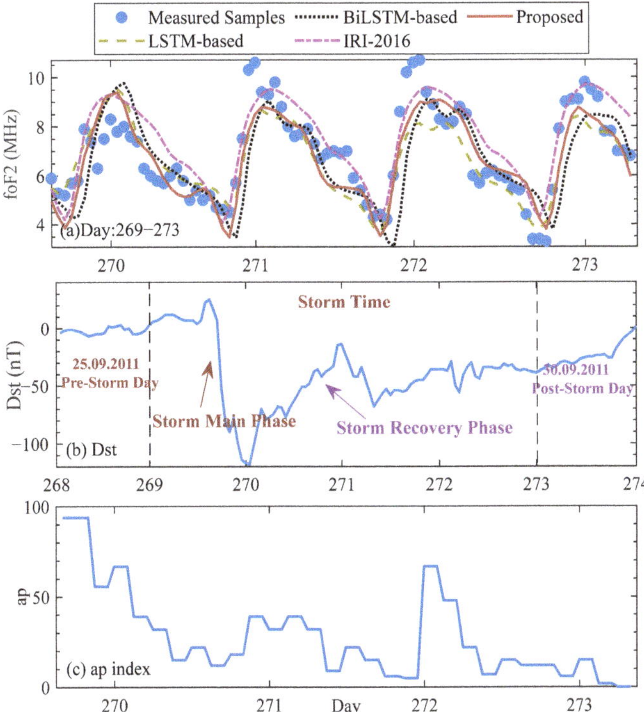

Figure 7. Variations of foF2 (**a**), Dst (**b**), and ap (**c**) during the 26 September 2011 geomagnetic storm.

On 26 September 2011, the Dst value turned positive to negative at 17:00 h UT 26 September and lower than -30 nT, representing the storm's beginning. It is noticed that the Dst value rapidly declined from 16:00 h UT [Figure 7b], and the maximum value of -118 nT was observed at 23:00 h UT. The above time indicates the crucial stage of the storms. A major part of the time from the 270th to 272nd days in 2011, was in the process of recovery from the storm event. The storms ended at 8:00 h UT on 30 September, and the Dst index completely recovered to normal. Significant deviations in foF2 were noticed on 26 September 2011 compared with the quiet periods, indicating the storm's effects on ionospheric conditions. The observation sequence from 16:00 h UT 26 to 8:00 h UT 30 September was used to evaluate different ionospheric foF2 prediction models.

The prediction results of the proposed method and other models are shown in Figure 7a. Similarly, the RMSE values of the proposed method and other models considered in the comparison are summarized in Table 12. The proposed method predicted foF2 features with RMSE of 0.966 MHz, and the prediction values are followed with the measured data during the whole storm [Figure 7a]. In addition, the BiLSTM-foF2 model performed better when compared with the IRI-2016, but the RMSE is significantly more than the proposed method. It is evident from the results that the proposed method performed well during the considered storm event.

Table 12. Performance evaluation of prediction model for 26 September 2011 storm event.

Model Configuration	RMSE (MHZ)	MAE (MHZ)	MAPE
IRI-2016	1.059	0.865	13.5%
LSTM-foF2	1.373	1.017	14.86%
BiLSTM-foF2	1.035	0.814	11.587%
Proposed method	0.966	0.784	11.181%

3.3.2. Scenario 4: The foF2 Prediction Analysis for the 15 July 2012 Geomagnetic Storm

Figure 8 depicts scenario 4 of the proposed model and other model results with measured data during the storm event of 15 July 2012. It is noticed that the storm took place in winter. On 15 July 2012, the Dst value turned positive to negative at 2:00 h UT, and a peak negative value of −139 nT was observed at 16:00 h UT on 15 July. The process indicated the main phase of the storm [Figure 8b]. The recovery phase of the storm begins at 19:00 h UT, and the Dst value had steadily risen [Figure 8b]. The end of the event is at 10:00 h UT on 18 July.

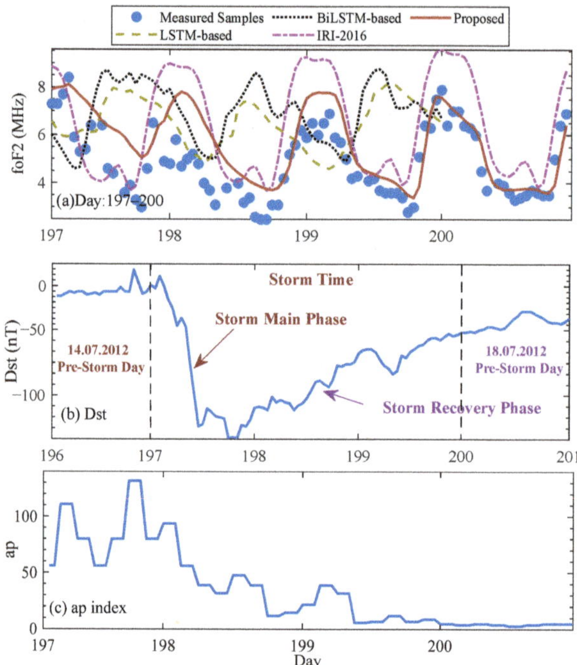

Figure 8. Variations of foF2 (**a**), Dst (**b**), and ap (**c**) during the 14 July 2012 geomagnetic storm.

Evaluation of the different models' performance is from 4:00 22 to 10:00 h UT 18 July during this storm event, and the prediction results of the proposed method and other models are shown in Figure 8b. Similarly, the RMSE, MAE, and MAPE of the proposed method and other models considered in the comparison are summarized in Table 13.

Table 13. Performance evaluation of prediction model for 15 July 2012 storm event.

Model Configuration	RMSE (MHZ)	MAE (MHZ)	MAPE
IRI-2016	2.057	1.760	37.427%
LSTM-foF2	1.61	1.306	31.802%
BiLSTM-foF2	1.496	1.207	27.364%
Proposed method	1.265	0.996	22.737%

The predicted results of the proposed method are RMSE of 1.265 MHz, MAE of 0.996 MHz, and MAPE of 22.73%, respectively. It is clear from Figure 8a that the proposed method performed well during this storm event. Similarly, Table 13 reveals that the proposed method predicted well when compared with other models during this storm period. The error is relatively high with the LSTM-based model compared with BiLSTM-based and the proposed method. Similarly, the BiLSTM-foF2 model predicts better when compared with the IRI-2016 model, but their RMSE is significantly more compared with the proposed model.

4. Discussion

4.1. Analysis of Prediction Error Results

In this section, we calculate the samples error from different foF2 prediction models under low solar activity year (scenario 1), and the results are presented in Figure 9.

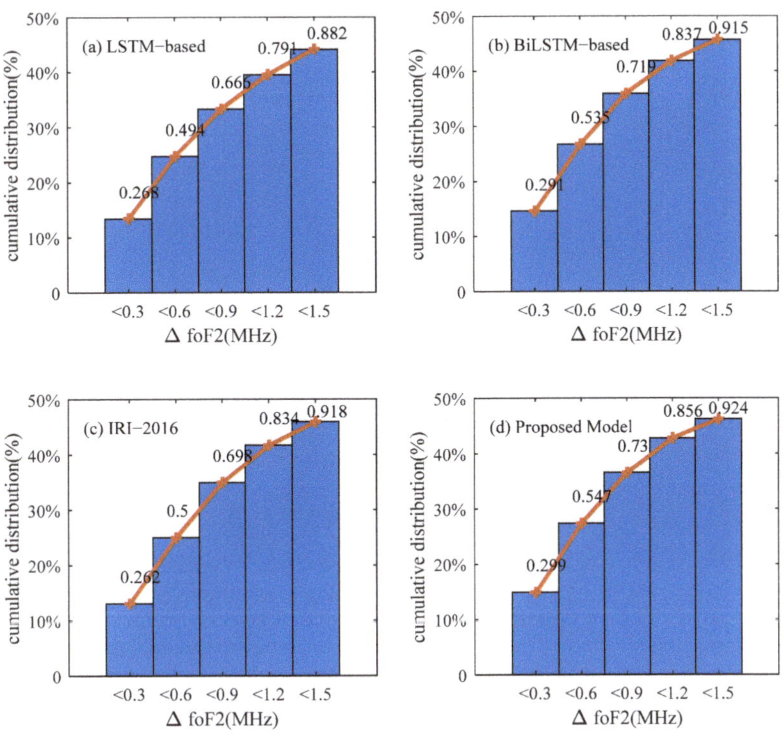

Figure 9. Comparison of cumulative distribution of prediction errors: (**a**) LSTM-foF2; (**b**) BiLSTM-foF2; (**c**) IRI-2016; (**d**) proposed model.

Figure 9d describes that the proposed method predicted well compared with the other models. The proposed model is performed well when the absolute error is less than 0.3 and 1.5 MHz. On the contrary, the LSTM-foF2 model has the worst correlation with measured samples. Similarly, the prediction errors are analyzed during the high solar activity year (scenario 2), as described in Figure 10.

Figure 10 elucidates the statistical and normal prediction error distribution. In addition, the proposed method presented in Figure 10d is crucially concentrated between −2 and 2 MHz, and the LSTM-based with the maximum error is between −4 and 4 MHz. Moreover, the LSTM-based, IRI-2016, BiLSTM-based, and proposed model predicted results have a maximum error of 1.816, 1.58, 1.808, and 1.262 MHz, respectively. The proposed method has a clear advantage compared with the above prediction models. Furthermore, Figure 10d reveals that the normal distribution curve of the proposed method is more well-fitted than the others, in which the normal distribution curve of the proposed model's error is prominently concentrated. At the same time, the LSTM-based is scattered, indicating the improved performance of the proposed model compared with the other models. The MAE in the IRI 2016 model (0.851 MHz) and proposed model (0.626 MHz) reveals that the proposed method is well correlated with the measured samples under the high solar activity year.

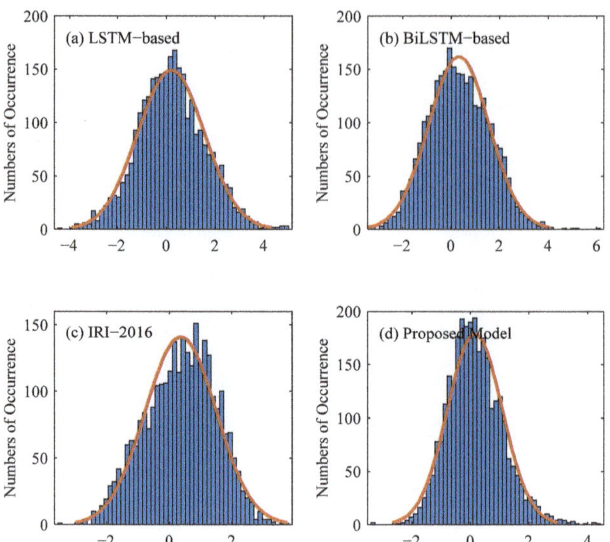

Figure 10. Histograms and the normal distribution curve of the prediction errors: (**a**) LSTM-foF2; (**b**) BiLSTM-foF2; (**c**) IRI-2016; (**d**) proposed model.

Figure 11 shows the two storm events foF2 prediction error box plots for the different foF2 prediction models. The box represents the foF2 prediction errors for space weather events, in which the red signs represent outlier values. The top and bottom black bars represent the maximum and minimum values of the prediction error, respectively. The upper and lower edges of the blue box represent the upper and lower quartiles, respectively, in which the middle red represents the median. It is noticed that the proposed foF2 prediction model performed well when comparing the boxes.

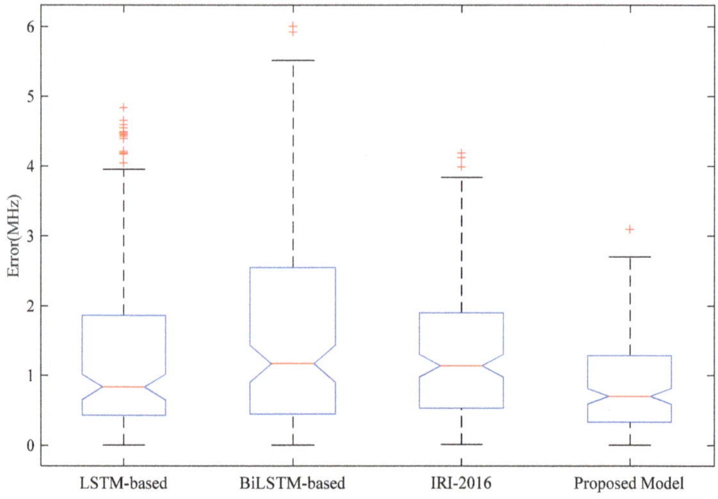

Figure 11. Statistical prediction error analyses for different prediction models.

4.2. Analysis of Quantile Regression Weight

Here, we further analyze the effect of quantile regression weight on the error of ionospheric foF2 prediction results. The month of February of the two solar years (2010 and

2014) was taken as an example for discussion. The weight is set to quartiles and deciles, and the configuration and prediction errors of the quartile's weight during 2010 and 2014 are summarized in Tables 14 and 15.

Table 14. Prediction errors for quartiles during February 2010.

Quartiles Weight	RMSE (MHZ)	MAE (MHZ)	MAPE
0.25	1.33	1.11	17.65%
0.5	0.98	0.77	12.77%
0.75	0.86	0.67	11.95%

Table 15. Prediction errors for quartiles during February 2014.

Quartiles Weight	RMSE (MHZ)	MAE (MHZ)	MAPE
0.25	1.4	1.13	12.89%
0.5	1.21	0.96	10.85%
0.75	1.05	0.81	9.56%

Tables 14 and 15 illustrate that the quantile weight in 0.75 has the smallest RMSE, MAE, and MAPE error in the case of quartiles during 2010 and 2014. In addition, the deciles are utilized to further discuss the prediction results. The RMSE values of prediction results under the different solar activity years are presented in Figure 12, and the detailed value is summarized in Table 16.

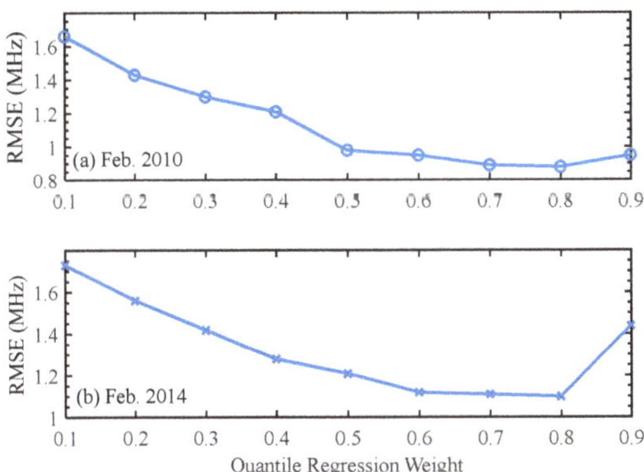

Figure 12. Prediction error analyses for deciles quantile regression weight: (**a**) February 2010 and (**b**) February 2014.

Table 16. The division of foF2 samples.

Input Time Series Length	Training Samples (Input)	Validation Samples	Test/Predict Samples (Output)
One month	December 2009	January 2010	February 2010
Three months	October–December 2009	January 2010	February 2010
Six months	July–December 2009	January 2010	February 2010
Nine months	April–December 2009	January 2010	February 2010
Full	January–December 2009	January 2010	February 2010

The quantile weight in 0.1 has a maximum RMSE value of 1.66 MHz and a minimum of 0.86 MHz in 0.75 during 2010. Similarly, the maximum and minimum RMSE values of 1.73 and 1.05 MHz are produced in quantile weights of 0.1 and 0.75, respectively, in 2014. In addition, Figure 12 describes that the ionospheric foF2 prediction error has significantly declined from a quantile weight of 0.2, and the foF2-RMSE value curve is relatively stable from a quantile weight of 0.6 to 0.8.

4.3. Analysis of Input and Output foF2 Time Series Length

In this section, we evaluate the limited time series input and different lengths of output for the ionospheric foF2 prediction model. The total foF2 samples are set from January 2009 to February 2010, and the division is presented in Table 16. We use hourly foF2 samples and set the complete training samples to include data from January to December 2009. Similarly, validation samples and test samples are set to January and February 2010, respectively. To analyze the limited time series, we set the input time lengths as one, three, six, and nine months, and the prediction results are presented in Table 17.

Table 17. Prediction error analyses for different time series lengths.

Input Time Series Length	RMSE (MHZ)	MAE (MHZ)	MAPE
One month	1.80	1.54	24.52%
Three months	1.49	1.23	19.84%
Six months	1.14	0.87	14.28%
Nine months	0.93	0.73	12.2%
Full	0.86	0.67	11.95%

It is clear from Table 17 that the complete training samples perform better than the other conditions. Generally, diurnal and seasonal variations are significant features of ionospheric foF2. With the input short time series, deep learning is hard to learn and capture features of foF2 variation. In addition, since deep learning-based ionospheric foF2 prediction models need sufficient training samples to learn features between input elements, we consider that input length (training samples) should include nine months (three seasons) at least to obtain better prediction performance (RMSE less than 1 MHz) for the input limited time series length.

In addition, the different lengths of outputs are analyzed. The total foF2 samples are consistent with the above, and we set training samples from January to December 2009, and validation samples are set for January 2010. A detailed explanation of the foF2 sample division is shown in Table 18. For the test/prediction set, we set one, three, six, nine, and twelve months to analyze performance, and the prediction results are presented in Table 19.

Table 18. The setting for output time series length.

Output Time Series Length	Training Samples (Input)	Validation Samples	Test/Predict Samples (Output)
One month	January–December 2009	January 2010	February 2010
Three months	January–December 2009	January 2010	February–April 2010
Six months	January–December 2009	January 2010	February–July 2010
Nine months	January–December 2009	January 2010	February–October 2010
Twelve months	January–December 2009	January 2010	February 2010–January 2011

Table 19. Prediction error analyses for different time series output lengths.

Output Time Series Length	RMSE (MHZ)	MAE (MHZ)	MAPE
One month	0.86	0.67	11.95%
Three months	0.88	0.68	13.16%
Six months	0.92	0.7	14.81%
Nine months	0.93	0.72	15.9%
Twelve months	1	0.79	16.05%

It is evident from Table 19 that the one-month condition performs better than others. The prediction performance in twelve months is the worst when compared with other conditions. In addition, the RMSE values are relatively stable from three-to-nine-month conditions.

5. Conclusions

This paper proposes a hybrid neural network with quantile regression mechanism, to predict the ionospheric foF2 variation over the low latitude region. The model input parameter (foF2) is considered from the ionosonde located at the Brisbane station (27°53′S, 152°92′E) along with space weather parameters F10.7, SSN, IMF-Bz, ap, and Dst from 2009 to 2014. We design four scenarios (high and low solar activity year and two space weather events) and use measured foF2 data to evaluate the performance of the proposed model. The prediction results are compared with IRI-2016, LSTM-based, and BiLSTM-based models' output values. In addition, we further analyzed and discussed the performance of different models in the considered storm scenarios (26 September 2011 and 15 July 2012). The RMSE, MAPE, MAE, and statistical error analysis demonstrate that the proposed model strikes a favorable performance compared with the other previous models during the high and low solar activity and geomagnetic storm periods.

The ionospheric foF2 determines the working frequency of the shortwave signal, and an accurate frequency prediction can reduce the complexity of frequency selection. In summary, the proposed model is ideally suited to predict the variations of ionospheric parameters, and the results are helpful in frequency selection and global navigation satellite system position. Future studies will constantly test the proposed model in foF2 measured data from other ionospheric observatories over low latitude regions, and discuss optimal space weather parameters for the deep network model's input during space weather events.

Author Contributions: Conceptualization, C.B. and T.Y.; methodology, C.B.; software, C.B.; validation, Z.X. and C.B.; formal analysis, C.B.; investigation, C.B. and T.Y.; writing—original draft preparation, C.B.; writing—review and editing, Y.Z. and P.R. All authors have read and agreed to the published version of the manuscript.

Funding: This research received no external funding.

Data Availability Statement: In this paper, we used Brisbane station ionospheric foF2 measured data from the Australian Bureau of Meteorology, Space Weather Services, and the solar cycle and geomagnetic information from the Goddard Space Flight Center, NASA (https://omniweb.gsfc.nasa.gov/ow.html (accessed on 31 December 2014)) and Helmholtz Centre Potsdam GFZ German Research Centre (https://www.gfz-potsdam.de/en (accessed on 31 December 2014)). The sunspot number (SSN) was used from the Royal Observatory of Belgium, Brussels (http://sidc.oma.be/silso/datafiles (accessed on 31 December 2014)).

Acknowledgments: The authors are grateful to the Australian Bureau of Meteorology, Space Weather Services for the provision of ionospheric data.

Conflicts of Interest: The authors declare no conflict of interest.

Abbreviations

The following abbreviations are used in this manuscript:

BiLSTM	Bidirectional long short-term memory
CNN	Convolutional neural network
LSTM	Long short-term memory
foF2	Ionospheric F2 layer Critical frequency
IRI	International Reference Ionosphere
TEC	Total electron content
SSN	Sunspot number
UT	Universal time
F10.7	Solar radio flux of 10.7 cm wavelength
IMF-Bz	Interplanetary magnetic field Bz component
Dst	Disturbance storm time
RNN	Recurrent neural network
QR	Quantile regression
RMSE	Root-mean-square error
MAE	Mean absolute error
MAPE	Mean absolute percentage error

References

1. Bilitza, D.; Obrou, O.; Adeniyi, J.; Oladipo, O. Variability of foF2 in the equatorial ionosphere. *Adv. Space Res.* **2004**, *34*, 1901–1906. [CrossRef]
2. Oyeyemi, E.O.; McKinnell, L.A.; Poole, A.W.V. Near-real time foF2 predictions using neural networks. *J. Atmos. Sol. Terr. Phys.* **2006**, *68*, 1807–1818. [CrossRef]
3. Bai, H.; Feng, F.; Wang, J. A Combination Prediction Model of Long-Term Ionospheric foF2 Based on Entropy Weight Method. *Entropy* **2020**, *22*, 442. [CrossRef] [PubMed]
4. Chen, C.; Wu, Z.S.; Xu, Z.W.; Sun, S.J.; Ding, Z.H.; Ban, P.P. Forecasting the local ionospheric foF2 parameter 1 hour ahead during disturbed geomagnetic conditions. *J. Geophys. Res. Space Phys.* **2010**, *115*, 135–146. [CrossRef]
5. Olga, M. The Influence of Space Weather on the Relationship Between the Parameters TEC and foF2 of the Ionosphere. *IEEE J. Radio Freq. Identif.* **2021**, *5*, 261–268.
6. Sun, W.; Long, X.; Xin, H. Forecasting of ionospheric vertical total electron content (TEC) using LSTM networks. In Proceedings of the 2017 International Conference on Machine Learning and Cybernetics (ICMLC), Ningbo, China, 9–12 July 2017 .
7. Suin, M.; Yong, H.K.; Jeong, H.; Young, K.; Jong, Y. Forecasting the Ionospheric F2 Parameters over Jeju Station (33.43°N, 126.30°E) by Using Long Short-Term Memory. *J. Korean Phys. Soc.* **2020**, *77*, 1265–1273.
8. Li, X.J.; Zhou, C.; Tang, Q.; Zhao, J.; Zhang, F.B.; Xia, G.C.; Liu, Y. Forecasting Ionospheric foF2 Based on Deep Learning Method. *Remote Sensing* **2021**, *13*, 3849. [CrossRef]
9. Kim, J.H.; Kwak, Y.S.; Kim, Y.H.; Moon, S.I.; Jeong, S.H.; Yun, J.Y. Potential of Regional Ionosphere Prediction Using a Long Short-Term Memory Deep-Learning Algorithm Specialized for Geomagnetic Storm Period. *Space Weather* **2021**, *19*, 1–20. [CrossRef]
10. Rao, T.V.; Sridhar, M.; Ratnam, D.V.; Harsha, P.B.S.; Srivani, I.A. Bidirectional Long Short-Term Memory-Based Ionospheric foF2 and hmF2 Models for a Single Station in the Low Latitude Region. *IEEE Geosci. Remote Sens. Lett.* **2021**, *19*, 1–5. [CrossRef]
11. Williscroft, L.A.; Poole, A.W. Neural networks, foF2, sunspot number and magnetic activity. *Geophys. Res. Lett.* **1996**, *23*, 3659–3662. [CrossRef]
12. McKinnell, L.A.; Poole, A.W.V. Ionospheric variability and electron density profile studies with neural networks. *Adv. Space Res.* **2001**, *27*, 83–90. [CrossRef]
13. Athieno, R.; Jayachandran, P.T.; Themens, D.R. A neural network based foF2 model for a single station in the polar cap. *Radio Sci.* **2017**, *52*, 784–796. [CrossRef]
14. Perna, L.; Pezzopane, M. foF2 vs solar indices for the Rome station: Looking for the best general relation which is able to describe the anomalous minimum between cycles 23 and 24. *J. Atmos. Sol. Terr. Phys.* **2016**, *148*, 13–21. [CrossRef]
15. Joshua, E.O.; Nzekwe, N.M. foF2 correlation studies with solar and geomagnetic indices for two equatorial stations. *J. Atmos. Sol. Terr. Phys.* **2012**, *80*, 312–322. [CrossRef]
16. Kane, R.P. Solar cycle variation of foF2. *J. Atmos. Sol. Terr. Phys.* **1994**, *54*, 1201–1205. [CrossRef]
17. Bai, H.M.; Feng, F.; Jian, W.; Wu, T.S. Nonlinear dependence study of ionospheric F2 layer critical frequency with respect to the solar activity indices using the mutual information method. *Adv. Space Res.* **2019**, *5*, 1085–1092. [CrossRef]
18. Campbell, W.H. Occurrence of AE and Dst geomagnetic index levels and the selection of the quietest days in a year. *J. Geophys. Res.* **1979**, *84*, 75–881.
19. Ergen, T.; Kozat, S.S. Online Training of LSTM Networks in Distributed Systems for Variable Length Data Sequences. *IEEE Trans. Neural. Netw. Learn. Syst.* **2018**, *29*, 5159–5165. [CrossRef]
20. Sherstinsky, A. Fundamentals of recurrent neural network (RNN) and long short-term memory (LSTM) network. *Phys. D Nonlinear Phenom.* **2020**, *404*, 132306. [CrossRef]
21. Mustaqeem Sajjad, M.; Kwon, S. Clustering-Based Speech Emotion Recognition by Incorporating Learned Features and Deep BiLSTM. *IEEE Access* **2020**, *8*, 79861–79875. [CrossRef]
22. Siami-Namini, S.; Tavakoli, N.; Namin, A.S. The performance of LSTM and BiLSTM in forecasting time series. In Proceedings of the 2019 IEEE International Conference on Big Data (Big Data), Los Angeles, CA, USA, 9 September 2019; IEEE: Piscataway, NJ, USA, 2019; pp. 3285–3292.
23. Kumar, S.D.; Subha, D. Prediction of Depression from EEG Signal Using Long Short Term Memory (LSTM). In Proceedings of the 2019 3rd International Conference on Trends in Electronics and Informatics (ICOEI), Tirunelveli, India, 23–25 April 2019; IEEE: Piscataway, NJ, USA, 2019; pp. 1248–1253.
24. Zhang, X.; Zhang, Y.; Lu, X.; Bai, L.; Chen, L.; Tao, J.; Wang, Z.; Zhu, L. Estimation of Lower-Stratosphere-to-Troposphere Ozone Profile Using Long Short-Term Memory (LSTM). *Remote Sens.* **2021**, *13*, 1374. [CrossRef]
25. Koenker, R.; Hallock, K. Quantile Regression-An introduction. *J. Econ. Perspect.* **2000**, *15*, 143–156. [CrossRef]
26. Zhang, W.; Quan, H.; Gandhi, O.; Rajagopal, R.; Tan, C.; Srinivasan, D. Improving Probabilistic Load Forecasting Using Quantile Regression NN With Skip Connections. *IEEE Trans. Smart Grid.* **2020**, *11*, 5442–5450. [CrossRef]
27. Taylor, W. A Quantile Regression Neural Network Approach to Estimating the Conditional Density of Multiperiod Returns. *J. Forecastin.* **2020**, *19*, 299–311. [CrossRef]

remote sensing

Article

Statistical Study of the Ionospheric Slab Thickness at Yakutsk High-Latitude Station

Jian Feng [1], Yuqiang Zhang [2,*], Na Xu [1], Bo Chen [2], Tong Xu [1], Zhensen Wu [3], Zhongxin Deng [1], Yi Liu [2], Zhuangkai Wang [2], Yufeng Zhou [4], Chen Zhou [2] and Zhengyu Zhao [2]

1   China Research Institute of Radiowave Propagation (CRIRP), Qingdao 266107, China
2   Department of Space Physics, School of Electronic Information, Wuhan University, Wuhan 430072, China
3   School of Physics, Xidian University, Xian 710126, China
4   Beijing Institute of Applied Meteorology, Beijing 100029, China
*   Correspondence: yqzhang_3@stu.xidian.edu.cn

Abstract: The ionospheric equivalent slab thickness (EST, also named τ) is defined as the ratio of the total electron content (TEC) to the F2-layer peak electron density (NmF2), and it is a significant parameter representative of the ionosphere. This paper presents a comprehensive statistical study of the ionospheric slab thickness at Yakutsk, located at the high latitude of East Asia, using the GPS-TEC and ionosonde NmF2 data for the years 2010–2017. The results show that the τ has different diurnal and seasonal variations in high- and low-solar-activity years, and the τ is greatest in the winter, followed by the equinox, and it is smallest in the summer in both high- and low-solar-activity years, except during the noontime of low-solar-activity years. Specifically, the τ in inter of high-solar-activity year shows an approximate single peak pattern with the peak around noon, while it displays a double-peak pattern with the pre-sunrise and sunset peaks in winter of the low-solar-activity years. Moreover, the τ in the summer and equinox have smaller diurnal variations, and there are peaks with different magnitudes during the sunrise and post-sunset periods. The mainly diurnal variation of τ in different seasons of high- and low-solar-activity years can be explained within the framework of relative variation of TEC and NmF2 during the corresponding period. By defining the disturbance index (DI), which can visually assess the relationship between instantaneous values and the median, we found that the geomagnetic storm would enhance the τ at Yakutsk. An example on 7 June 2013 is also presented to analyze the physical mechanism. It should be due to the intense particle precipitation and expanded plasma convection electric field during the storm at high-latitude Yakutsk station. The results would improve the current understanding of climatological and storm-time behavior of τ at high latitudes in East Asia.

Keywords: ionospheric slab thickness; high latitude; geomagnetic storm; East Asia

Citation: Feng, J.; Zhang, Y.; Xu, N.; Chen, B.; Xu, T.; Wu, Z.; Deng, Z.; Liu, Y.; Wang, Z.; Zhou, Y.; et al. Statistical Study of the Ionospheric Slab Thickness at Yakutsk High-Latitude Station. *Remote Sens.* **2022**, 14, 5309. https://doi.org/10.3390/rs14215309

Academic Editor: Fabio Giannattasio

Received: 29 September 2022
Accepted: 21 October 2022
Published: 24 October 2022

Publisher's Note: MDPI stays neutral with regard to jurisdictional claims in published maps and institutional affiliations.

1. Introduction

The ionospheric equivalent slab thickness, τ, is defined as the ratio of TEC (el/m^2) to NmF2 (el/m^3); the τ is thus expressed in meters, and it represents the equivalent depth of the ionosphere, which has a uniform electron density of NmF2. As it contains information on TEC and NmF2, it can be used to investigate the vertical distribution of plasma in the ionosphere–plasmasphere system and the vertical electron density profile. It is also related to the important ionosphere and thermosphere parameters, such as plasma scale height and neutral temperature [1–9]. Moreover, it can find applications in other fields, such as ionospheric modeling and data-assimilation methodologies, as it can convert the TEC to NmF2, and vice versa, according to its definition [10–14]. Therefore, numerous statistical and modeling studies have been performed since the 1960s when TEC measurements from geostationary satellites became available [15–26].

In previous studies on τ, researchers have found that the τ has various morphologies in different regions, as it has a special dependence on the local time, season, solar cycle, and geomagnetic activity according to the location [27–31]. Moreover, the effect of geomagnetic activity on τ is still an open question since the effect is strongly influenced by other factors such as location, local time, and solar activity [19,32–34]. In a recent survey, we found that the τ usually tends to increase during storm time, except for the sunrise period at Guam equatorial station by defining the disturbance index DI of τ [35].

As the installation and maintaining ionospheric facilities are difficult at equatorial and high latitudes, the studies of τ in high latitude are therefore relatively rare when compared with the studies of τ at low latitude and midlatitude [34]. Jayachandran et al. [36] studied the climatology of τ in Goosebay, which is located at high latitude, and they found that the nighttime mean values of τ are always larger than daytime mean values during both solar minimum and solar maximum. In addition, they discovered that the highest and lowest daytime mean values of τ are in summer and winter, respectively, at Goosebay for both solar phases. As for the effect of geomagnetic activity on the τ, they conclude that the geomagnetic activity would enhance τ values in high-solar-activity years at Goosebay. Using the data of Casey station, which is located at high latitudes in the southern hemisphere, Yadav and Bhawre [37] studied the variation of τ during the high-solar-activity year 2005. They also found that the nighttime mean values of τ are larger than daytime mean values, while they showed that the mean values of τ are largest during the equinox and smallest in the summer, demonstrating a different feature from the τ at Goosebay. In a recent study about the climatological behavior of τ, Pignalberi et al. [34] selected Tromso station (69.6°N 19.2°E, 66.5°N QD latitude) as a representative, and they found that the largest τ appeared during the nighttime and dawn hours in winter and equinox. Moreover, the pre-sunrise peak is more evident in the low-solar-activity period and mid-solar-activity period, except for the summer, and the post-sunset peak is only clearly seen in winter, except for the high-solar-activity period.

Yakutsk is located at the high latitude of East Asia, where the ionospheric instruments are relatively few. As previous studies described, there is still much room for improvement in τ global modeling, especially for the τ at high latitudes, where the pattern of τ is completely different from those at the other latitudes. In previous studies of τ at high latitudes, the main focus is on North American, the Europe sector, or the Southern hemisphere; there are no specific studies investigating the τ in the high latitude of East Asia (best to our knowledge). To better understand the variation of τ in the high latitudes of East Asia and improve the global empirical modeling of τ, this paper investigates the climatology of τ at the Yakutsk high-latitude station. In addition, we investigated the effects of magnetic activity disturbances on τ by using the disturbance index DI of τ. This paper is organized as follows: The data and methodology are briefly described in Section 2. Section 3 presents the results, especially for the variation of τ during geomagnetic storms. The interpretation of the derived results is given in Section 4. Section 5 sums up the paper.

2. Data and Methods of Analysis

The original receiver-independent exchange data (RINEX format) used to deduce TEC are obtained from UNAVCO (University NAVSTAR Consortium) database (http://www.unavco.org, (accessed on 20 October 2022)). After downloading the RINEX format data, the GPS-TEC software Gopi developed by Seemala is used to derive the TEC values (https://seemala.blogspot.com/, (accessed on 20 October 2022)). Moreover, the elevation angle cutoff is set to 30° to eliminate the multiple-path effects. Previous studies described, in detail, how the software works and compared it with other techniques, and the results showed that the software is capable of TEC computation [38,39]. Noteworthy is that this GPS-TEC software has been extensively used in previous studies to derive TEC in the ionosphere [40–43]. Additionally, the regional kriging interpolation method which converts STEC to VTEC has been applied in this paper to derive more accurate VTEC values after obtaining STEC from the software, rather than using the average values of VTEC data calculated by the software directly.

The foF2 data used in this paper were downloaded from the GIRO (Global Ionospheric Radio Observation) database (https://giro.uml.edu/didbase (accessed on 20 October 2022)) [44]. Only the ionosonde data with a Confidence Score (CS value) larger than 80 are selected in this paper to calculate τ values. The NmF2 are computed by the formula NmF2 = 1.24 * 1010 (foF2)^2. Since the resolution of foF2 is 15 min, the TEC data are also calculated with a resolution of 15 min, and the τ is obtained by the following equation:

$$\tau = TEC/NMF2 = TEC/(1.24 \times 10^{-6}(foF2)^2) \tag{1}$$

where TEC is measured with electrons per m^2, and foF2 is measured with the critical frequency of F2 layer, with foF2 given in MHz and τ in meters. In the following section, this paper uses kilometers to measure τ for simplicity.

As the solar EUV radiation is the main source for the neutral gas to ionize and it can be inferred by F10.7, this paper adopts the F10.7 index which is downloaded from NOAA (ftp://ftp.swpc.noaa.gov/pub/indices/old_indices/, (accessed on 20 October 2022)) to represent the solar activity. Figure 1 displays the overall variation of the solar F10.7 index from 2010 to 2017. As shown in Figure 1, the F10.7 was the greatest in 2014 among these 8 years, with an average value of 145.9 SFU. The F10.7 was small in 2010, 2016, and 2017, with mean values being 80SFU, 8.7 SFU, and 77.3 SFU, respectively. It is of great importance to note that only the first half of foF2 data in 2016 and 2017 are available for the Yakutsk station, while only the second half of foF2 data are available for 2010. Therefore, this paper used 2014 for the high-solar-activity periods (represented with blue line in Figure 1) and used 2010/2016/2017 for the low solar activity periods (represented with red line in Figure 1) to study τ variation in low-solar-activity years at Yakutsk. Moreover, this paper defined summer (May to August), winter (January, February, November, and December), and equinox (March, April, September, and October) according to the classification of Lloyd season.

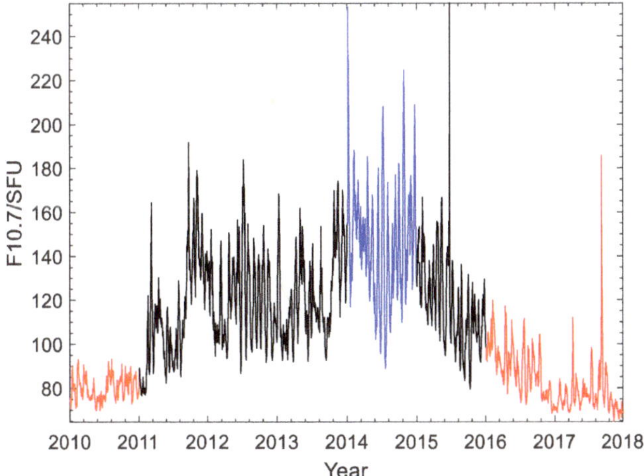

Figure 1. The F10.7 variation from the year 2010 to 2017.

The effects of geomagnetic storms on τ have been studied for several decades, and it is found that the effects depend on location, solar cycle, and other factors [35,36,45]. To study the climatology of τ under geomagnetic quiet conditions, this study excluded the data during geomagnetic storms, satisfying Dst min $<$ -30, and the Dst data were downloaded from World Data Center (WDC, http://wdc.kugi.kyoto-u.ac.jp/, (accessed on 20 October 2022)). The selected dataset was then binned into monthly/local-time hour grids according to the 0–23LT and Jan–Dec division. Recently, Pignalberi et al. [26] applied a similar methodology to analyze the main global features of τ during the geomagnetic quiet period. It is important to note that the median of τ rather than the mean of τ is used in this paper to study the climatological of τ at Yakutsk, as it could cut off the outliers of the distribution, as Pignalberi et al. [26] suggested. Moreover, adopting the same methodology as our last studies [35], this paper used the disturbance index (DI) to study the effect of geomagnetic storms on τ. The DI index is defined as follows:

$$DI = \tau_s / \tau_m \tag{2}$$

where τ_S and τ_m are the storm-time τ, and corresponding monthly median τ, respectively. Moreover, the monthly median is the median of a 27-day running window centering for the observed day. Figure 2 displays an example during the period 6 June 6 12 UT–8 June 12 UT in 2013 at Yakutsk. Figure 2 shows the Dst variation during this period (top panel), and it can be seen from the figure that the main phase onset of the geomagnetic storm is at 18:00 UT on June 6; the middle panel shows the measured τ (blue line) and corresponding τ median (red line) during 6–8 June; the bottom panel presents the DI index of τ according to Equation (1), using the measured τ and its median value. It is important to note that DI indexes are classified into storm-time DI index and quiet-time DI index to study their characteristics during the geomagnetic storm and geomagnetic quiet conditions, respectively. For example, the DI index in the storm-time period (colored in red) is used to study the features of DI index during geomagnetic storms, while DI index in the quiet-time period (colored in purple) is used to study the one at geomagnetic quiet conditions.

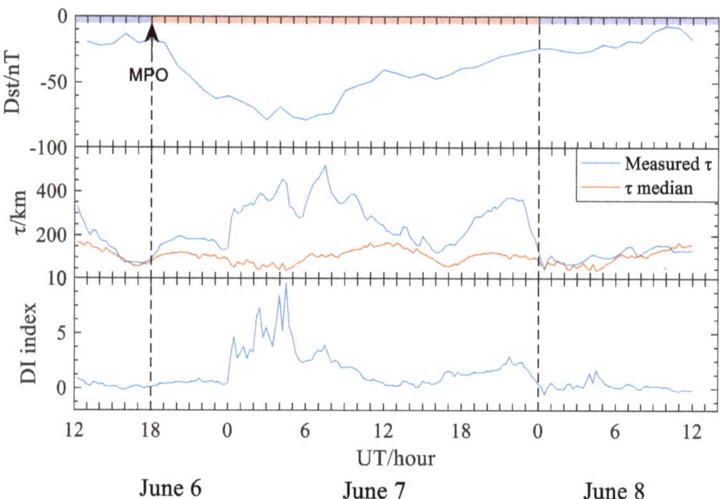

Figure 2. An example for the calculation of DI index during the period 6 June 12 UT−8 June 12 UT in 2013 at Yakutsk.

3. Results

According to the 0–23 LT and January–December grid division, Figure 3 displays the median and standard deviation of τ during the geomagnetic quiet period in the high- and low-solar-activity years at Yakutsk. It is important to note that the foF2 data are not available in November of 2010/2016/2017; the mean and standard deviation of τ in November are therefore white colored in the figures of low-solar-activity years. In addition, the sunrise (represented by yellow line) and sunset (represented by black line) time are shown in Figure 3. It can be seen from the figure that the τ is greatest in winter, followed by equinox, and smallest in summer in both high- and low-solar-activity years, except in the noontime of low-solar-activity years, as shown in Figure 3c. Moreover, the τ shows a completely different diurnal variation in winter of high- and low-solar-activity years. Specifically, it reaches its maximum during the noon to sunset hours and keeps relatively small during the nighttime in high-solar-activity years, while it stays large during 16 LT–06 LT, and it is the smallest at noon in low-solar-activity years. In addition, the τ has the highest variability in the winter, as shown in in the standard deviation of τ in Figure 3b,d. The figure also demonstrates that the τ in summer and equinox have smaller diurnal variations when compared to the τ in winter, and there are peaks with different magnitudes in the sunrise and post-sunset periods. Moreover, the τ in the summer and equinox seems to have a larger standard deviation in low-solar-activity years than in high-solar-activity years.

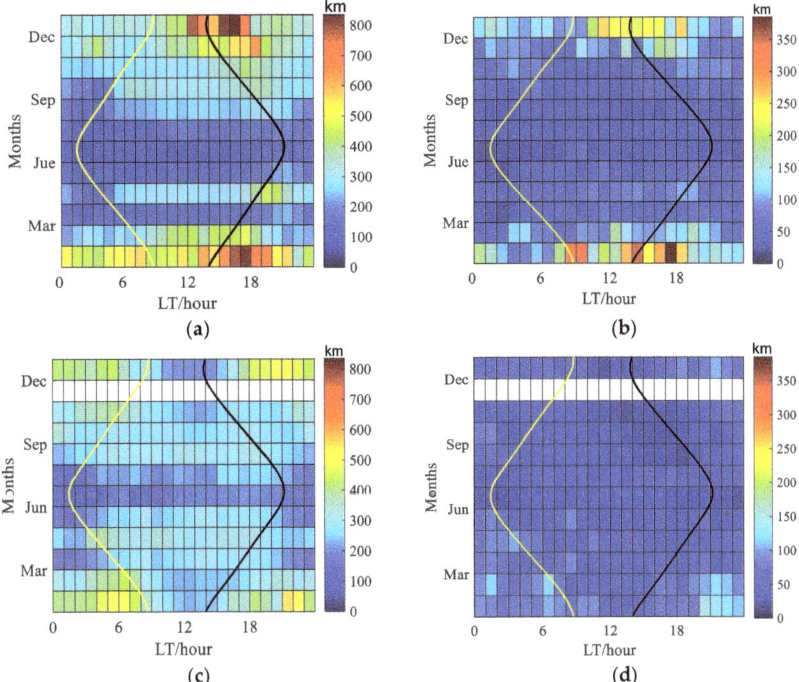

Figure 3. The median and standard deviation of τ during geomagnetic quiet period. (**a**) Median of τ in high-solar-activity year 2014. (**b**) Standard deviation of τ in high-solar-activity year 2014. (**c**) Median of τ in low-solar-activity years. (**d**) Standard deviation of τ in low-solar-activity years.

In order to more accurately demonstrates the diurnal variation of τ in different seasons, Figure 4 gives the diurnal variation of the median and standard deviation of τ in winter. It can be seen from the figure that the ionospheric slab thickness has different statistical characteristics in different seasons and solar-activity years:

(1)   In winter, the τ in the high-solar-activity year shows an approximate single-peak pattern, while it displays a double-peak pattern in low-solar-activity years. Specifically speaking, the τ during the daytime is far larger than that in the nighttime in high-solar-activity years, whereas the opposite situation applies for low-solar-activity years. In the winter of the high-solar-activity year, the τ increases continuously after sunrise, reaching its first peak of 675 km at 14 LT, and then it decreases to 579 km at 15 LT and keeps increasing to the maximum of 746 km at 17 LT. After that, the τ decreases to its minimum 325 km at midnight 0 LT. In addition, it changed little during the midnight-to-sunrise period. On the other hand, the τ in the winter of low-solar-activity years showed a totally different pattern. It decreases to the minimum of 242 km in 12 LT during the pre-noon hours and continuously increases to the peak of 445 km at 22 LT.

Moreover, it starts increasing at 3 LT and reaches its maximum of 480 km at 6 LT, as shown in Figure 3c.

(2) In the summer, the τ has a similar variation both in the high- and low-solar-activity years, except that the τ has a post-sunset peak in high-solar-activity years, and τ in the high-solar-activity years is smaller than that in the low-solar-activity years during all periods, except in the evening period (20–22 LT). Specifically, in the high-solar-activity years, it stays relatively steady after sunrise, remaining in the 160 ± 10 km range. From 17 LT, it keeps increasing until it reaches the maximum of 225 km at 21 LT, and then it decreases to the minimum of 93 km at 3 LT. After that, the τ continuous increases to the pre-sunrise peak of 160 km at 5 LT. In the low-solar-activity years, it changed little during the sunrise-to-sunset period, remaining in the 235 ± 15 km range. Moreover, it begins to decrease after sunset, until it reaches a minimum of 161 km in 0 LT, and then it increases until the peak of 240 km at 6 LT.

(3) In equinox, the τ in low-solar-activity years has a small range of diurnal variation, while in high-solar-activity years shows, it more variability, with τ having a maximum/minimum during the post-sunset/pre-sunrise period. Specifically speaking, it increases rapidly from 3 LT to its first peak of 264 km at 6 LT in high-solar-activity years. It remains stable (264–268 km) during the post-sunrise to afternoon (6–14 LT) period, and then it increases to the maximum of 357 km at 19 LT. From the evening to the post-midnight period (19–2 LT), the slab thickness decreases continuously to its minimum of 166 km at 2 LT. Compared with the τ in high-solar-activity years, the τ in low-solar-activity years shows less variability, especially during the pre-sunrise and post-sunset period, for which the τ does not have an apparent peak.

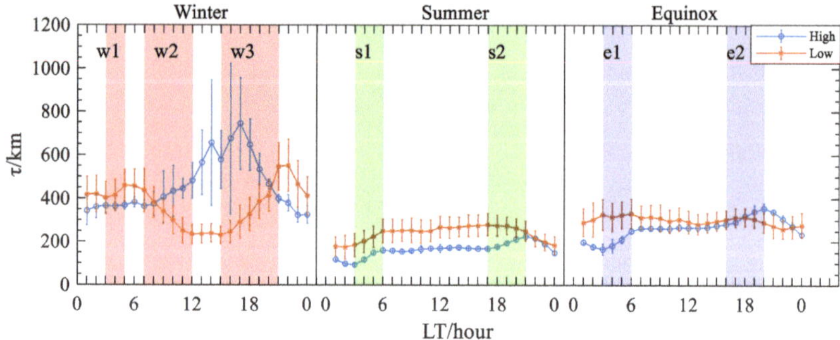

Figure 4. The diurnal variation of the mean and standard deviation of τ in different seasons of high and low-solar-activity years.

This paper uses the DI index to evaluate the geomagnetic activity effect on the slab thickness, as described above. To demonstrate the difference of τ between storm time and quiet time, we first calculated the DI index during quiet time. According to whether the DI is positive or negative, the daily variation of the mean and standard deviation of the DI index are shown separately. The positive DI variation is represented by blue curves, and the negative DI variation is represented by red curves. As shown in Figure 5a, the magnitude of the ionospheric slab thickness perturbation at Yakutsk station is relatively stable at all times of the day, with the DI index fluctuating between -0.18 and 0.29. In addition, the disturbance is more intense during the sunset time compared with other periods, and the positive/negative DI index is at least greater than 0.21/less than -0.15 within 16:00–20:00 LT, with the DI index reaching a maximum/minimum of 0.29/-0.18, respectively, at 17:30 LT.

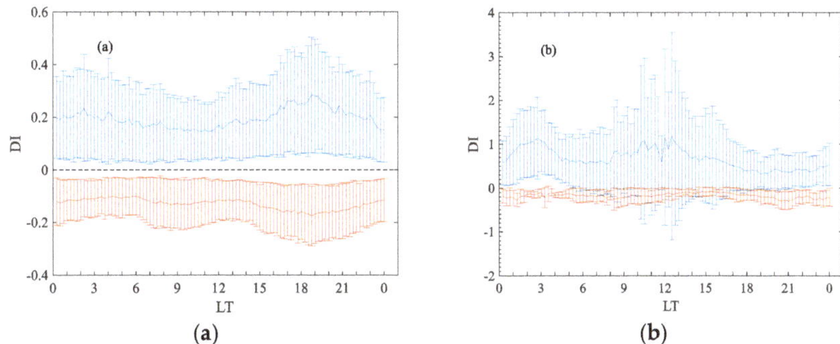

Figure 5. The diurnal variation of the mean and standard deviation of DI index during the (**a**) geomagnetically quiet periods and (**b**) geomagnetic storm periods.

Figure 5b shows the variation of DI index during storm time. The variation of the storm-time DI index differs significantly from the variation of quiet-time DI index. The positive DI index in the storm time is far larger than that in the quiet time, with the smallest DI index at 0.32, which is greater than the largest positive DI index at 0.28 in quiet time. In the period from sunset to evening (17:00–23:00 LT), the storm-time positive DI index is less than 0.5. The DI index reached a peak of 1.14 at 2:30 LT after midnight, and another peak of 1.16 is also present at 12:00 LT. Overall, the most intense disturbances in τ and the corresponding largest DI indices were observed before midnight and around the midday hours. In contrast, during the sunset hours, when the DI index is the largest and the disturbance is the most intense on quiet days, the storm time disturbance is the weakest, and the DI index is the smallest. The negative DI index in storm time also shows a different morphology from that in the quiet time. It is relatively stable in the 11–17 LT period, with a value greater than -0.16. The negative perturbation is intense in the period after sunset to before midnight, with the minimum value of -0.29 at 21:00 LT. The negative DI index also reaches -0.27 at 1:00 LT after midnight and -0.25 at 8:00 LT after sunrise. Overall, the intensity of the positive and negative disturbance varies dramatically. The magnitude of the positive disturbance is much larger than the magnitude of the negative disturbance, and the local time variation of the positive/negative disturbance is also different.

To better show the geomagnetic effect on τ, Figure 6a presents the comparison of the quiet-time DI index and storm-time DI index. It is of great importance to note that the DI index in Figure 6a is the same as the mean DI index in Figure 5. As can be seen from the figure, the storm-time DI index has a greater range of variation than the quiet-time DI index. The storm-time positive DI index is larger than the quiet-time positive DI index in all periods, and it is particularly significant in the period from midnight to sunrise and around noon. The storm-time negative DI index has a more variable pattern compared to the quiet-time negative DI index. For most of the time, the storm-time negative DI index is smaller than the quiet-time negative DI index, except for the afternoon period from 12:00 to 15:00 LT when the storm-time negative DI index is slightly larger than the quiet-time negative DI index. Figure 6b shows the difference between the storm-time DI index and quiet-time DI index according to the sign of the DI index. As can be seen from the figure, the difference in the positive DI index is relatively small during the pre-sunset-to-evening period (17–23 LT), with the difference below 0.21. The difference is greater than 0.2 in all other time periods, reaching 0.93/1.02 at 02:45 LT/12:30 LT after midnight/around noon, respectively. On the other hand, the negative DI index difference is greater than -0.2 in all time periods, and even greater than 0 in the afternoon period, 12:00–15:00 LT, indicating that the negative disturbance during the storm is very weak. Therefore, we preliminarily conclude that the geomagnetic storm would enhance the τ at Yakutsk, especially for the τ in post-midnight-to-sunrise hours and noontime.

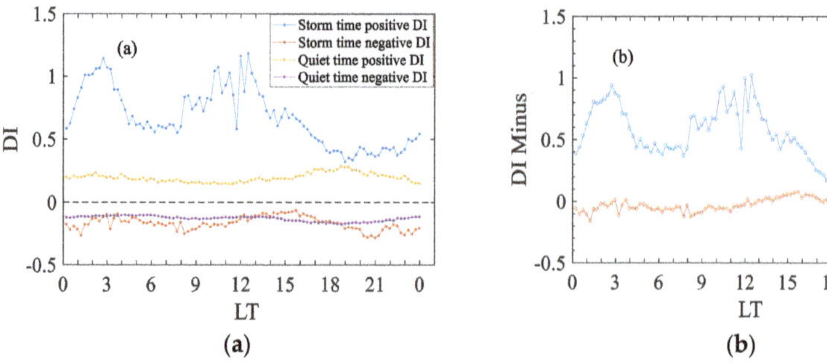

Figure 6. (**a**) Comparison of the quiet-time DI index and storm-time DI index at Yakutsk (the DI index in Figure 6a is the same as the mean DI index in Figure 5). (**b**) The minus between the storm-time DI and quiet-time DI according to the sign of the DI index.

4. Discussion

Yakutsk is located at a high latitude in East Asia; the most important factors affecting the ionospheric climatological behavior in this region are solar radiation, solar zenith angle, and background atmospheric composition, which are associated with summer-to-winter prevailing circulation [46–48]. At the same time, the nighttime downward plasma influx from the plasmasphere and conjugate hemisphere can also significantly influence the ionospheric condition at high latitudes [49–51]. In addition, the geomagnetic activity could result in positive or negative ionospheric storms, which severely affect the ionosphere states, according to the local time, season, and other ionosphere–thermosphere system background information when the geomagnetic storms occur [52–54]. Consequently, the τ at Yakutsk shows a complicated pattern. Since the geomagnetic activity is the main contributor to the relative ionospheric variability, we first discussed the climatology of τ, followed by the variation of τ in storm time.

To explain the diurnal variation of τ, the TEC and NmF2 during the same geomagnetic quiet period as the period of τ shown in Figure 4 are selected, and we displayed them in Figure 7, according to the definition of τ. It can be seen from the figure that TEC and NmF2 have different morphology in different seasons at high latitudes. The reversal of the meridional wind to equatorward and continuous strong solar radiation in the summer leads to a pronounced increase in TEC and NmF2 during post-sunset period [51,55–57]. Moreover, the TEC and NmF2 in Yakutsk display different magnitude 'winter anomaly' which referred to the TEC and NmF2 around noontime in winter being larger than those in summer [47,58]. Specifically, the "winter anomaly" of TEC is more evident in the high-solar-activity year, while the "winter anomaly" of NmF2 is more significant in the low solar activity years, consistent with previous studies [48]. It is due to the ionosphere strongly depending on the solar zenith angle in the winter, and the increasing [O/N2] is less/more important than the strong/weak solar radiation in high- and low-solar-activity years, respectively. Therefore, the TEC around noontime is greatest in winter of high-solar-activity years, and it caused the maximum of τ, while the large NmF2 caused the small τ in winter of low-solar-activity years, as shown in Figure 7. Moreover, the Yakutsk Anomaly (YA) is similar to the Weddell Sea Anomaly (WSA), which is characterized by greater nighttime electron density than daytime electron density in summer, as also seen in Figure 6, consistent with previous sub-auroral summer longitudinal anomalies study [59–61].

Moreover, we could also calculate the variation rates of τ, which are determined by the variation rates of TEC and NmF2 as follows (deduced from Equation (1)):

$$\frac{d\tau}{dt} = \frac{1}{NmF2}\frac{dTEC}{dt} - \frac{TEC}{NmF2^2}\frac{dNmF2}{dt} \tag{3}$$

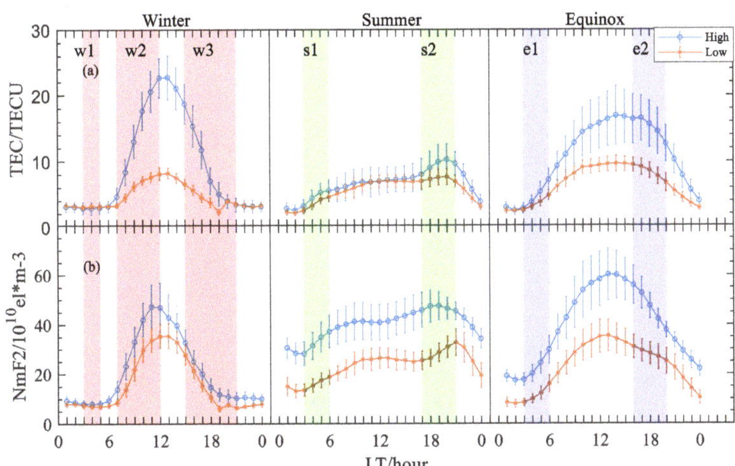

Figure 7. (**a**) The diurnal variation of median and standard deviation of TEC in winter, summer, and equinox of high- and low-solar-activity years. (**b**) The diurnal variation of median and standard deviation of NmF2 in winter, summer, and equinox of high- and low-solar-activity years.

Figure 8 presents the temporal variation rates of TEC and NmF2. According to Equation (3): The increase in variation rates of τ can be classified into the following groups (1) TEC increases and NmF2 drops; (2) TEC and NmF2 rise, but the variation rates of TEC are larger than that of NmF2; and (3) TEC and NmF2 fall, but the variation rates of NmF2 is smaller than that of TEC. The opposite situation applies to the decrease in τ.

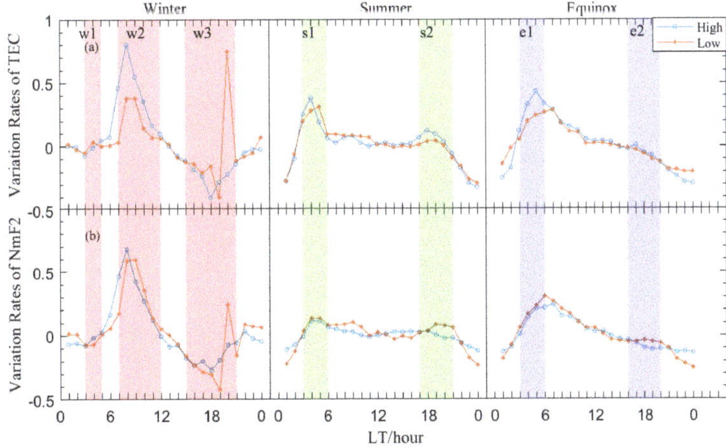

Figure 8. (**a**) Variation rates in TEC for winter, summer, and equinox in high- and low-solar-activity years, respectively. (**b**) Variation rates in NmF2 for winter, summer, and equinox in high- and low-solar-activity years, respectively.

As the τ has a totally different diurnal variation in the winter of the high- and low-solar-activity year, we discussed the physical mechanism related to the phenomenon first. As the Yakutsk is located at a high latitude, the solar zenith angle is small and the duration of solar radiation is short in the winter. Therefore, the electron density is highly sensitive to the solar zenith angle in this region [48]. In the high solar activity year, the variation rates of TEC are larger than that of NmF2 due to the solar zenith angle increases, although the summer-to-winter circulation would attenuate it (the increase of NmF2 caused by circulation is greater than that of TEC as the main area affected by summer to winter circulation is near the peak height of F2 layer) [47,62]. In other words, the solar zenith angle is more important than the prevailing wind circulation to TEC and NmF2 in high latitudes. Therefore, in the w2 period of the high-solar-activity year, the variation rate of TEC is greater than that of NmF2, and τ is increased. On the other hand, the variation rates of TEC and NmF2 are small in the w2 period of low-solar-activity years due to the low solar radiation in the low-solar-activity year. At the same time, the prevailing summer-to-winter circulation also influences the ionosphere significantly. As the summer-to-winter circulation mainly affects the area near the peak height of the F2 layer, it would result in the NmF2 increase faster than TEC, as shown in Figure 7. Therefore, the τ decreases during the w2 period in the low solar activity. In addition, the opposite situations could explain the increases of τ in low-solar-activity years. As the prevailing circulation effect attenuate, it would result in the decrease in rates of TEC smaller than that of NmF2, and the τ would therefore increase in the first half of the W3 period.

Around sunset hours, the decrease of the O+-H+ conversion height (i.e., the increase of the downward influx of the plasmasphere) also led to the increase of the τ in the winter of the low solar activity year (second half of the W3 period). The weak solar radiation around sunset in winter will not cause an obvious TEC increase, and the increase of TEC and NmF2 after sunset in low solar activity years shown in the figure is due to the lower O+-H+ conversion height. Previous studies also suggested that the decrease of the O+-H+ conversion height has a negative correlation with solar activity, as with the lower recombination rates in low-solar-activity years [63]. As the recombination rates are lower in high altitudes, in the process of downward influx from the plasmasphere and conjugate hemisphere, the increase of TEC is more than that of NmF2, resulting in the increase of τ around sunset in winter of low-solar-activity years. It can also be seen from Figure 4 that there is an increase of τ in the w1 period in low-solar-activity years, and it should also be attributed to the plasma transport from the plasmasphere and conjugate hemisphere. As the w1 period is 2–4 LT before sunrise in winter, the meridional wind is still equator-ward during the period, and the plasmasphere is not exposed to solar radiation. Therefore, NmF2 does not decrease rapidly, and TEC does not increase significantly, as shown in Figure 7, and the peak of the W1 time period is due to the decrease of the O+-H+ transition height, i.e., the transport of plasma flux. As the transport process is downward, and the ionosphere at higher heights than hmF2 has a recombination rate, the variation rates of TEC are, therefore, larger than those of NmF2, resulting in the increases of τ in the w1 period.

As for the τ variation in the summer and equinox, it can be seen from Figure 7 that TEC and NmF2 increase in the summer s1 period, as well as in the equinox e1 period, while it is not a nighttime increase. Because the increases in TEC and NmF2 are due to the change of solar zenith angel during this period at high latitudes, as indicated by Figure 7 the TEC and NmF2 increase continuously even after the s1 and e1 periods, and the increases in τ during the s1 and e1 periods are caused by the greater effect of the solar zenith angle on TEC than on NmF2 after sunrise.

Figure 7 also shows that TEC, as well as NmF2 is still increasing after 18 LT in summer (s2 period), which is due to the change of wind direction from poleward to equatorward around dusk hours and the continuous solar radiation after sunset [64,65]. On the one hand, the equatorward wind would raise the ionosphere to higher altitudes, where the recombination rates of electrons are lower. On the other hand, the topside ionosphere (along with the plasmasphere) is exposed to solar radiation for a longer time than the bottomside ionosphere, and the solar radiation is strong in the high-solar-activity year, so the increase rate of TEC is larger than that of NmF2, causing the τ increases in the s2 time period in high-solar-activity year. Due to the lack of sufficient solar radiation intensity in the equinox, the equatorward wind at sunset could not cause an increase in TEC and NmF2; however, they would cause the rate of decrease in TEC to be less than the rate of decrease in NmF2, which in turn leads to an increase in τ during e2 period. Additionally, as the wind reversal mainly influences the height around hmF2 and the solar irradiance is low in the low-solar-activity years, the increase/decrease rate of NmF2 is more intense/weak than the increase/decrease rate of TEC during the dusk period in the summer low-solar-activity years, resulting the decrease of τ in s2/e2 period in low-solar-activity years.

The geomagnetic storm is an important factor affecting the ionosphere, as it can severely disturb the ionosphere–thermosphere system through various process. Previous studies have found that the correlation between the τ and geomagnetic activity seems to be different for varying location, local time, season, solar activity, and magnetic activity intensity [27,33,34]. As for the effect of geomagnetic storms on the τ in high latitudes, Jayachandran et al. [36] found that the τ would increase throughout the day at Goose Bay, which is located in high latitudes during both the solar minimum year (1985) and solar maximum (1981) by comparing the τ in geomagnetically quiet days (Ap < 10) and geomagnetically disturbed days. In this paper, we also found that the storm-time positive DI index is larger than the quiet-time positive DI index throughout the day, suggesting that the τ tends to increase during geomagnetic storms. By contrast, the negative DI index of τ during storm time varies little from that of the quiet-time negative DI index, which is larger than -0.2 during both daytime and nighttime, indicating that the negative disturbance of τ during magnetic storms is weak.

During geomagnetic storms, there are several processes affecting the τ, and these processes include the following: (1) prompt penetration of the electric field due to the imbalance between the R1 (region1) and R2 (region 2) field-aligned currents FAC; (2) disturbance dynamo electric field due to the storm-time equatorward winds; (3) equatorward neutral wind result from the particle precipitation and Joule heating at high latitude, sometimes accompanied with traveling atmosphere disturbance (TAD); and (4) composition changes caused by the expansion of the neutral atmosphere at high latitudes and transport by the equatorward winds. In addition, there are two important processes that could affect the ionosphere at high latitudes during geomagnetic storms [55]. The particle precipitation gets stronger as there are more energetic particles input to the ionosphere through the nearly perpendicular field line during storms. At the same time, the cross-polar-cap potential drop increases significantly, resulting in the intensification and expansion of the plasma convection pattern. In the following, we present an example on 7 June 2013 to analyze why the τ would increase during the geomagnetic storms. Note that the Dst variation during the geomagnetic storm was presented in Figure 2.

Figure 9 shows the variations of the DI index of TEC, NmF2, and τ during the period from 6 June 12 UT to 8 June 0 UT. Correspondingly, the TEC/NmF2 observations and their monthly median were also presented in Figure 9. It can be seen from the figure that there is an interrupted TEC increase in 0 UT on 7 June, resulting in a TEC positive disturbance that lasts for 12 h (S1 period). Meanwhile, the NmF2 observations are smaller than the corresponding monthly median, and there are negative disturbances of NmF2. Therefore, the τ increases markedly, and there are positive disturbances of τ during the s1 period.

It is generally accepted that the PPEF, DDEF, equatorward neutral wind, and composition changes could influence the ionosphere of different altitudes in a similar way [40,55]. In addition, the ionospheric background and dominant physical process are different for varying altitudes. Therefore, the responses of TEC and NmF2 to geomagnetic storms often behave in a similar way but with different magnitudes for geomagnetic storm periods. Figure 9 shows that there is a contrasting behavior between the TEC and NmF2 during the s1 period, resulting in the positive disturbance of τ. Hence, there might have been other processes affecting the ionosphere at Yakutsk during the storm time. The particle precipitation and convection electric field are two important processes influencing the high-latitude ionosphere. As previous studies suggested, the particle precipitation mainly occurred in the cusp and auroral E region, which are a source of substantial ionization [66,67]. During geomagnetic storms, there is intense particle precipitation in the E region, and the high-energy precipitating particles collide with the neutral composition, resulting the high electron density in the E region. Meanwhile, the plasma convection intensifies and expands to lower latitudes during the storm. Therefore, the increased electron density in the E region could circulate to Yakutsk station through the storm-time, thus expanding plasma convection electric field and resulting in the electron density enhancement in the E region at Yakutsk. On the other hand, the intensified convection electric field would cause the recombination rate of the electron density in the F region to become stronger and electron density decreases. Consequently, the electron density in the E region could be larger than that in the F region. Previous studies have also found that the total electron content in the E region (TEC_E) could be a major component of the total electron content (TEC) during the geomagnetic storms [68,69]. Therefore, we preliminarily infer that it is the intense particle precipitation and expanding magnetospheric convection electric field during the storm that causes the TEC to increase even as NmF2 decreases during the s1 period. More observations and modeling work will be conducted in the future to study the physical mechanism of the storm-time τ increases at Yakutsk.

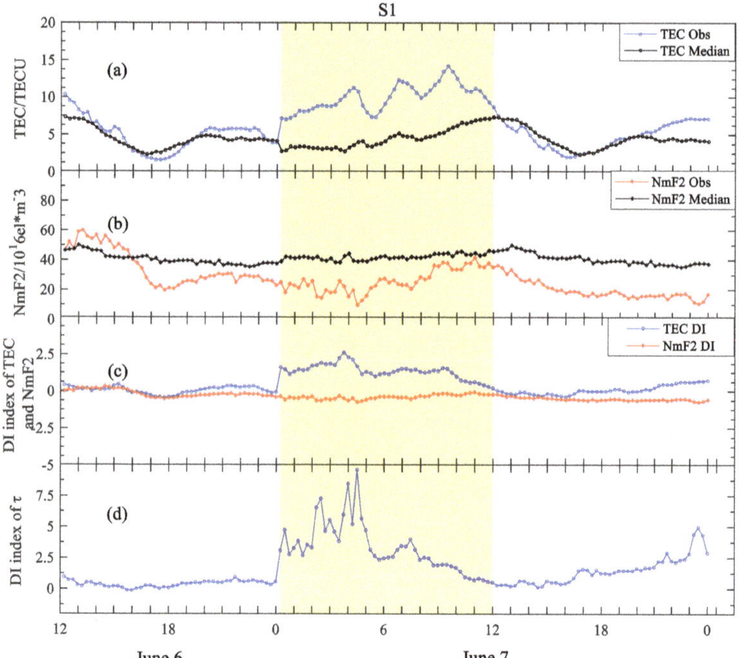

Figure 9. The variation of ionospheric parameters during the geomagnetic storm. (**a**) The TEC observation and its corresponding monthly median during the period. (**b**) The NmF2 observation and its monthly median during the period. (**c**) The DI index of TEC and NmF2 calculated from their observations and monthly medians. (**d**) The DI index of τ during the period.

5. Conclusions

This paper presents a statistical analysis of the ionospheric slab thickness τ at Yakutsk which is located in the high latitude of East Asia, using the TEC and NmF2 data during the years 2010–2017. The results show that τ has great local time, seasonal and solar activity variation, and the climatology of the τ is distinct from that in other high latitude regions. In addition, the geomagnetic storm seems to have a positive effect on the τ, consistent with previous study [36]. The main conclusions of this study are summarized as follows:

(1)   The τ is greatest in the winter, followed by the equinox; and it is smallest in summer in both high- and low-solar-activity years, except in the noontime of low-solar-activity years. It is due to the ionosphere strongly depending on the solar zenith angle in

the winter at Yakutsk, and the increasing [O/N2] is less/more important than the strong/weak solar radiation in high- and low-solar-activity years, respectively.

(2) In the winter, the τ in high-solar-activity years shows an approximate single-peak pattern around noontime, while it displays a double peak in pre-sunrise and post-sunset periods in low-solar-activity years. In the high-solar-activity years, the noontime peak was caused by the fact that the solar zenith angle is more important than the prevailing wind circulation to t TEC and NmF2 in Yakutsk. In the low solar activity years, the post-sunset and pre-sunrise peaks were caused by the downward plasma influx from the plasmasphere and conjugate hemisphere.

(3) In the summer and equinox, there is an increase during the forenoon period due to the greater effect of the solar zenith angle on TEC than on NmF2 in the period. In addition, there are post-sunset peaks in summer and equinox of high-solar-activity years, and they were caused by the equatorward neutral wind and continuous strong solar radiation in summer and equinox.

(4) Geomagnetic storms seem would enhance τ during the storm period, and this effect should be associated with intense particle precipitation and expanded plasma convection electric field during the storm time.

Author Contributions: Conceptualization, Y.Z. (Yuqiang Zhang), J.F. and N.X.; methodology, Z.D.; investigation, Y.L. and C.Z.; validation, J.F. and B.C.; formal analysis, Y.Z. (Yuqiang Zhang) and Z.W. (Zhuangkai Wang); resources, Y.Z. (Yufeng Zhou) and Z.Z.; visualization, J.F. and Z.W. (Zhensen Wu); funding acquisition, Y.L. and T.X. All authors have read and agreed to the published version of the manuscript.

Funding: This work was supported by the foundation of National Key Laboratory of Electromagnetic Environment (Grant No. 20200101), the Fundamental Research Funds for the Central Universities (Grant No. 2042021kf0022), the Stable-Support Scientific Project of China Research Institute of Radiowave Propagation (Grant No. A132102W06), the National Natural Science Foundation of China (Grant Nos. 42074187 and 42204161), the National Key R&D Program of China (Grant No. 2018YFC1503506), the Excellent Youth Foundation of Hubei Provincial Natural Science Foundation (Grant No. 2019CFA054), and Stable Support Project of Basic Scientific Research Institutes (Grant No. A131901W10).

Data Availability Statement: The use of ionosonde data provided by GIRO (Global Ionospheric Radio Observation) database (https://giro.uml.edu/didbase, (accessed on 20 October 2022)). The use of GPS data provided by UNAVCO database (University NAVSTAR Consortium) (http://www.unavco.org/, (accessed on 20 October 2022)). The use of Dst data provided by WDC (World Data Center, http://wdc.kugi.kyoto-u.ac.jp/, (accessed on 20 October 2022)). The use of F10.7 data provided by NOAA (ftp://ftp.swpc.noaa.gov/pub/indices/old_indices/, (accessed on 20 October 2022)).

Acknowledgments: We acknowledge the use of ionosonde data provided by GIRO (Global Ionospheric Radio Observation) database (https://giro.uml.edu/didbase, (accessed on 20 October 2022)), GPS data provided by UNAVCO database (University NAVSTAR Consortium) (http://www.unavco.org/, (accessed on 20 October 2022)), Dst data provided by WDC (http://wdc.kugi.kyoto-u.ac.jp/, (accessed on 20 October 2022)), and F10.7 data provided by NOAA (ftp://ftp.swpc.noaa.gov/pub/indices/old_indices/, (accessed on 20 October 2022)).

Conflicts of Interest: The authors declare no conflict of interest.

References

1. Wright, J.W. A model of the F-region above hmaxF2. *J. Geophys. Res.* **1960**, *65*, 185–191. [CrossRef]
2. Rishbeth, H.; Garriott, O. *Introduction to Ionospheric Physics*; International Geophysics Series; Academic Press: New York, NY, USA, 1969; Volume 14.
3. Titheridge, J.E. The ionospheric slab thickness of the mid-latitude ionosphere. *Planet Space Sci.* **1973**, *21*, 1775–1793. [CrossRef]
4. Jakowski, N.; Putz, E.; Spalla, P. Ionospheric storm characteristics deduced from satellite radio beacon observations at three European stations. *Ann. Geophys.* **1990**, *8*, 343–352.
5. Jakowski, N.; Mielich, J.; Hoque, M.M.; Danielides, M. Equivalent ionospheric slab thickness at the mid-latitude ionosphere during solar cycle 23. In Proceedings of the 38th COSPAR Scientific Assembly, Bremen, Germany, 18–25 July 2010.

6. Jakowski, N.; Hoque, M.M. Global equivalent ionospheric slab thickness model of the Earth's ionosphere. *J. Space Weather Space Clim.* **2021**, *11*, 10. [CrossRef]
7. Pignalberi, A.; Pezzopane, M.; Rizzi, R. Modeling the lower part of the topside ionospheric vertical electron density profile over the European region by means of Swarm satellites data and IRI UP method. *Space Weather* **2018**, *16*, 304–320. [CrossRef]
8. Pignalberi, A.; Pezzopane, M.; Themens, D.R.; Haralambous, H.; Nava, B.; Coïsson, P. On the analytical description of the topside ionosphere by NeQuick: Modeling the scale height through COSMIC/FORMOSAT-3 selected data. *IEEE J. Sel. Top. Appl. Earth Obs. Remote Sens.* **2020**, *13*, 1867–1878. [CrossRef]
9. Mendillo, M.; Papagiannis, M.D.; Klobuchar, J.A. Average behavior of the midlatitude F-region parameters NT, Nmax and τ during geomagnetic storms. *J. Geophys. Res.* **1972**, *77*, 4891–4895. [CrossRef]
10. Krankowski, A.; Shagimuratov, I.I.; Baran, L.W. Mapping of foF2 over Europe based on GPS-derived TEC data. *Adv. Space Res.* **2007**, *39*, 651–660. [CrossRef]
11. Gerzen, T.; Jakowski, N.; Wilken, V.; Hoque, M.M. Reconstruction of F2 layer peak electron density based on operational vertical total electron content maps. *Ann. Geophys.* **2013**, *31*, 1241–1249. [CrossRef]
12. Maltseva, O.A.; Mozhaeva, N.S.; Nikitenko, T.V. Validation of the Neustrelitz Global Model according to the low latitude ionosphere. *Adv. Space Res.* **2014**, *54*, 463–472. [CrossRef]
13. Frón, A.; Galkin, I.; Krankowski, A.; Bilitza, D.; Hernández-Pajares, M.; Reinisch, B.; Li, Z.; Kotulak, K.; Zakharenkova, I.; Cherniak, I.; et al. Towards Cooperative Global Mapping of the Ionosphere: Fusion Feasibility for IGS and IRI with Global Climate VTEC Maps. *Remote Sens.* **2020**, *12*, 3531. [CrossRef]
14. Galkin, I.; Frón, A.; Reinisch, B.; Hernández-Pajares, M.; Krankowski, A.; Nava, B.; Bilitza, D.; Kotulak, K.; Flisek, P.; Li, Z.; et al. Global monitoring of ionospheric weather by GIRO and GNSS data fusion. *Atmosphere* **2022**, *13*, 371. [CrossRef]
15. Bhonsle, R.V.; Da Rosa, A.V.; Garriott, O.K. Measurement of Total Electron Content and the Equivalent ionospheric slab thickness of the Mid latitude Ionosphere. *Radio Sci.* **1965**, *69*, 929–937.
16. Huang, Y.N. Some results of ionospheric slab thickness observations at Lunping. *J. Geophys. Res.* **1983**, *88*, 5517–5522. [CrossRef]
17. Davies, K.; Liu, X.M. Ionospheric slab thickness in middle and low latitudes. *Radio Sci.* **1991**, *26*, 997–1005. [CrossRef]
18. Fox, M.W.; Mendillo, M.; Klobuchar, J.A. Ionospheric equivalent ionospheric slab thickness and its modeling applications. *Radio Sci.* **1991**, *26*, 429–438. [CrossRef]
19. Chuo, Y.J.; Lee, C.C.; Chen, W.S. Comparison of ionospheric equivalent slab thickness with bottomside digisonde profile over Wuhan. *J. Atmos. Sol.-Terr. Phys.* **2010**, *72*, 528–533. [CrossRef]
20. Jin, S.; Cho, J.-H.; Park, J.-U. Ionospheric slab thickness and its seasonal variations observed by GPS. *J. Atmos. Sol. Terr. Phys.* **2007**, *69*, 1864–1870. [CrossRef]
21. Stankov, S.M.; Warnant, R. Ionospheric slab thickness—Analysis, modelling and monitoring. *Adv. Space Res.* **2009**, *44*, 1295–1303. [CrossRef]
22. Guo, P.; Xu, X.; Zhang, G.X. Analysis of the ionospheric equivalent ionospheric slab thickness based on ground-based GPS/TEC and GPS/COSMIC RO. *J. Atmos. Sol. Terr. Phys.* **2011**, *73*, 839–846. [CrossRef]
23. Huang, H.; Liu, L.; Chen, Y.; Le, H.; Wan, W. A global picture of ionospheric slab thickness derived from GIM TEC and COSMIC radio occultation observations. J. Geophys. *Res. Space Phys.* **2016**, *121*, 867–880. [CrossRef]
24. Odeyemi, O.O.; Adeniyi, J.O.; Oladipo, O.A.; Olawepo, A.O.; Adimu, I.A.; Oyeyemi, E.O. Ionospheric slab thickness investigation on slab-thickness and B0 over an equatorial station in Africa and comparison with IRI model. *J. Atmos. Sol. Terr. Phys.* **2018**, *179*, 293–306. [CrossRef]
25. Jakowski, N.; Hoque, M.M.; Mielich, J.; Hall, C. Equivalent ionospheric slab thickness of the ionosphere over Europe as an indicator of long-term temperature changes in the thermosphere. *J. Atmos. Terr. Phys.* **2017**, *163*, 92–101.
26. Pignalberi, A.; Nava, B.; Pietrella, M.; Cesaroni, C.; Pezzopane, M. Mid-latitude climatology of the ionospheric equivalent slab thickness over two solar cycles. *J. Geod.* **2021**, *95*, 124. [CrossRef]
27. Kersley, L.; Hajeb-Hosseinieh, H. Dependence of ionospheric slab thickness on geomagnetic activity. *J. Atmos. Terr. Phys.* **1976**, *38*, 1357–1360. [CrossRef]
28. Venkatesh, K.; Rama Rao, P.V.S.; Prasad, D.S.V.V.D.; Niranjan, K.; Saranya, P.L. Study of TEC, slab thickness and neutral temperature of the thermosphere in the Indian low latitude sector. *Ann. Geophys.* **2011**, *29*, 1635–1645. [CrossRef]
29. Minakoshi, H.; Nishimuta, I. Ionospheric electron content and equivalent ionospheric slab thickness at lower mid-latitudes in the Japanese zon. In Proceedings of the Beacon Satellite Symposium (IBSS), Aberystwyth, UK, 11–15 July 1994; University of Wales: Wales, UK, 1994; Volume 144.
30. Gulyaeva, T.L.; Jayachandran, B.; Krishnankutty, T.N. Latitudinal variation of slab thickness. *Adv. Space Res.* **2004**, *33*, 862–865. [CrossRef]
31. Huang, Z.; Yuan, H. Climatology of the ionospheric slab thickness along the longitude of 120°E in China and its adjacent region during the solar minimum years of 2007–2009. *Ann. Geophys.* **2015**, *33*, 1311–1319. [CrossRef]
32. Balan, N.; Iyer, N. Ionospheric slab thickness during geomagnetic storm. *Indian J. Radio Space Phys.* **1978**, *7*, 238–241.
33. Breed, A.M.; Goodwin, G.L.; Vandenberg, A.-M.; Essex, E.A.; Lynn, K.J.W.; Silby, J.H. Ionospheric total electron content and ionospheric slab thickness determined in Australia. *Radio Sci.* **1997**, *62*, 1635–1643. [CrossRef]
34. Pignalberi, A.; Pietrella, M.; Pezzopane, M.; Nava, B.; Cesaroni, C. The Ionospheric Equivalent Slab Thickness: A Review Supported by a Global Climatological Study Over Two Solar Cycles. *Space Sci. Rev.* **2022**, *218*, 37. [CrossRef]

35. Zhang, Y.; Wu, Z.; Feng, J.; Xu, T.; Deng, Z.; Ou, M.; Xiong, W.; Zhen, W. Statistical study of ionospheric equivalent slab thickness at Guam magnetic equatorial location. *Remote Sens.* **2021**, *13*, 5175. [CrossRef]
36. Jayachandran, B.; Krishnankutty, T.; Gulyaeva, T. Climatology of ionospheric slab thickness. *Ann. Geophys.* **2004**, *22*, 25–33. [CrossRef]
37. Yadav, R.; Bhawre, P. Ionospheric slab thickness over high latitude Antarctica during the maxima of solar cycle 23rd. *Int. J. Curr. Res.* **2020**, *12*, 10041–10046. [CrossRef]
38. Abe, O.; Villamide, X.O.; Paparini, C.; Radicella, S.; Nava, B.; Rodríguez Bouza, M. Performance evaluation of GNSS-tec ionospheric slab thickness estimation techniques at the grid point in middle and low latitudes during different geomagnetic conditions. *J. Geod.* **2017**, *91*, 409–417. [CrossRef]
39. Uwamahoro, J.C.; Giday, N.M.; Habarulema, J.B.; Katamzi-Joseph, Z.T.; Seemala, G.K. Reconstruction of storm-time total electron content using ionospheric tomography and artificial neural networks: A comparative study over the African region. *Radio Sci.* **2018**, *53*, 1328–1345. [CrossRef]
40. Seemala, G.; Valladares, C. Statistics of total electron content depletions observed over the South American continent for the year 2008. *Radio Sci.* **2011**, *46*, RS5019. [CrossRef]
41. Olwendo, O.; Baki, P.; Mito, C.; Doherty, P. Characterization of ionospheric GPS Total Electron content (GPS TEC) in low latitude zone over the Kenyan region during a very low solar activity phase. *J. Atmos. Sol. Terr. Phys.* **2012**, *84*, 52–61. [CrossRef]
42. Matamba, T.M.; Habarulema, J.B.; McKinnell, L.-A. Statistical analysis of the ionospheric response during geomagnetic storm conditions over South Africa using ionosonde and GPS data. *Space Weather* **2015**, *13*, 536–547. [CrossRef]
43. De Dieu Nibigira, J.; Sivavaraprasad, G.; Ratnam, D.V. Performance analysis of IRI-2016 model TEC predictions over Northern and Southern Hemispheric IGS stations during descending phase of solar cycle 24. *Acta Geophys.* **2021**, *69*, 1509–1527. [CrossRef]
44. Reinisch, B.W.; Galkin, T.A. Global ionospheric radio observatory (GIRO). *Earth Planets Space* **2011**, *63*, 377–381. [CrossRef]
45. Gulyaeva, T.; Stanislawska, I. Night-day imprints of ionospheric slab thickness during geomagnetic storm. *J. Atmos. Sol. Terr. Phys.* **2005**, *67*, 1307–1314. [CrossRef]
46. Fuller Rowell, T.J.; Rees, D. Derivation of a conservation equation for mean molecular weight for a two constituent gas within a three dimensional time-dependent model of the thermosphere. *Planet. Space Sci.* **1983**, *31*, 1209–1222. [CrossRef]
47. Rishbeth, H. How the thermospheric circulation affects the ionosphere F2-layer. *J. Atmos. Terr. Phys.* **1998**, *60*, 1385–1402. [CrossRef]
48. Ma, R.; Xu, J.; Wang, W.; Yuan, W. Seasonal and latitudinal differences of the saturation effect between ionospheric NmF2and solar activity indices. *J. Geophys. Res.* **2009**, *114*, A10303.
49. Bailey, G.J.; Sellek, R.; Balan, N. The effect of interhemispheric coupling on nighttime enhancements in ionospheric total electron content during winter at solar minimum. *Ann. Geophys.* **1991**, *9*, 738–747.
50. Jakowski, N.; Förster, M. About the nature of the Night-time Winter Anomaly effect (NWA) in the F-region of the ionosphere. *Planet. Space Sci.* **1995**, *43*, 603–612. [CrossRef]
51. Chen, Y.; Liu, L.; Le, H.; Wan, W.; Zhang, H. The global distribution of the dusk-to-nighttime enhancement of summer NmF2 at solar minimum. *J. Geophys. Res. Space Res.* **2016**, *121*, 7914–7922. [CrossRef]
52. Fuller-Rowell, T.J.; Codrescu, M.V.; Moffett, R.J.; Quegan, S. Response of the thermosphere and ionosphere to geomagnetic storm. *J. Geophys. Res. Atmos.* **1994**, *99*, 3893–3914. [CrossRef]
53. Buonsanto, M.J. Ionospheric storms—A review. *Space Sci. Rev.* **1999**, *88*, 563–601. [CrossRef]
54. Mendillo, M. Storms in the ionosphere: Patterns and processes for total electron content. *Rev. Geophys.* **2006**, *44*, 335–360. [CrossRef]
55. Bellchambers, W.; Piggott, W. Ionospheric measurements made at Halley Bay. *Nature* **1958**, *182*, 1596–1597. [CrossRef]
56. Horvath, I.; Essex, E.A. The Weddell sea anomaly observed with the Topex satellite data. *J. Atomos. Sol. Terr. Phys.* **2003**, *65*, 693–706. [CrossRef]
57. Lin, C.H.; Liu, J.Y.; Cheng, C.Z.; Chen, C.H.; Liu, C.H.; Wang, W.; Burns, A.G.; Lei, J. Three-dimensional ionospheric electron density structure of the Weddell Sea Anomaly. *J. Geophys. Res. Space Res.* **2009**, *114*, A02312. [CrossRef]
58. Duncan, R.A. F-region seasonal and magnetic-storm behavior. *J. Atmos. Terr. Phys.* **1969**, *31*, 59–70. [CrossRef]
59. Mamrukov, A.P. Evening anomalous enhancement of ionization in F region. *Geomagn. Aeron.* **1971**, *21*, 984–988.
60. Maxim, K.; Vladimir, K.; Alexander, K.; Konstantin, R. Sub-auroral longitudinal anomalies in ionosphere-protonosphere system according to GSM TIP model and IK-19 satellite and ground-based observation. In Proceedings of the 2014 XXXIth URSI General Assembly and Scientific Symposium (URSI GASS), Beijing, China, 16–23 August 2014; pp. 1–4. [CrossRef]
61. Richards, P.G.; Meier, R.R.; Chen, S.; Dandenault, P. Investigation of the Causes of the Longitudinal and Solar Cycle Variation of the Electron Density in the Bering Sea and Weddell Sea Anomalies. *J. Geophys. Res. Space Res.* **2018**, *123*, 7825–7842. [CrossRef]
62. Torr, D.G.; Torr, M.R.; Richards, P.G. Causes of the F region winter anomaly. *Geophys. Res. Lett.* **1980**, *7*, 301–304. [CrossRef]
63. Jakowski, N.; Hoque, M.M.; Kriegel, M.; Patidar, V. The persistence of the NWA effect during the low solar activity period 2007–2009. *J. Geophys. Res. Space Phys.* **2015**, *120*, 9148–9160. [CrossRef]
64. He, M.; Liu, L.; Wan, W.; Ning, B.; Zhao, B.; Wen, J.; Yue, X.; Le, H. A study of the Weddell Sea Anomaly observed by FORMOSAT-3/COSMIC. *J. Geophys. Res. Space Res.* **2009**, *114*, A1230. [CrossRef]
65. Xiong, C.; Lühr, H. The Midlatitude Summer Night Anomaly as observed by CHAMP and GRACE: Interpreted as tidal features. *J. Geophys. Res. Space Res.* **2014**, *119*, 4905–4915. [CrossRef]

66. Roble, R.G. The Polar Lower Thermosphere. *Planet. Space Sci.* **1992**, *40*, 271–297. [CrossRef]
67. Robel, R.G.; Rees, M.H. Time-dependent studies of the aurora: Effects of particle precipitation on the dynamic morphology of ionospheric and atmospheric properties. *Planet. Space Sci.* **1977**, *25*, 991–1010. [CrossRef]
68. Wu, Y.; Liu, R.; Zhang, B.; Wu, Z.; Hu, H.; Zhang, S.; Zhang, Q.; Liu, J.; Honary, F. Multi-instrument observations of plasma features in the Arctic ionosphere during the main phase of a geomagnetic storm in December 2006. *J. Atmos. Sol. Terr. Phys.* **2013**, *105–106*, 358–366. [CrossRef]
69. Yang, S.; Zhang, B.; Fang, H.; Liu, J.; Zhang, Q.; Hu, H.; Liu, R.; Li, C. F-lacuna at cusp latitude and its associated TEC variation. *J. Geophys. Res. Space Phys.* **2014**, *119*, 10384–10396. [CrossRef]

MDPI AG

Grosspeteranlage 5

4052 Basel

Switzerland

Tel.: +41 61 683 77 34

Remote Sensing Editorial Office

E-mail: remotesensing@mdpi.com

www.mdpi.com/journal/remotesensing